Principles of Hormone/Behavior Relations

Principles of Hormone/ Behavior Relations

Donald W. Pfaff
Professor
Laboratory of Neurobiology and Behavior
The Rockefeller University
New York, New York

M. Ian Phillips
Vice-President for Research
University of South Florida
College of Medicine
Tampa, Florida

Robert T. Rubin
Professor
Drexel University College of Medicine
Allegheny General Hospital Campus
Pittsburgh, Pennsylvania

Amsterdam Boston Heidelberg London New York Oxford
Paris San Diego San Francisco Singapore Sydney Tokyo

Elsevier Academic Press
200 Wheeler Road, 6th Floor, Burlington, MA 01803, USA
525 B Street, Suite 1900, San Diego, California 92101-4495, USA
84 Theobald's Road, London WC1X 8RR, UK

This book is printed on acid-free paper. ♾

Library of Congress Cataloging-in-Publication Data
Application submitted

British Library Cataloguing in Publication Data
A catalogue record for this book is available from the British Library

ISBN: 0-12-553149-4

For all information on all Academic Press publications
visit our Web site at www.academicpress.com

Printed in the United States of America
04 05 06 07 08 09 9 8 7 6 5 4 3 2 1

ACKNOWLEDGMENTS

Creation and Management
Of Illustrations by
Par Parekh,
The Rockefeller University

The authors are pleased to thank David Rubinow, M.D., of the National Institute of Mental Health, Bethesda, MD for helpful criticisms of the text. We also thank Dr. Jasna Markovac for raising the notion of a small text to follow the reference source Hormones, Brain and Behavior (HBB).

CONTENTS

SECTION II

History: Hormone Effects Can Depend on Family, Gender, and Development

SECTION III

Time: Hormonal Effects on Behavior Depend on Temporal Parameters

SECTION IV

Space: Spatial Aspects of Hormone Administration and Impact are Important

SECTION V

Mechanisms: Molecular and Biophysical Mechanisms of Hormone Actions Give Clues to Future Therapeutic Strategies

SECTION VI

Environment: Environmental Variables Influence Hormone/Behavior Relations

SECTION VII

Evolution

INTRODUCTION

The mechanisms by which hormones affect behaviors of all mammals, including humans, have achieved a primary spot in 21st-century neuroscience. Terrific progress in this field has been due to several factors. Stimuli and responses are relatively simple. Behaviors in their natural form can be evoked in the laboratory. Importantly, we can make use of the tremendous bodies of knowledge in hormone chemistry and pharmacology—especially for steroid hormones—to enrich and extend our ability to manipulate and measure neuronal events as they cause behavior. Steroid hormone receptors, discovered in brain, turn out to be nuclear proteins which are ligand-activated transcription factors, thus allowing us, *pari passu,* to do state-of-the-art molecular biology in the most complex organ in the body, the central nervous system (CNS). Therefore, it is timely to try to encompass our hard-won findings in an orderly fashion. It is also important to do so in order to make available to the medical student and resident some of the broadest generalizations that have emerged from the animal neurobiology literature. Included among these is the amazing development of proven causal linkages between specific molecular events and behavioral results.

Hormone/behavior relations always serve either homeostasis or reproduction. To preserve homeostasis, a wide variety of hormone actions bring behavioral responses into play in order to maintain body caloric and fluid balance and body temperature, to reduce pain and stress, etc. On other occasions, hormone-related behaviors cannot be argued to be protecting the internal environment, but rather to be supporting reproduction through courtship, mating, and parental response sequences. The axioms proposed in this book include both types of hormone actions.

WHAT THIS TEXT IS

This book represents the first attempt to state some of the major truths of the general field of hormone/behavior relations and their mechanisms in a systematic fashion. It is intended to be logically orderly in the fashion of a high-school geometry text. We put forward and illustrate a number of simple statements, "principles" of hormone/behavior relationships supported by the literature to date. Each statement (*i.e.*, principle) is the title of a chapter. Then, that general principle of hormone/behavior relations is exemplified in two ways—from basic scientific work in the laboratory and from clinical experience.

These principles of hormone/behavior relations are grouped into six sections of related statements. In turn, the sections themselves have been arranged in a logical order. First, in Section I, the very existence of powerful hormone/behavior phenomena is claimed and, in several chapters, partially characterized. This includes, for example, showing the dependence of hormone effectiveness on combinatorial actions. Likewise, as shown in Section II, the strength of a hormone action on a given behavior can depend upon family background, gender, and early developmental events. Hormone actions can be grouped and illustrated according to parameters of time (Section III) and space (Section IV). For the latter, it is important (1) to demonstrate the orchestration of neurobiological mechanisms with physiological regulatory mechanisms throughout the body, and (2) to consider different parts of the neuraxis itself. Section V addresses the question of how, exactly, hormones act on behavior. The study of cellular and molecular mechanisms of hormone action is emerging as one of the most dynamic fields of biological science; here, a few interesting generalizations are listed as chapter heads. The important pair of chapters in Section VI offers the broadest perspective on hormone/behavior relations. Especially in humans, the details of the effectiveness of a hormone in altering behavior cannot be understood without knowledge of the environmental context of the individual in whose brain the hormone is acting. Finally, in Section VII, the beginning of an evolutionary perspective is offered.

Each chapter provides examples for further reading, listed chapter by chapter at the end of the book. To help the reader with very specific interests, the index is intended to be detailed and comprehensive.

Teachers will recognize that many applications of each principle have been left out of this short text. They are encouraged to add examples from their own scholarship and research. Further, because this textbook represents a first attempt at a logically systematic overview of the field, the authors hope that interested students and medical residents will e-mail their interesting perspectives (pfaff@mail.rockefeller.edu), and, especially, will challenge the concepts and examples presented.

WHAT THIS TEXT IS NOT

In attempting to keep the book reasonably short and affordable, we have not tried to fill in background details from all the fields that border the integrative field of neuroendocrinology. Therefore, this text does not stand in for a steroid chemistry treatise, a neuroanatomy or neurophysiology primer, or a molecular biology handbook. A major bibliographic reference source has recently been published (*Hormones, Brain, and Behavior*, Academic Press, 2002), and compared to that source the examples in this text can neither be exhaustive nor complete. They are intended simply to be physiologically clear and to represent a variety of mammalian (including human) endocrine systems and central nervous system (CNS) or behavioral endpoints.

BRIEF OVERVIEW OF NEUROENDOCRINE ANATOMY AND PHYSIOLOGY AS THEY APPLY TO BEHAVIORAL AND MOLECULAR MECHANISMS

The brain has been referred to by some as the "largest gland in the body." While this may be hyperbole, it serves to highlight the close functional relationships between the brain and endocrine systems. The pituitary, the so-called master gland, is closely regulated by the brain, being anatomically connected to the hypothalamic area by the pituitary stalk. The hypothalamus is a major integrating center for many other areas of the brain, and, through specialized secretions, it provides the primary functional regulation of the anterior and posterior pituitary gland. The hormones secreted by the pituitary, in turn, regulate the output of other endocrine glands throughout the body, as well as having direct metabolic effects themselves. Most of the hormones secreted by the hypothalamus, the anterior and posterior pituitary, and the peripheral endocrine glands in turn can profoundly affect brain function. Thus, there is a full reciprocity in the concept of hormone–behavior relations.

To develop an appreciation of the magnitude and complexity of neuroendocrine anatomy and physiology, it is important to understand the basic organization of the brain itself. Throughout evolution, parts of the brain that were present in earlier forms of life have been overlaid with newer parts, culminating (at least at this stage of evolution) in the mammalian, primate, and uniquely human brains. Older brain areas regulate functions necessary for survival and reproduction. In particular, the hypothalamus, at the base of the brain, regulates five basic physiological functions: (1) blood pressure and electrolyte composition (vasomotor tone, thirst, salt appetite); (2) body temperature (metabolic thermogenesis—shivering, seeking a different environment); (3) energy metabolism (feeding, digestion, metabolic rate); (4) reproduction (hormonal control of mating, pregnancy, lactation); and (5) emergency responses to stress (blood flow to muscle, hormonal and immunological changes). It does this by influencing the autonomic nervous system, the endocrine system, and the immunological system, all of which act together to produce coordinated physiological changes throughout the body.

The hypothalamus has many afferent and efferent connections with other parts of the brain, particularly evolutionarily older structures. These include the amygdala and hippocampus in the temporal lobe, the septal nuclei, the mammillary bodies, the thalamus, the cingulate cortex, and the orbitofrontal cortex—structures that can be grouped functionally as the limbic system. Fig. 1 portrays a sagittal (midline) section of the human brain: The pituitary is shown at the base of the brain, connected to the hypothalamus by the pituitary stalk. The limbic structures lie around the central area of the brain and are darkly shaded in Fig. 1.

Some of the connections of the limbic structures of the brain are schematically portrayed in the top part of Fig. 2. The hypothalamus receives

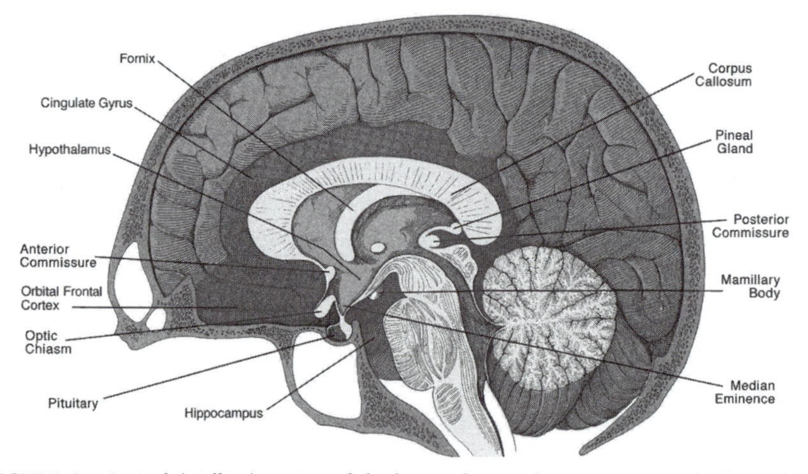

FIGURE 1. Sagittal (midline) section of the human brain. The pituitary is at the base of the brain, connected to the median eminence of the hypothalamus by the pituitary stalk. The limbic structures (darkly shaded) lie around the corpus callosum and other central brain structures. (Modified from Krieger, D.T., in *Neuroendocrinology*, Krieger, D.T. and Hughes, J.C., Eds., HP Publishing, New York, 1980, p. 4; Kandel, E.R. *et al.*, *Principles of Neural Science*, 4th ed., McGraw-Hill, New York, 2000, p. 987.)

inputs from several of these structures. Within the hypothalamus are a number of nuclei, and within these nuclei are specialized neuroendocrine cells that secrete several different hormones. As portrayed in the lower part of Fig. 2, vasopressin, important in blood pressure control and water conservation by the kidney, and oxytocin, important in uterine contractions during childbirth and milk ejection during nursing, are produced in neurosecretory cells in the paraventricular and supraoptic nuclei and are carried into the posterior pituitary directly by the long axons of these cells, where they are secreted into the general circulation.

Other neurosecretory cells produce releasing and inhibiting hormones, which regulate anterior pituitary secretion. As also indicated in Fig. 2, the axons of these cells terminate in the external layer of the median eminence, in proximity to one capillary bed of the pituitary portal blood vessels. Following their secretion, the releasing and inhibiting hormones are carried by this portal system down the pituitary stalk to the anterior pituitary, where a second capillary bed distributes them to the pituitary cells. In response, these cells secrete their hormone products into the general circulation.

A few examples, shown in Fig. 2, will illustrate the organization of the anterior pituitary hormonal cascade: The hypothalamo–pituitary–gonadal (HPG) system is regulated primarily by gonadotropin-releasing hormone (GnRH) from the hypothalamus. GnRH stimulates secretion of the gonadotropins, luteinizing hormone (LH), and follicle-stimulating hormone (FSH), from the anterior pituitary which in turn: (1) stimulate follicular growth and estrogen production (FSH) and corpus luteum formation and progesterone production (LH) in the ovary in the female, and (2) stimulate testosterone production (LH) and sperm production (FSH) in the testis in the male.

The hypothalamo–pituitary–thyroid (HPT) system is regulated primarily by thyrotropin-releasing hormone (TRH) from the hypothalamus. TRH stimulates thyrotropin (thyroid-stimulating hormone; TSH) from the anterior pituitary, which in turn stimulates thyroxine and triiodothyronine secretion from the thyroid gland; these thyroid hormones have widespread effects on metabolic rate. A third example is the hypothalamo–pituitary–adrenal cortical (HPA) system, which is regulated primarily by corticotropin-releasing hormone (CRH) from the hypothalamus. CRH stimulates corticotropin (adrenocorticotropic hormone; ACTH) secretion from the anterior pituitary, which in turn stimulates cortisol (in primates) or corticosterone (in other species) and aldosterone secretion from the adrenal cortex; these adrenal hormones have widespread effects on glucose metabolism and salt retention.

Other anterior pituitary hormones—for example, growth hormone (which is regulated primarily by a hypothalamic inhibitory hormone, somatostatin) and prolactin (which is regulated primarily by an inhibitory neurotransmitter, dopamine)—are additional examples. And, regulating the hypothalamic

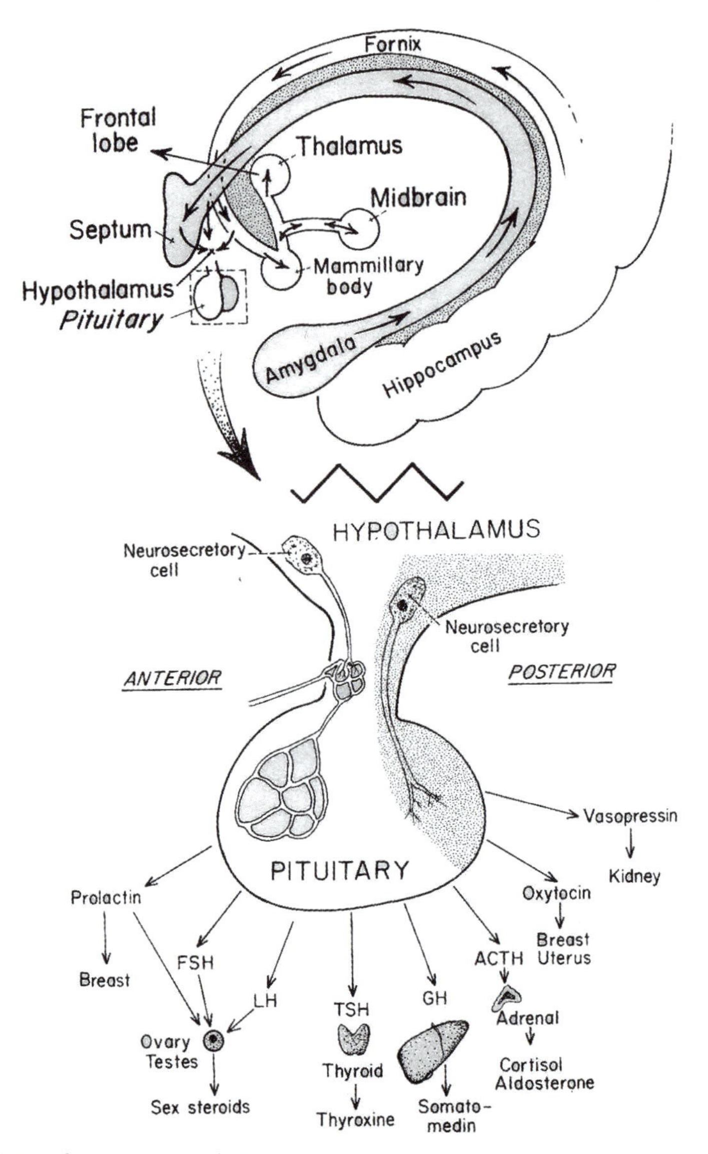

FIGURE 2. The upper part of the figure illustrates particular limbic structures and their connections, including inputs into the hypothalamus. The lower part of the figure is an expanded portrayal of the pituitary and its connection via the pituitary stalk to the median eminence of the hypothalamus. Also illustrated are the neurosceretory cells that secrete releasing and inhibiting factors into the pituitary portal circulation, from where they influence the cells of the anterior pituitary; the neurosecretory cells that carry their hormone products directly into the posterior pituitary; and several anterior and posterior pituitary hormones and the peripheral glands and tissues that they, in turn, influence.

releasing and inhibiting hormones are neurotransmitters and neuromodulators from the limbic and other neuronal circuits that feed into the hypothalamus itself.

The above hormonal cascades are oversimplified, in that other hormones participate in the regulatory steps, as do the autonomic nervous system and immune system. For example, in the HPA system, vasopressin, in addition to CRH, stimulates ACTH secretion from the anterior pituitary. Autonomic nervous system regulation of blood flow to the adrenal gland can modify its response to ACTH stimulation. And, inflammatory cytokines such as interleukin-1 (IL-1) can be powerful stimulants of the HPA system. Some of these influences will be considered in greater detail in later chapters.

Most hormones are secreted episodically, in bursts. Of particular importance, negative feedback loops at several levels in the hypothalamo–pitiuitary–target endocrine gland cascade serve to provide a certain constancy to episodic hormone secretion, just as a thermostat regulates temperature within a specified range. For example, hormones from the thyroid gland feed back to the pituitary and hypothalamus to regulate their release within fairly narrow limits, by suppressing TRH and TSH secretion. Thus, the secretion of thyroid hormones over the 24 hours is reasonably constant. Many hormone systems, however, are controlled more by CNS driving (so-called open-loop mechanisms) than by closed-loop, negative feedback. A notable example of open-loop regulation is, again, in the HPA axis: ACTH and cortisol have very prominent, similar circadian (24-hour) rhythms, with a peak that is 2 to 3 times greater than the low point (nadir).

The importance of CNS–hypothalamic regulation of pituitary hormone secretion can be seen experimentally following complete transection (severing) of the pituitary stalk. The basal secretion of pituitary hormones is lessened, and their circadian rhythms are abolished. The secretion of prolactin, however, is increased, because its inhibitory neurotransmitter, dopamine, cannot reach the pituitary lactotrophs (prolactin-secreting cells). In later chapters, the circadian rhythms of specific hormones and their pathological dysregulation will be presented in detail.

At the cellular level, hormones (ligands) bind to and activate receptors, which then regulate intracellular metabolic events. There are primarily two types of hormone receptors: those spanning the cell membrane and those within the cytoplasm. The largest family of cell-membrane receptors are coupled by G-proteins to their intracellular second-messenger systems, usually adenylate cyclase or phosphoinositol. They are comprised of loops of glycosylated amino acids that span the width of the cell membrane, with an extracellular binding domain and an intracellular tail. Many polypeptide hormones bind to this type of receptor.

A second family of cell-membrane receptors has the effector activity (*e.g.*, the enzyme tyrosine kinase) as part of the receptor structure. These receptors contain an extracellular ligand-binding domain, a single transmembrane-spanning amino acid sequence, and an intracellular kinase domain. Insulin, insulin growth factor I (IGF-I), and several other growth factors bind to this type of receptor.

A third family of cell-membrane receptors is comprised of ligand-gated ion channels. The effector mechanism for these receptors is a change in conformation of the channel which, depending on the receptor, allows the cations sodium, potassium, or calcium to traverse the cell membrane. These receptors have four or six transmembrane-spanning domains. Their ligands are primarily neurotransmitters such as acetylcholine (nicotinic cholinergic receptors) and γ-aminobutyric acid (GABA). Some hormones may act as allosteric (conformational) modulators of ligand-gated ion channels (*e.g.*, progesterone and other steroid hormone modulation of the GABA receptor).

Other membrane receptors have a structure similar to those with intrinsic effector activity but do not contain such an effector; their mechanism of action is not yet understood. Growth hormone, prolactin, and several cytokines bind to this type of receptor.

Intracellular receptors are in the cytoplasm and nucleus and bind their ligands after the ligands traverse the cell membrane. After ligand binding, the cytosolic ligand–receptor complex is translocated into the cell nucleus. Within the nucleus, the ligand–receptor complex binds to regulatory elements of the genome, resulting usually in gene activation and new protein synthesis. Steroid hormones, such as those produced by the adrenal glands and gonads, bind primarily to intracellular receptors. Because of the processes required between gene activation and protein synthesis (transcription, translation), the cascade of metabolic events resulting from hormone activation of intracellular receptors is longer (a few hours) than the effects of hormone activation of cell membrane receptors.

Because some steroid hormone actions occur in a matter of minutes, cell-membrane receptors for steroids also have been postulated, but they have not yet been clearly demonstrated. And, because steroid hormones are highly lipid (fat) soluble, the cell membrane is a lipid bilayer, and steroid hormones partition with high affinity into the cell membrane, direct effects of steroid hormones on cell-membrane width and other physicochemical characteristics have been suggested.

To build upon this brief overview, the reader may want to consult Knobil and Neill's *The Physiology of Reproduction* (Academic Press, 1994/2004) and DeGroot's *Endocrinology* (Saunders, 1995) and do targeted searches in *Molecular Endocrinology* and *Endocrine Reviews*.

SECTION I

Characterizing the Phenomena: Hormone Effects Are Strong and Reliable

CHAPTER 1

Hormones Can Both Facilitate and Repress Behavioral Responses

I. BASIC EXPERIMENTAL EXAMPLES

Hormones have powerful influences on behavior, acting over many time courses, and can act in either direction. We highlight these very briefly and then illustrate the principle of this chapter with major bodies of data.

Power. One of the most important hormone/behavior causal links is the complex set of mechanisms controlling aggression. Among mammals, males are almost always more aggressive than females. Simon (2002) reviews the large body of work explaining hormone effects on offensive intermale aggression. Some neural controls over such aggression are turned on by estrogens and some by androgens, while some require a synergy between estrogens and androgens. The relative importance of these three hormonal routes depends on the species studied. Further, during adulthood, the sensitivity of a given hormonal trigger for aggression depends on hormone exposures early in development (Fig. 1.1).

Clearly, steroid sex hormones facilitate mating behaviors in a wide variety of experimental animals. Estrogens given over a long period will turn on

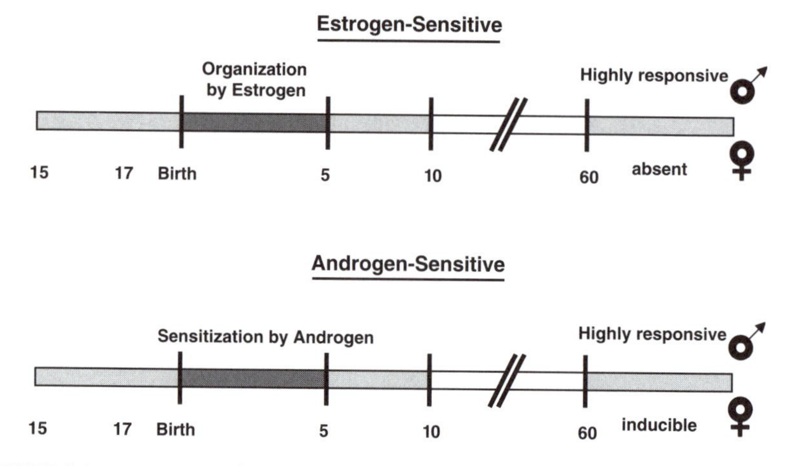

FIGURE 1.1. A summary of the major hormonal events and their timing in the establishment of androgen- and estrogen-sensitive regulatory pathways for offensive male-typical aggressive behavior in the mouse. Time is marked in days. The development of each pathway depends on exposure to specific testosterone metabolites during a restricted period shortly after birth. (From Simon, N. *et al.* Development and expression of hormonal systems regulating aggression. *Ann. NY Acad. Sci.* 794, 8–17. ©1996 New York Academy of Science. With permission.)

females' courtship and copulatory behaviors, especially if the estrogenic effect is amplified by subsequent progesterone treatment. In the male, testosterone administered peripherally enters the brain readily. It leads to heightened sexual motivation and mating behaviors, acting both in the chemical form of testosterone and in the form of one of its metabolites, estradiol (see Chapter 4).

Time Courses. Many actions of steroid hormones on behavior take a long time. Androgenic facilitation of male aggressive behavior requires many days of testosterone exposure. An exception is the striking elevation of salt hunger by the mineralocorticoid hormone aldosterone. In a wide variety of species, aldosterone or desoxycorticosterone, both mineralocorticoids produced in the zona glomerulosa cells of the adrendal cortex, will dramatically increase sodium ingestion. When aldosterone is delivered to the central nervous system (CNS), salt intake can rise within one hour. Also, laboratory animals will learn and perform arbitrarily chosen operant behavioral responses to obtain salt, thus indicating a hormone-altered motivational state.

Bidirectionality. The principle that hormones are capable of either increasing or decreasing certain behaviors can be extended to the molecular mechanisms in forebrain neurons underlying such behavior. Watts, at the University of Southern California, reported, for example, that the adrenal hormone corticosterone can inhibit gene expression for corticotropin-releasing hormone (CRH) by working through glucocorticoid receptors in certain hypothalamic neurons, whereas it can facilitate CRH messenger RNA (mRNA) levels, through mineralocorticoid receptors (MRs), in the face of sustained stress.

A. Ingestive Behaviors

Many of the best understood mechanisms for hormone effects on behavior have to do with sex and with stress. These examples are rife in the chapters to follow. Precisely for that reason, we here emphasize hormonal relations with ingestive behaviors—less frequently considered as part of traditional neuroendocrinology. Ingestive behaviors have been the subject of a wide variety of neuroscientific methods of analysis. They illustrate not only the direct hormone/behavior causal relations that are the subject of this chapter, but also the convergence of several hormonal influences, which will be the subject of Chapter 3.

1. Feeding and Hunger

Since the 1940s, lesion studies in the hypothalamus have suggested that certain areas are involved in feeding and other areas involved in inhibition of feeding. Electrolytic lesioning of the ventromedial hypothalamic nucleus (VMH) produced voracious, aggressive rats that became very fat. Lesioning the adjacent lateral hypothalamic nucleus (LH) induced an anorexic-like state where animals had to be forcefully fed to prevent self-starvation. Lesioning, however, is a crude and often confusing technique. You never know if you have destroyed a center or a pathway or if the brain is compensating. More evidence has shown that there are hormones that stimulate feeding and hormones that inhibit feeding. A seminal discovery at the University of Florida by Kalra (Clark *et al.*, 1984), showed that neuropeptide Y (NPY) is a powerful stimulator of eating and food intake in rats when injected directly into the brain. That by itself is not proof that under normal physiological circumstances NPY is active, but further studies showed that in food-deprived animals, levels of NPY mRNA were very high. Also, in diabetic mice which became obese and in a genetic strain of obese mice (*Ob/Ob*), both of which show hyperphagia, NPY levels are high. The source of the NPY in the brain involved in hunger was located to the arcuate nucleus (ARC) (Fig. 1.2).

Thus, NPY initiates the complex behavior of feeding. While this experimental approach revealed the importance of NPY in stimulating the

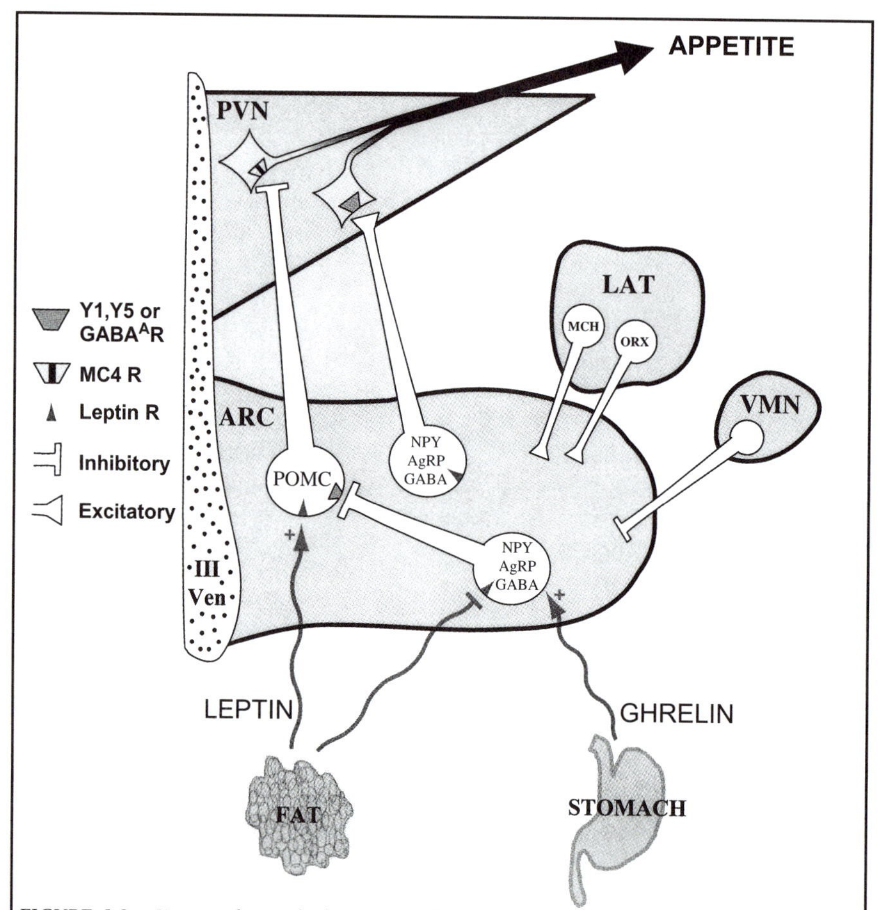

FIGURE 1.2. Hormonal signals from stomach and fat impact neurons of specific chemical identities in the arcuate nucleus of the hypothalamus (ARC), lying next to the third ventricle (III Ven). These neurons, in turn, influence downstream neurons in the paraventricular nucleus (PVN) that facilitate food intake. A network of neuropeptide-expressing and neuropeptide-receiving neurons in the hypothalamus regulates appetitive behavior. The orexigenic peptides, neuropeptide Y (NPY) and agouti-related peptide (AgRP) along with γ-aminobutyric acid (GABA) are coproduced by neurons in the arcuate nucleus (ARC) of the hypothalamus. Interspersed among these neurons are neurons that produce anorexigenic peptides, primarily α-melanocyte-stimulating hormone (α-MSH) derived from proopiomelanocortin (POMC). Both these neuronal populations project to the paraventricular nucleus (PVN), where α-MSH binds to MC-4 receptors to inhibit appetite. NPY and GABA released in the PVN activate the Y1/Y5 and GABA-A receptors, respectively, to stimulate appetite. Additionally, NPY released within the ARC nucleus binds to Y1 receptors on POMC neurons to reduce α-MSH synthesis. Thus, NPY stimulates appetite by direct action in the PVN and indirectly by suppressing POMC neuronal activity in the ARC. AgRP, coproduced with NPY, stimulates appetite by binding to MC-4 neurons in the PVN and preventing α-MSH action. Additional neuropeptides such as melanin-concentrating hormone (MCH) and orexins (ORX) from the lateral hypothalamus (LH) enhance

urge to eat, it is just one of many hormones involved in normal everyday food intake. A startling discovery by Friedman and colleagues at Rockefeller University (Halaas *et al.*, 1994) showed that fat cells produce a hormone, leptin, which is essential to maintaining normal body weight. When leptin is completely absent, obesity develops because of overfeeding. Mice without leptin, or the leptin receptor, appear to have a ravenous appetite. Leptin is a true endocrine hormone, as it circulates in the blood and at some point enters the brain. (Because the brain has a blood–brain barrier, the uptake is either by a transport mechanism or through circumventricular organs which have no blood–brain barrier.) Leptin inhibits feeding by inhibiting the release of neuropeptide Y. Leptin acts on specific leptin receptors in the hypothalamus. Leptin receptors are particularly concentrated in the arcuate nucleus of the hypothalamus, which is where the NPY-containing cells are also found (Fig. 1.2). Thus, leptin is a hormone that monitors adipose tissue mass. Adipose cells increase in size with accumulation of fat. When there is too much fat, these cells secrete leptin into the blood which inhibits NPY in the brain and reduces feeding. Under normal circumstances, reduced food intake decreases the amount of fat. Leptin is a 167-amino-acid peptide product of the obesity gene (*Lep*). Leptin secretion also is stimulated by insulin and glucocorticoids and inhibited by testosterone and beta-adrenergic agonists.

Since that important discovery, many other peptides and hormones have been discovered to have a role in feeding (Fig. 1.2). Those peptides that increase feeding are known as *orexigenic* peptides. These include NPY, ghrelin, melanocortin, melanin-concentrating hormone (MCH), the receptors of MCH (MCHR), $PYY_{3\text{-}36}$, galanin, and agouti-regulated protein (AGRP). Peptides that inhibit feeding are *anorectic* or *anorexigenic*. The anorexigenic peptides include cholecystokinin (CCK), melatonin, neurotensin, alpha-melanocyte-stimulating hormone (α-MSH), and peptides that bind to DNA and are called cocaine- and amphetamine-regulated transcripts (CART). In this chapter and several later chapters, we will come back to these hormones.

Figure 1.2. (*Continued.*)
appetite by upregulating the NPY network, and other unidentified signals from the ventromedial nucleus (VMN) inhibit this network. In sum, the appetite regulating axis in the ARC and PVN is modulated by afferent hormonal signals that cross the blood–brain barrier and exert opposing effects. Food intake is suppressed by the adipocyte hormone leptin that conveys signals to the brain regarding the body's energy stores. Under conditions of positive energy balance, leptin binds to its receptors on NPY neurons to suppress NPY synthesis and release, and concomitantly it activates receptors on POMC neurons to enhance α-MSH production. Conversely, hunger is signaled by ghrelin, a hormone secreted by oxyntic cells in the stomach. Blood ghrelin levels rise in conditions of negative energy balance and enhance appetite by activating NPY neuronal activity in the ARC. (Courtesy of P. and S. Kalra, University of Florida.) (See Morton and Schwartz, 2001; Schwartz, 2001.)

2. α-MSH

α-Melanocyte-stimulating hormone is a component of the proopiomelanocortin (POMC) molecule (see Chapter 4), which contains other peptides, including beta-endorphins and adrenocorticotropic hormone (ACTH). The molecule is produced in the arcuate nucleus (ARC) and released from the anterior pituitary gland. α-MSH is a 13-amino-acid peptide that inhibits food intake by activating melanocortin receptors (Mc3R and Mc4R). When these receptors are stimulated, adenylate cyclase is activated to increase intracellular concentrations of cyclic adenosine monophosphate (cAMP). The effect is to dramatically decrease food intake. Agouti-regulated protein (AGRP) is the major antagonist of melanocortin receptors (McRs). Thus, in cells in the ARC, AGRP acting on McR prevents body weight loss. The POMC molecule is found in neurons in the arcuate that also contain the cocaine- and amphetamine-regulated transcript (CART).

3. CART

Cocaine- and amphetamine-regulated transcript (CART) encodes a neuropeptide that also inhibits food intake and promotes weight loss. It has been suggested that the POMC/CART neurons in the arcuate nucleus are activated by hormones that reflect body fat stores, such as leptin and insulin. In rats that are starved, POMC and CART mRNA levels are downregulated. When animals are overfeeding, the levels of these mRNAs are upregulated. Therefore, it has been hypothesized that POMC/CART neurons respond to excess energy and try to reduce it and maintain fat levels. Separate cells in the ARC that synthesize NPY also contain AGRP. This is an orexigenic peptide, and it has been suggested that AGRP/NPY neurons primarily sense energy deficits (Morton and Schwartz, 2001).

4. AGRP

As we have just noted, agouti-related protein (AGRP) is a neuropeptide that binds to Mc3R and Mc4R, melanocortin receptors, but it inhibits them and stimulates food intake and excess weight gain. The name is derived from agouti protein, which is produced in the skin. The peptide causes melanocytes to synthesize a yellow pigment in the hair follicles known as pheomelanin. Mice with this peptide in the skin are the yellow-colored agouti strain. Agouti-related protein is normally expressed in hypothalamus but not in the skin, and it binds to the melanocortin receptors Mc3R and Mc4R on arcuate neurons, where it antagonizes them. Therefore overexpression of AGRP or a lack of Mc4R caused by a mutation leads to increased food intake and to obesity (Nijenhusiu *et al.*, 2003).

5. MCH

Melanin-concentrating hormone (MCH) is a neuropeptide produced in the lateral hypothalamus that receives input directly from the arcuate nucleus. Note that MCH increases feeding, while α-MSH, which is structurally different, inhibits feeding. MCH, when administered centrally, provokes an increase in food intake. MCH mRNA expression is higher in genetically obese (*Ob/Ob*) than in fasting mice. Overexpression of MCH in transgenic mice leads to obesity. An MCH receptor, MCH-R1, is expressed in various brain regions, with the highest concentrations being found in the hypothalamus. MCH-R2 is also localized in the brain and more widely distributed in the cerebral cortex and hippocampus, in addition to the hypothalamus. As feeding is a complex set of behaviors involving both social factors and memory, the involvement of the hippocampus is not surprising.

In addition to the peptides, cytokines such as interleukin-1 (IL-1) and interleukin-6 (IL-6) and tumor necrosis factor (TNF), plus the neurotransmitters serotonin (5HT) and dopamine, are all involved in feeding. The question is how and when do these peptides and hormones work.

6. LepR

The receptors for leptin (LepR) are found on both AGRP and NPY neurons and on POMC/CART neurons. LepR is a cytokine receptor that modulates gene expression through the intracellular signaling pathway known as the Jak/STAT pathway. Jak stands for *Janus kinases*, which phosphorylate signaling proteins, and STAT stands for *signal transducer and activator of transcription*. The STAT pathway (STAT3) leads to gene expression within the cells. The Jak pathway activates an insulin receptor, which in turn stimulates a protein that activates phosphatidylinositol-3-kinase (PI-3K). This enzyme activates adenosine triphosphate (ATP)-sensitive potassium channels in the neuron plasma membrane. This is important because, through the Jak/STAT pathway, leptin is linked with insulin, and they have become known as *adiposity hormones*. That is, they detect when there is too much fat or too little and help maintain a constant body weight for many years.

7. Insulin

This is the peptide released from the pancreas when glucose levels increase during feeding. The primary role of insulin is to stimulate uptake of glucose from the blood into various tissues. In type 1 diabetes mellitus, insulin is absent because Islet cells in the pancreas which synthesize insulin have been destroyed, probably by an autoimmune action. In type 2 diabetes, insulin is present and frequently levels are high, but patients have developed a resistance to insulin (*i.e.*, they do not respond to it). The various explanations for this resistance include dysfunctional receptors, fatty acids that interfere with binding, or the absence of adiponectin. The insulin receptor is different from the leptin receptor, but it activates tyrosine kinase, which phosphorylates the insulin receptor substrate (IRS) protein. As stated earlier, the link between leptin and insulin is that phosphorylation of the IRS protein activates PI-3K, which in turn activates the ATP-sensitive K channel (K_{ATP}). This results in a hyperpolarization and causes an inactivation of the neuron. Because insulin is released when glucose levels are increased, insulin is a sensor for energy. Thus, when energy levels are high, insulin use and the AGRP/NPY neurons in the ARC are inhibited. The connection of insulin and feeding has been further established by Barsh and Schwartz (2002), who have proposed that reduced PI-3K signaling is a key factor that underlies insulin resistance and might be the key to the relationship between type 2 diabetes and obesity. The theory would be that the lack of inhibition of the K_{ATP} channel in the AGRP/NPY cell by insulin resistance and/or leptin resistance would induce more feeding behavior, thus leading to obesity.

8. Other Factors

Feeding behaviors not only respond to energy expenditure or metabolism (Fig. 1.3), energy storage, or the amount of fat in cells but also invoke the motor functions to forage or hunt for food and try to activate the reward system that motivates the behavior. A simple but brilliant experiment by Olds and Milner in 1961 showed that rats will work to directly stimulate their brains electrically through implanted electrodes. In their study, the rat's work involved pressing levers or running down alleys in a maze for a half-second buzz of electricity in the hypothalamus. What was impressive about this finding was how compelling the behavior was. Hungry rats would ignore food in order to receive this stimulation. They would even run over an electrified grid. Olds and Milner suggested that the electrodes tapped into the very substrate that gives feeding, and all other positive behaviors, reward or satisfaction. At first, researchers sought out a magic "pleasure center" in the brain. Numerous experiments eventually revealed the presence of a neuronal circuit, which, when stimulated, is dopaminergic, involving the hypothalamic nuclei and the nucleus accumbens. Feeding is obviously rewarding; therefore, we might expect that dopamine is involved in the pleasure of feeding. Some evidence comes from Palmiter and colleagues (the creators of gene knockout mice lacking the rate-limiting enzyme for

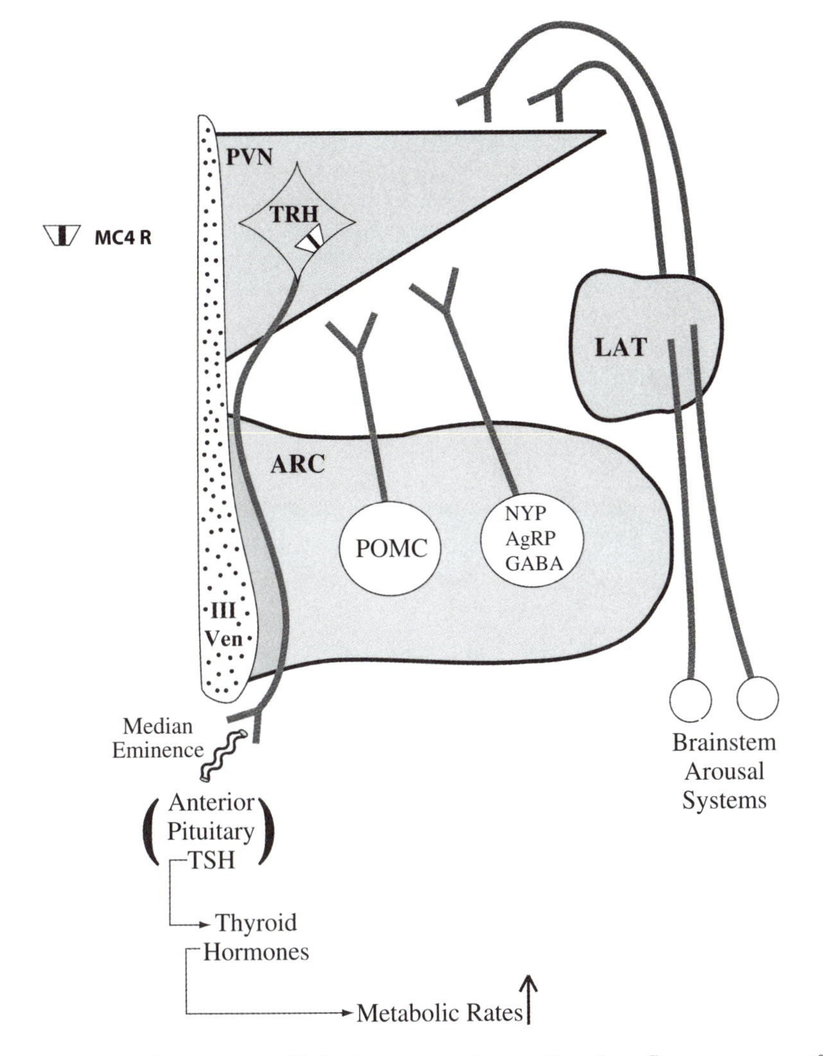

FIGURE 1.3. The activation of behavior, in general, as well as the influences over specific behaviors related to food depend not only on the *input* of metabolic energy but also on the *expenditure* of metabolic energy. Shown here are some of the neural inputs to the paraventricular nucleus (PVN) that influence nerve cells which secrete TRH (thyroid-stimulating hormone-releasing hormone). In addition to interesting electrophysiological actions on other neurons (not shown here), the major action of TRH is in the pituitary, where it releases TSH and activates the thyroid gland. That is, TRH is released in the median eminence to travel down the portal vessels and cause the release of TSH. Throughout the body, thyroid hormones increase metabolic rates by both genomic (see Chapter 18) and non-genomic (see Fig. 16.4) mechanisms. In doing so, they prepare the body for the initiation of vigorous behaviors. (Modified from Barsh, G.S. and Schwartz, M.W., *Nat. Rev. Genet.*, 3(8): 589–600, 2002.)

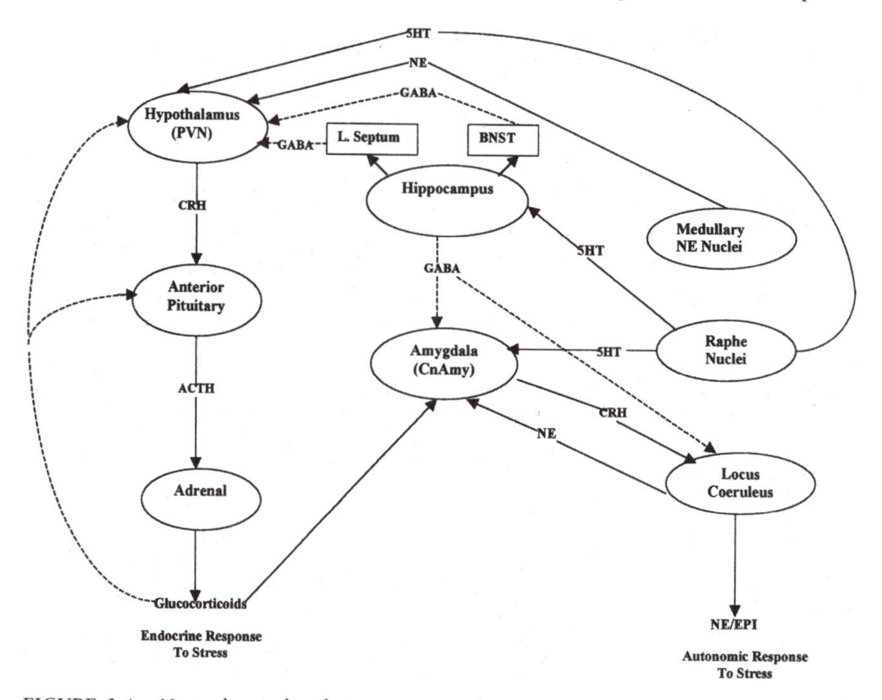

FIGURE 1.4. Neurochemical and neuroanatomical systems involved in stress responses. Key are the pathways involving corticotrophic-releasing hormone (CRH) originating in the paraventricular nucleus of the hypothalamus (PVN) and the central nucleus of the amygdala (CnAmy). Notice the two main outputs: endocrine responses and autonomic nervous system responses. See Section II, Clinical Examples. (Modified from Kaufman, J. *et al.*, *Biol. Psychiatry*, 48: 778–790, 2000.)

making dopamine), who have suggested that human patients with Parkinson's disease, caused by loss of dopamine neurons in the substantia nigra, suffer from a lack of motivation to eat. The patients lose weight and are thin. Experiments with dopamine-deficient mice have shown that hypoactivity and reduced feeding could be reversed with the dopamine precursor, l-DOPA, which can be ingested and then enters the brain. It is the normal treatment for patients with Parkinson's disease in the early stages. Zhou and Palmiter (1995) hypothesized that the dopamine effects on feeding were due to a different location than the dopamine effects on motor activity. They suggested the nucleus accumbens, which we have noted is the area that studies on the reward system have indicated is the major source of dopamine for reinforcing behavior. If motivation or reward is lacking, then there is no incentive to eat. Dopamine-deficient mice will starve to death.

B. Motivation For Feeding and Drinking

All of these data raise an underlying issue that will receive scant attention in a textbook that primarily treats concrete, objectively defined, specific, hormone-influenced behaviors; that is, the ability of hormones to change the frequency and intensity of a behavioral response when all other conditions are held constant proves the existence of an underlying variable: hormone-dependent motivation. Phillips and Olds proposed a theory based on the electrical recording of single cells in the brains of freely moving rats. The rats were conditioned to remain motionless when a light came on. This would be followed two seconds later by either a signal telling the animal it would get food or a signal telling the animal it would not get food. When rats were hungry, they responded to the signal informing them that they would get food with increased neuronal activity in the thalamus. If, however, the signals were changed so that the food signal now became a signal for water, the response was reduced or absent. When the animals were thirsty and not hungry, then they responded to the signal for water with increased neuronal activity in the thalamus. Thus, the significance of the sound in the brain depended on the motivational state of the animal. This experiment was done before we had knowledge of the thirst-inducing effects of angiotensin and the feeding-inducing effects of NPY and ghrelin. Nevertheless, the model for this behavior is clear. Hormones change the motivational state. As we have seen, many of these effects occur in the hypothalamus. Signals (smell, taste, auditory, or light) are received through sensory pathways that analyze them only as sensory modalities. In the thalamus, however, the signals become integrated with the activity developed by hormones in the hypothalamus. In this way, a neutral 10-Hz frequency sound becomes a signal for either food or water. Depending on the animal's state and, therefore, the hypothalamic neuronal activity responding to that state through certain hormonal levels, the signal takes on meaning. A signal with significance or meaning is essential to conscious thought. Only the significant signals are integrated in this way and responded to in hormone-driven behavior.

C. Meal Size

Feeding behavior is not constant. We are conditioned to expect meals at certain times of the day whether we really need them or not. As that time gets closer, cells in the stomach secrete ghrelin. Ghrelin is a 28-amino-acid peptide produced in the stomach that induces the sensation of hunger by activating NPY/AGRP neurons in the arcuate nucleus (see Chapter 3). When food is eaten and reaches the stomach, the levels of ghrelin go down and appetite is suppressed. With feeding, glucose in the blood increases and signals the release of insulin from islet cells of the pancreas. Insulin, through an action on insulin receptors, stimulates cells to take up glucose and the liver to convert excess glucose into glucagon and fat. The fat is stored in the adipose cells. Once food reaches the small intestine, cholecystokinin (CCK) is released into the blood and inhibits eating. A relative of NPY, PYY_{3-36} is released by cells in the digestive tract in response to food. This peptide also inhibits feeding. Again, the mechanism for these satiety signals is a feedback on the hypothalamic peptides in the brain. Thus, ghrelin, CCK, and PYY_{3-36} regulate meal feeding; however, their effects are either short term or long term. Over the long term, energy balance involves leptin and insulin. When energy levels are low and leptin levels are low, there is an increase in NPY, galanin, and AGRP release in the hypothalamus to induce hunger and food-seeking behavior. When the body has too much fat and stores of glucose are readily available, there is an increase of α-MSH, CART, and neurotensin to reduce appetite. For the short term, appetite is increased, NPY/ARGP induces feeding, and the food releases CCK from the gut. CCK is released from the duodenum and small bowel in response to lipids, and it acts on the stomach by causing neural signals to activate the vagus nerve. The vagus nerve sends action potentials to the nucleus tractus solitarius (NTS) to inhibit feeding.

To maintain energy balance, all of these hormones must work together. Energy balance has a simple formula: The number of calories taken in should equal the number of calories burned. People can either eat more calories and work or exercise those calories away, or they can reduce their intake. However, in our society, we have a much greater tendency to take in more calories than we burn so the incidence of obesity is rising. One might think that an effective approach to reducing obesity would be to give leptin to reduce appetite; however, doing so has not proven to be effective. One problem is that leptin appears to develop a resistance. Leptin in the plasma becomes ineffective in overweight individuals. This situation is very similar to the insulin resistance of type 2 diabetes. Although the mechanism for resistance is not fully understood, it probably relates to a decrease in receptor number and receptor signal transduction and to one or more of the following: fatty acids, TNFα, resistin, and adiponectin. The resistance to leptin explains why it is so difficult to maintain weight loss. High levels of leptin in overweight individuals become constant without reducing the urge to eat or increasing energy output; consequently, the orexigenic peptides are not inhibited. Even in normal-weight individuals on diets, the orexigenic peptide are not inhibited.

In summary, ingestive behaviors show clearly how a variety of hormones can influence several parameters related to food intake in experimental animals. The potential importance of these phenomena to modern medicine is quite obvious.

II. CLINICAL EXAMPLES

It is time to consider here just a few examples of the hormone/behavior phenomena already shown to play a part in clinical practice.

A. CRH can Promote Anxiety-Related Behaviors

Corticotropin-releasing hormone (CRH), produced in the hypothalamus and released into the pituitary portal circulation, is the main stimulus to ACTH secretion by the anterior pituitary. This, however, is not its only function in the brain (Fig. 1.4). For example, CRH also is synthesized by neurons with cell bodies in the amygdala and axons that project to the locus ceruleus in the pons. The locus ceruleus contains about half the norepinephrine (NE)-secreting neurons in the entire CNS, and CRH can activate these neurons. The projections of these NE-secreting neurons are widespread—to the hypothalamus, thalamus, limbic system, and cerebral cortex—and they are involved in orienting and attentional responses (see Chapter 2).

Abnormally increased and prolonged activation of the NE-releasing neurons of the locus ceruleus has been implicated in the production of anxiety states, a component of which may be inappropriately heightened arousal and vigilance. CRH acts in other brain areas as well, so that it is likely that the anxiogenic effects of CRH administration to experimental animals involve multiple sites of action. The relative specificity of CRH in anxiety production can be demonstrated in laboratory rats by the anxiolytic effects of CRH receptor antagonists under several conditions: (1) when exogenous CRH is administered, (2) when rats are subjected to experimental stress, and (3) when rats are bred to show innate heightened anxiety-like behaviors and to have a hyperactive hypothalamo–pituitary–adrenal cortical (HPA) axis.

Because CRH has anxiogenic effects, in addition to its hormonal effects, because 30 to 50% of psychiatric patients with major depression have increased HPA axis activity, and because about a third of such patients also have anxiety syndromes, CRH has been proposed as an etiological hormone in the pathogenesis of major depression. Consequently, drug development efforts for the treatment of depressed patients have focused on CRH antagonists. As with most new drugs, the effectiveness of CRH antagonists in the treatment of major depression so far has been modest, and some troubling side effects have led to the discontinuation of trials. Efforts toward the development of safer and more effective CRH antagonists for use in depressed patients are continuing.

B. Oxytocin Promotes Affiliative Behaviors

In humans, as in laboratory animals, social tendencies include a wide variety of courtship and reproductive behaviors, which further the survival of the species; aggressive behaviors, especially for food acquisition and protection of offspring; communicative responses, which convey emotional and behavioral intent; and pure affiliation, which provides emotional support and promotes friendly synergy within a group. Pair bonding is a complex, affiliative behavior that serves several purposes, including reproduction, food acquisition, protection of offspring, and, certainly in humans and likely in other species, emotional support. The antecedents of pair bonding take many forms in different species, from elaborate plumage and nest displays in birds to arranged marriages and the "social scene" in humans. The higher on the evolutionary ladder the species is, the greater the role societal factors play in pair bonding.

Many hormones help establish the physiological substrates of affiliative behavior. For example, the two posterior pituitary peptides, oxytocin (OT) and vasopressin, are secreted during social and sexual interactions, situations that can result in pair bonding. In laboratory rats, oxytocin has been shown to facilitate exploration and approach to novelty, positive social behaviors, onset of pair bonding in adults, and mother–infant attachment (see Chapter 2). Also, it decreases responses to stress and pain, which also facilitates affiliation. The behavioral effects of oxytocin are supported by many other hormones, from those that promote general metabolic balance, (*e.g.*, thyroid hormones) to those that contribute specifically to pair-bonding behaviors (*e.g.*, the gonadal [sex] steroids).

Pair bonding may be extremely strong in some subprimate species in which the influence of hormones predominates—for example, prairie voles, which form tight families comprised of a breeding pair and their offspring. An interesting sex difference in the effects of the posterior pituitary hormones can be found in these animals: In males, vasopressin, but not oxytocin, administration will induce pair bonding in the absence of mating, and in females oxytocin, but not vasopressin, administration will induce pair bonding. Incest and other threats to the communal family are avoided by sexual suppression of male offspring, which must leave the nest to reproduce, and by the onset of estrus in females only after a pair bond is formed. In contrast, close relatives of the prairie vole, montane voles and meadow voles, show little pair bonding. Males and females live in separate nests and tend to be polygamous. One of many articles extending these considerations to normal human social behaviors has been presented by Taylor and Klein and their colleagues in *Psychological Reviews* (2001).

C. Neuroactive Steroids can Both Antagonize and Promote Anxiety-Like Behaviors

Well-estabiished functions of steroids are to bind to intracellular receptors and influence genomic activity (see Chapters 16 to 18). As well, in the CNS some steroids also bind to and modulate the activity of traditional neurotransmitter receptors. These hormones can be synthesized elsewhere in the body and transported to the CNS (neuroactive steroids) and/or synthesized within the CNS itself (neurosteroids). CNS receptors affected by neuroactive steroids include the γ-aminobutyric acid (GABA) receptor family, and *N*-methyl-D-aspartate (NMDA), glycine, nicotinic cholinergic, and serotonin-3 (5HT3) receptors.

Progesterone, an ovarian steroid produced primarily during late proestrus and estrus in the rodent and during the luteal phase of the primate menstrual cycle, has prominent inhibitory effects on CNS activity. Behaviorally, depending on hormone concentration, these include suppression of anxiety-related behavior, anticonvulsant activity, sedation, analgesia, and anesthesia. These properties of progesterone are due to one of its metabolites, allopregnanolone, which does not bind to progesterone receptors but does bind to subunits of the GABA-A receptor, changing its conformation (allosteric modulation) and thereby influencing its function. GABA is a major inhibitory neurotransmitter in the mature CNS. The GABA-A receptor is a ligand-gated chloride ion channel, and allopregnanolone enhances its function by prolonging the duration of channel opening induced by GABA.

The physiological effects of allopregnanolone in the CNS are similar to those of barbiturates and benzodiazepines and are manifest under normal circumstances. For example, during the rat estrus cycle and during pregnancy, when allopregnanolone levels are high, responses to anxiogenic stimuli are reduced. Conversely, during the late luteal phase in women, when progesterone and allopregnanolone levels are decreasing, anxiety and dysphoria can increase, resulting in premenstrual syndrome in certain individuals. In patients with epilepsy, increased CNS irritability in the late luteal phase also can result in an increased susceptibility to seizures prior to menses.

Neuroactive steroids can modulate not only the inhibitory GABA-A receptor, but excitatory receptors as well. The excitatory NMDA receptor is gated by glutamate; is permeable to sodium, potassium, and calcium; and contributes to establishing long-term potentiation of neuronal circuits in the cerebral cortex and hippocampus. The sulfated metabolite of allopregnanolone modulates the NMDA receptor by inhibiting receptor activity, which is synergistic with the enhancing effect of allopregnanolone itself on the inhibitory GABA-A receptor.

In contrast, the neurosteroid pregnenolone sulfate has opposite effects: It enhances excitatory NMDA-channel activity and attenuates both GABA-receptor and glycine-receptor activity, glycine being another major inhibitory neurotransmitter in the CNS. In rodents, pregnenolone sulfate is behaviorally activating, enhancing cognition and improving learning and memory.

The distinction between neuroactive steroids and neurosteroids is not just semantic; rather, it has important physiological implications. Neuroactive steroids, produced outside the CNS but readily crossing the blood–brain barrier because of their lipid solubility, are distributed to a similar degree throughout the CNS. Neurosteroids, on the other hand, because they are locally synthesized within the CNS, can be concentrated in certain brain areas, potentially having a considerably greater specificity of targeted activity. Allopregnanolone, for example, is both a neuroactive steroid and a neurosteroid. The net modulating effects of neuroactive steroids and neurosteroids on CNS excitatory and inhibitory neurotransmission, therefore, are both hormone-dependent and site-dependent.

D. Thyroid Hormones

Effects of thyroid hormones on human behavior have been noted for more than a century. As reviewed by Whybrow (2002), high levels of thyroxine cause nervousness, irritability, impaired concentration, and insomnia. In contrast, levels that are too low are associated with depression and cognitive impairments (Fig. 1.5). We conclude that normal levels of thyroid hormone support widespread activation of neural circuits important for broad categories of both emotional and cognitive functions. The actions of thyroid versus sex hormones are examined in Chapter 3, thyroid hormone metabolites are discussed in Chapter 4, and the clinical implications of hyper- and hypothyroid states are covered more extensively in Chapter 5.

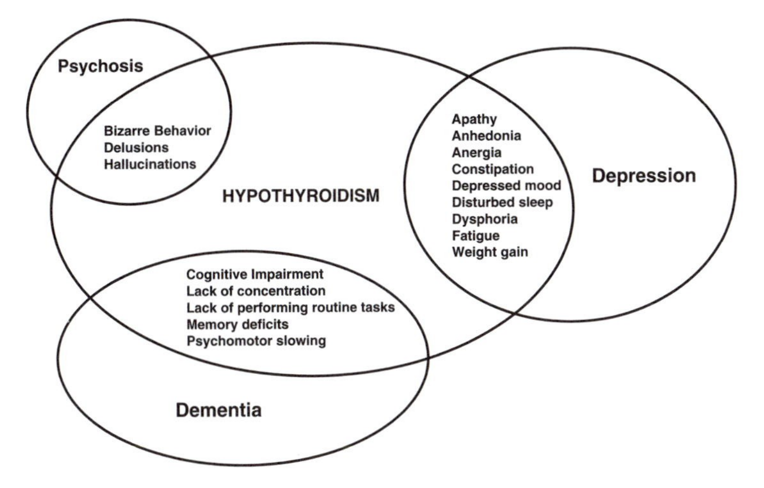

FIGURE 1.5. Most common psychiatric syndromes in hypothyroidism. (From Bauer, M. and Whybrow, P.C. (2002). Thyroid Hormone, Brain, and Behavior. In: Hormones, Brain, and Behavior, Vol. 2, ch. 21, Pfaff, D.W. *et al.*, ed. San Diego: Elsevier, pp. 245. With permission.)

E. Differential CNS Sensitivity to Changes in Sex Steroids May Underlie Premenstrual Dysphoria in Women

In some women, negative (dysphoric) emotions and negative behaviors and social interactions can occur for 7 to 10 days prior to menses and during menstruation itself. There also can be painful physical changes, including breast swelling, uterine cramping, and fluid retention. If this occurs with most menstrual cycles for at least a year and involves a variety of symptoms including depression, anxiety, mood swings, irritability, fatigability, appetite and sleep changes, and physical changes, it can be formally diagnosed as *premenstrual dysphoric disorder (PMDD)*, and treatment is recommended. The assumption has been that, because the ovary undergoes profound changes in its production of estrogen and progesterone across the menstrual cycle, these hormones must be intimately involved in the etiology of this syndrome.

Many studies have examined basal and stimulated activity levels of the hypothalamo–pituitary–gonadal (HPG) axis in women affected by PMDD. Studies of circulating estrogen and progesterone have been quite variable, in sum indicating that there is no excessive or deficient secretion of either hormone or testosterone (converted in women in small amounts from adrenal androgens) in patients compared to normal women. There may, however, be a positive correlation between severity of premenstrual symptoms and circulating female sex steroids (although they are still within the normal range).

More recent studies have considered other possibilities. The rate of change of gonadal steroids in the late luteal phase (*e.g.*, progesterone decline) may be important—depressive symptoms become more frequent in girls than in boys at the time of puberty, when major sex hormone changes occur, and there is a predominance of cyclic mood disorders in women throughout adult life. Studies in which similar hormone changes have been experimentally produced in women with and without premenstrual dysphoria indicate that those without antecedent premenstrual complaints do not develop such complaints when their sex steroids are experimentally altered, whereas women who do have antecedent premenstrual dysphoria develop similar complaints with experimental alteration of their hormones. These findings suggest a heightened sensitivity of the CNS to changes in circulating hormones in women with premenstrual dysphoria.

Another consideration, and fitting with the differential CNS sensitivity hypothesis, is the role of neuroactive steroids, given that they have a predominantly inhibitory effect on CNS excitability via GABA-A receptor modulation, as discussed above. Studies of circulating concentrations of, for example, allopregnanolone, however, have been conflicting, with both higher and lower concentrations reported in premenstrually dysphoric versus normal women. Again, the rate of change may be an important factor. And, a potentially important influence on CNS sensitivity to gonadal hormone changes may be CNS-produced neurosteroids, which, as indicated above, may be concentrated differently in local CNS areas important in the mediation of affect and mood.

Finally, the role of sociocultural factors must be considered in premenstrual dysphoria. Some symptoms, such as sleep and appetite disturbances, may be relatively environmentally independent, whereas others, such as depression and irritability, can be provoked by interpersonal and other environmental stressors. Given that treatment of this major cause of discomfort in women remains empirical, attention to social as well as to biological factors during treatment is quite important.

F. Obesity

Returning to the subject of food intake, as introduced earlier with experimental physiological material, mutation in the leptin gene, or the POMC gene, leads to severe or early onset of obesity in humans. The mutation results in the perception of starvation; therefore, the patients have a ravenous appetite even though their body mass index (BMI) clearly indicates they do not need the food. A proportion of obese people gain their weight by periodic binges. While this has been thought of as a loss of will power for those on a diet, evidence now suggests that it could be due to a mutation of the melanocortin-4 receptor (Mc4R). In a study of 469 obese patients 5% had the mutated gene for Mc4R, and all were binge eaters (Branson *et al.*, 2003).

Many drugs have been developed as anti-appetite drugs. Amphetamines have been in use for a long time. These act by increasing wakefulness and inducing hyperactivity by elevating catecholamines. Catecholamines, such as norepinephrine or epinephrine, increase tissue metabolism; therefore, amphetamines burn up more fuel but do not produce long-term control of feeding. Drugs that increase the activity of serotonin (5-hydroxyitryptamine; 5HT) were once widely prescribed. Millions of people used d-fenfluramine (d-Fen) or phentermine, or a combination of the two (Phen-Fen), to achieve weight loss until Phen-Fen was withdrawn in 1997 by the Food and Drug Administration because of reports of cardiac complications. These drugs increased 5HT, by blocking its reuptake and stimulating its release. The target neurons for d-Fen action are those that receive direct input from the dorsal raphe nucleus, the source of 5HT in the brain. Sibutamine, a 5HT uptake inhibitor, is still used to treat obesity and is used in combination with orlistat, which inhibits fat absorption in the gut.

The World Health Organization lists obesity among the top 10 global health risks. Worldwide, one billion people are overweight or obese, including 22 million children younger than the age of 5. In the United States, 65.5% of adults and 15% of children ages 6 to 19 are overweight. While dieting seems like an obvious solution, most people who try dieting lose weight but gain it back. There are many temptations in our society, but the reason why people cannot stop eating is because eating is a fundamental human drive due to the hormones involved.

Obesity in children generally leads to obesity in adulthood. In most cases, children are overfed high-fat, high-sugar diets. They get into a vicious cycle of eating without exercising because being overweight makes exercise and sports more difficult. There are also genetic syndromes and medical conditions that lead to obesity. Examples of obesity induced by genetic syndromes include Prader–Willi syndrome (PWS) and Bardet–Biedl syndrome, where early obesity is seen in children. PWS is a genetic disorder occurring in 1 out of 10,000 live births. It is characterized by excessive appetite and progressive obesity. It also affects behavior in that it leads to mental retardation and, generally, the patients have delayed puberty and short stature. The characteristic hormone deficiency is in growth hormone (GH) and hypogonadotropic hormones, including luteinizing hormone (LH), follicle-stimulating hormone (FSH), testosterone, and estrogen. Ghrelin is elevated in subjects with PWS. Del Parigi *et al.* (2002) showed that ghrelin levels in PWS was greater than in normal subjects, fasted for 36 hours. Endocrine disorders that contribute to obesity are seen in Cushing's disease and in hypothyroidism. Many new genes, called obesity genes, have been recently discovered in mice, but the role they play in humans has yet to be defined.

Obesity is becoming more and more predominant in our society, and worldwide. Obesity is a risk factor for type 1 diabetes mellitus, heart disease, stroke, and hypertension. One reason why it is so difficult to control is that it is a hormonal behavioral problem. Over 25 endocrine pathways have been identified as playing a role in the regulation of energy and weight homeostasis.

While the theory of leptin inhibiting NPY in the hypothalamus to reduce eating would suggest that obese people lack leptin, they in fact have abundant leptin in the blood. Leptin synthesized in adipose cells should switch off when the fat stores are reduced. In dieting, this is what happens, and that leads to more feeding stimulated by the NPY in ghrelin systems. The reason why high levels of leptin are still apparent in obese patients without an inhibition of feeding is because of the resistance to leptin. Thus, in obesity when leptin levels are high, the body develops resistance to leptin and it has no effect on feeding. When leptin levels are low, the feeding mechanisms are released from inhibition, leading to more hunger and eating. In addition, ghrelin levels rise in dieters who lose weight and then try to keep it off. Ghrelin is produced in the stomach, in the oxyntic cells; and induces hunger in one treatment for obesity, stomach stapling, appetites sharply decline, which would suggest that stapling the stomach reduces the amount of ghrelin produced, either by isolating the cells from signals or preventing secretion.

G. Anorexia Nervosa

Anorexia nervosa is an extreme form of weight loss without any apparent organic disorder (see Rubin [2004] for review). It is a psychiatric disorder involving self-starvation and must be differentiated diagnostically from hyperthyroidism, which can cause weight loss due to a high metabolic rate, diabetes mellitus type 1 (where the weight reduction is due to muscle wasting), or other diseases such as inflammatory bowel syndrome, brain tumor, hypothalamic lesions, and depression. In anorexia nervosa, weight loss leads to hypogonadism and primary and secondary amenorrhea. Young women show this syndrome much more frequently than men. Patients have low estradiol and low progesterone levels. Anorexic patients are characterized by continuous efforts to avoid food and a false perception of body size. Patients believe they are fatter than they are or that they are becoming fatter.

The condition is most frequent in adolescent females. The consequences of continuous self-starvation can be fatal.

Anorexia nervosa is not necessarily the result of a loss of appetite. Patients have increased plasma levels of ghrelin; consequently, anorexic patients will sometimes feed voraciously (bulimia nervosa) but then make efforts to prevent food absorption by vomiting or excessive use of laxatives. Because leptin is an anorexic hormone, studies are ongoing to establish a connection between leptin and anorexia nervosa. It is unlikely that the source of the leptin is fat cells, as the patients are so thin, but leptin may be produced in the hypothalamus. Numerous studies have implicated obsessive–compulsive disorder (OCD) with anorexia nervosa. The link between these two morbidities is serotonin. High levels of serotonin have been correlated with OCD (Barbarich, 2002). Anorexia nervosa is accompanied frequently by osteoporosis. This can be explained in terms of hormonal changes, particularly the reduction of estrogen. The lack of nutrition may stimulate a vicious cycle of low vitamin D intake, low calcium, and low estrogen with hypercortisolemia. Some patients are literally "dying to be thin."

III. OUTSTANDING NEW BASIC OR CLINICAL QUESTIONS

As the above examples indicate, individual hormones can have facilitating or inhibiting effects on behavior, depending on their concentrations and locations in the CNS and, especially, on the overall hormonal milieu in which they exert their actions. And, as pointed out below, most groups of hormones with related metabolic effects are secreted in an orderly fashion, particularly with respect to their time courses. Of the many remaining questions, the questions provided below could be mentioned. Students may want to consider these questions during extra reading for term papers. The authors welcome hearing about additional questions from teachers and students as they occur.

1. Which characteristics of environmental stimulation, as well as environmental stress (*e.g.*, their type, intensity, duration, etc.), influence the balance achieved between hormonal activation of a given behavior versus its repression?
2. How do stimulation and stress alter hormonal activity and hormonal metabolism in discrete areas of the CNS? (The same question applies for hormone receptors.)
3. In those discrete areas of the CNS where a given hormone can excite or repress a given behavior, is the balance of effects on inhibitory and

excitatory neurotransmitters similar or different? What local characteristics (*e.g.*, receptor densities, receptor subunit compositions, receptor sensitivities to specific neurosteroid concentrations) influence this functional balance?

4. How is the sex difference in posterior pituitary hormone (oxytocin, vasopressin)-induced pair bonding mediated? Is it by sexual dimorphism of hormone receptors resulting from early hormonal influences on receptor development? Is it by modulation of hormone receptors by different concentrations of sex steroids in the adult animal?
5. Is the coordinated response to stress of early secretion of catabolic (energy-providing) hormones, followed by later secretion of anabolic (tissue rebuilding) hormones, similar across a wide range of stress circumstances? Does the overlap of the secretion of these two sets of metabolically influential hormones expand as a stress becomes less acute and more chronic?
6. How can the contributions of sex steroids and neuroactive steroids to premenstrual dysphoria be assessed in individual women in order to tailor specific therapeutic hormonal interventions to a given patient?
7. More generally, what determines the individuality of human CNS sensitivity to hormone changes ? For example, why do some individuals develop severe mental symptoms with a certain hormone change and others do not?
8. What is the interplay between extrahypothalamic pituitary-hormone-releasing and -inhibiting factor circuits in the CNS with their classical releasing factor effects on downstream hormones (*e.g.*, CRH and anxiety or depression compared to CRH releasing ACTH)?

One Hormone Can have Many Effects: A Single Hormone Can Affect Complex Behaviors

I. BASIC SCIENTIFIC EXAMPLES

Hormones related to stress responses provide a superb system for discussing the multiple effects of a given compound. Consider corticotropin-releasing hormone (CRH), which among other actions, causes the release of adreno-corticotropic hormone (ACTH) from the pituitary gland, which, in turn, among other actions causes the release of glucocorticoids such as cortisol from the adrenal cortex. Not only does CRH have a neuroendocrine effect, acting through the median eminence of the ventral hypothalamus, but it also potentiates a variety of anxiety-related behaviors (see below) and even affects the autonomic nervous system. The effects include increased blood pressure and heart rate, decreased gastric acid secretion and increased colonic motility (reviewed by Valentino and van Bockstaele, *Hormones, Brain and Behavior*, 2002).

Adrenocorticotropic hormone not only has the stated endocrine effect on the cortex of the adrenal gland but also affects fear learning. Most amazing are the steroid hormones themselves, glucocorticoid hormones such as cortisol and corticosterone, which affect body metabolism under stressful,

TABLE 2.1 Effects of Glucocorticoid Hormones

Short-term adaptation	Long-term disruption
Inhibition of sexual motivation	Inhibition of reproduction
Regulate immune system	Suppress immune system
Increase glucogenesis	Promote protein loss
Increase foraging behavior	Suppress growth

Adapted from Wingfield and Romero.

"flight or fight" situations. As summarized by Wingfield and Romero, steroid hormones have different properties according to whether their levels are high briefly or high chronically (Table 2.1). They clearly point the animal's behavior away from activities such as sex and toward activities that will provide energy for metabolic support.

Still another consideration is how different glucocorticoids can elicit responses that are not identical to each other. From a single glucocorticoid receptor gene, different promoter and different mRNA splicing combinations yield proteins that not only would have different affinities for various glucocorticoid ligands but also different DNA response element affinities and different relations with cell nuclear co-activators (see Chapters 16 to 18). Such molecular variations could account not only for differences between natural and synthetic glucocorticoids but also for a variety of responses among individuals. Along these lines, we note that some patients are resistant to cortisol treatment, and this must be explained.

Thyroid hormones, as well, can be cited as having several effects. Besides widespread activation of metabolic rate acting through signal transduction pathways, directly on cell mitochondria and through the cell nucleus, thyroxine has effects on the central nervous system (CNS) and behavior. T4 heightens mood, increases activity (especially fidgeting), can make a person irritable, and decreases food intake.

A. Angiotensin

Another example of a single hormone having many effects is angiotensin II. Angiotensin is an octapeptide; that is, it has eight amino acids. It is found in the blood and in tissues, including the brain. When injected directly into the brain, angiotensin induces drinking behavior. This phenomenon, which was first discovered by Fitzsimmons and colleagues in 1969, is one of the most compelling examples of a hormone-inducing behavior. Within a few seconds of being injected with nanogram amounts of angiotensin in the brain, rats will immediately seek out the waterspout and begin drinking

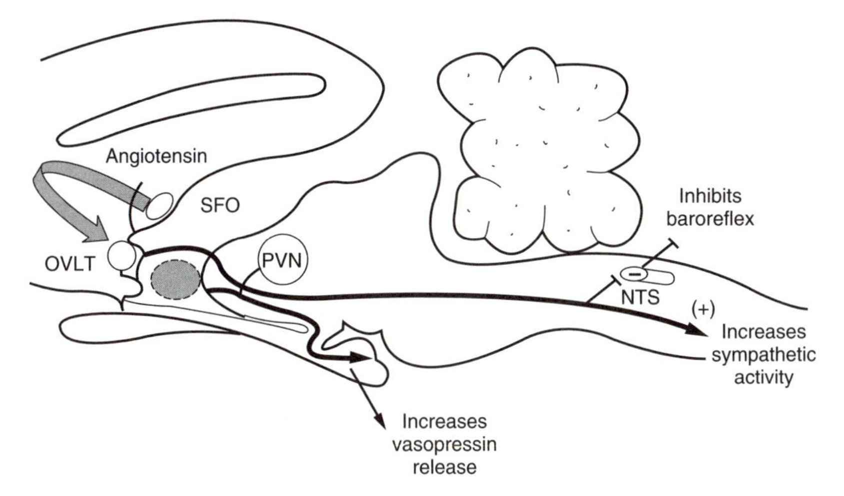

FIGURE 2.1. Angiotensin II has many effects. It acts on the subfornical organ (SFO), on the medial preoptic area (MPO), on the organum vasculosum of the lamina terminalis (OVLT), on the magnocellular nuclei of the anterior hypothalamus, in the paraventricular nucleus (PVN), and in the supraoptic nucleus (SO). As a result, vasopressin is released from the posterior pituitary gland, sympathetic autonomic nervous system activity is increased, and blood pressure rises because of inhibition of the baroreceptor reflex from the nucleus of the tractus solitarius (NTS). (From Phillips, M.I., *Ann. Rev. Physiol.*, 49: 413–435, 1987. With permission.)

continuously for about 15 minutes. In addition to this dramatic form of behavior, the same dose of angiotensin II that induces drinking also stimulates the release of vasopressin, an increase of sympathetic activity, and release of other hormones, including ACTH (Fig. 2.1). Drinking behavior is quite complex. First, there is the coordination of motor movements required to drink, from locomotion and positioning of the body to the source of water to catching the water with the tongue and swallowing it. In addition, the behavior requires memory of a source of water. The initial action of angiotensin is to motivate the animal to drink (*i.e.*, making the animal thirsty). Once the level of motivation is raised, information coming into the brain is converted into significant or nonsignificant information, based on the level of motivation. Angiotensin II stimulates angiotensin receptors in the hypothalamic region, and particularly the areas around the anterior ventral wall of the third ventricle (Phillips, 1987). These receptors are connected by neurons to circuits within the brain that stimulate all of these complex activities leading to thirst and drinking behavior. Attempts to dissect these circuits have been made with lesions, stimulation, microinjections, and even cream plugs to prevent access to sites on the ventricular wall. Lesioning has revealed two areas of significance: the subfornical organ (SFO) and the organum vasculosum of the laminae terminalis (OVLT). Lesioning either or both of these regions inhibits or abolishes the drinking response to central injections of angiotensin II. Electrical stimulation has shown that the SFO is connected neurally to the paraventricular nucleus and the supraoptic nuclei (Ferguson). Both of these nuclei contain vasopressin. When stimulated, these nuclei release vasopressin through neuronal axons coursing down to the posterior pituitary (the neurohypophysis), where vasopressin is released into the blood. Further studies have shown that the sites excited by angiotensin II are also connected to brainstem sites, for example the rostral ventral lateral medulla (RVLM) where sympathetic nerves are stimulated and affect the entire body. The effect is to increase blood pressure by vasoconstricting blood vessels with norepinephrine.

In addition to central injections of angiotensin II, thirst can be stimulated by large injections via the blood (see Fig. 2.2a). However, these concentrations are so large it has always been dubious whether the effects are due to peripheral action or to opening up the blood–brain barrier by increasing blood pressure excessively and allowing angiotensin to enter the brain. In addition, angiotensin can reach the circumventricular organs (CVOs), which have no blood–brain barrier. These include the SFO and OVLT.

During hemorrhage, blood pressure falls and volume falls. The fall in blood pressure increases the release of renin produced by the juxtaglomerular cells of kidney cortex. A cascade occurs in which angiotensin is synthesized by the cleavage of angiotensinogen in the liver to angiotensin II, a decapeptide; also, in the lungs, angiotensin-converting enzymes (ACE) cleaves angiotensin I to angiotensin II, the octapeptide, and from there on various aminopeptidases can break down angiotensin into active metabolites, including angiotensin III and angiotensin IV. One of the behavioral characteristics of hemorrhage is that eventually the patient becomes very thirsty. Whether this is due to high levels of angiotensin circulating in the blood entering the brain directly or to activating angiotensin receptors in the CVOs has not been fully resolved. High levels of circulating angiotensin result in high blood pressure due to its vasoconstrictive effects on blood vessels. High blood pressure could force angiotensin II into the brain by opening the tight junctions in blood vessels that normally keep proteins out. Nevertheless, one can see that all the effects of angiotensin II, whether they are peripheral or central, are related to fluid balance; therefore, we will now try to understand some of these hormone/behavior mechanisms in the context of fluid balance.

B. Fluid Balance

Fluid balance depends on two major principles: maintenance of fluid by osmotic pressure and maintenance of fluid by volume (Fig. 2.3). To

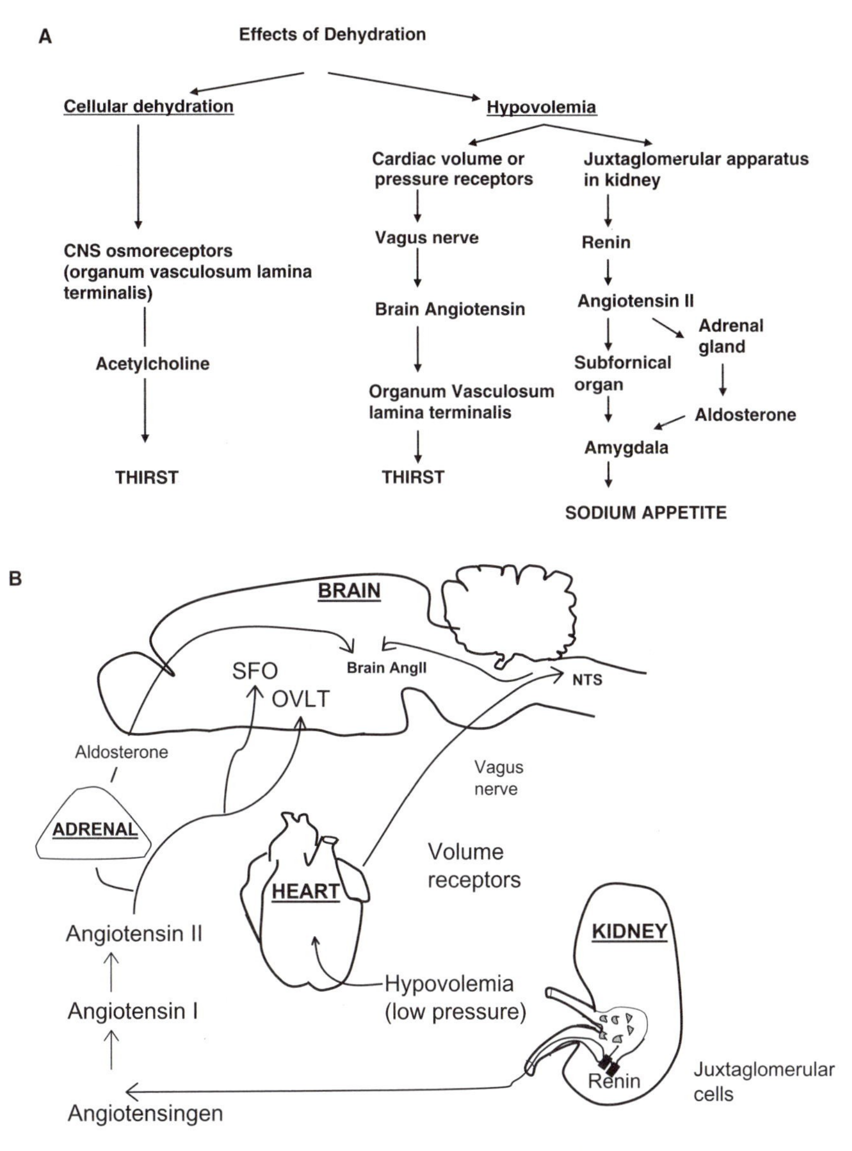

FIGURE 2.2. Hormonal involvement in water deprivation and thirst. (a) A logical chart of causal routes that could stimulate drinking following water deprivation. Diagram shows different major pathways involved for cellular dehydration leading to thirst and drinking, and extracellular dehydration (reduced fluid volume) leading to thirst and sodium appetite. (b) The renin/angiotensin system has many routes of action to increase extracellular thirst and to stimulate salt hunger (Na appetite) following hypovolemia (decreased blood volume; see Fig. 2.3) sensed eventually by the kidneys. Hypovolemia (low volume) leads to increased renin/angiotensin, which has central and peripheral routes of action to increase thirst and to stimulate (salt appetite). The low pressure of reduced blood volume is sensed by the kidneys and by the brain.

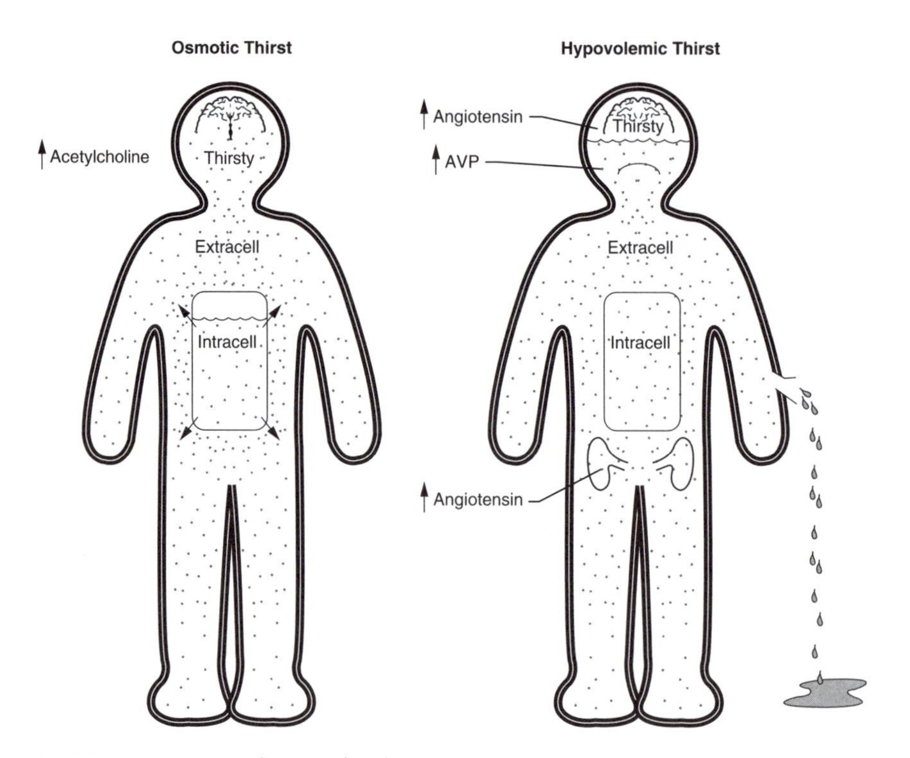

FIGURE 2.3. Two mechanisms for thirst: *Osmotic thirst* results from cellular dehydration without change in volume. Excess salt in the extracellular compartment draws water from the intracellular compartment by osmosis. Acetylcholine in the brain stimulates thirst. *Hypovolemic thirst* results from reduced blood volume (*e.g.*, hemorrhage) and activates renin release (see Fig.2.2).

understand this, imagine a cell surrounded by a high salt solution. By osmosis water will be drawn from the cell into the surrounding compartment and the cell will shrink. When sodium levels are high in the plasma, thirst is induced. Osmotic thirst appears to involve a cholinergic mechanism in which acetylcholine is the principal mediator. Just as with angiotensin, an injection of carbachol (an acetylcholine analog) induces thirst. Thirst leads to drinking and dilution of the high osmolality of the plasma. The kidney will eventually get rid of the high plasma sodium and restore volume balance. According to the volume hypothesis, the osmotic balance between the inside of the cell and the outside compartment is equal. However, blood, which represents some of the fluid outside the cell, can be reduced in volume without shrinking the cell (see Fig. 2.3). When the extracellular volume decreases, renin is released from the kidneys, and the fall in blood pressure causes a second neuropathway to stimulate renin via activation of beta-1 receptors on the juxta glomerular (JG) cells of the kidney. Angiotensin is also activated within the brain due to a reflex action of the baroreceptors in the heart that detect the decreased blood pressure. The pathway into the brain through the vagus nerve reaches its first synapse at the nucleus tractus solitarius (NTS). From this nucleus, afferent input stimulates the paraventricular nucleus (PVN) and supraoptic nucleus (SON) which contain brain angiotensin. Brain angiotensin then stimulates thirst and vasopressin release and activates the sympathetic nervous system. The role of vasopressin is to inhibit the loss of water by the kidney, and the role of angiotensin is to increase water intake by drinking and to increase blood pressure by sympathetic nerve activation and overriding the baroreflex. The baroreflex slows the heart when blood pressure goes up.

Another effect of angiotensin is to induce salt appetite. This again is related to the response to hypovolemia. As we noted above, the main role of angiotensin is to conserve and replenish volume. In addition to drinking, vasopressin release, and vasoconstriction, ingesting sodium would increase the osmotic pressure and pull water out of cells.

In addition to drinking and salt appetite, angiotensin also has been associated with memory and cognitive functions. These may relate to the complexity of drinking behavior and salt appetite behavior. Locating and memorizing sources of water and salt are functions of the hippocampal neurons and their mediation of spatial memory. In line with this, angiotensin is a neuromodulator of hippocampal neuronal firing. (See Chapters 4, 11, 15.)

C. Salt Appetite

The need for salt is very basic to human physiology. Salt (NaCl) is essential for neuronal activity, for contraction of heart and skeletal muscle, and for maintaining body fluid volume. In modern society, the intake of salt is extremely high, far above physiological needs. In societies where salt is absent from the diet or when salt is lost through sweat at a high rate, salt replacement is required. This has led to salt being a valued commodity in the course of human history. In fact, the English word "salary" comes from the word for salt. Physiologically, sodium hunger is also seen in animals, and experiments show that sodium appetite is due to hormones motivating the behavior. When the adrenal glands are removed, the mineralocorticoid hormone, aldosterone, is absent. Normally, aldosterone maintains sodium reabsorption from the kidney. This stops the loss of sodium, but when it is absent, the sodium levels in the blood are greatly reduced, resulting in hyponatremia (see Chapter 3).

The symptoms of hyponatremia include headaches, nausea, apathy, and a hunger for salt. The hormone most associated with the appetite and the behavior of looking for salt is angiotensin II (see Fig. 2.2). When injected into the brain, angiotensin II increases fluid intake (see previous discussion), and, if sodium solutions are available, it will increase sodium intake. If offered a solution high in salt (3–8%), rats injected with angiotensin II will avidly drink the salt solution (Buggy, 1984). These experiments showed that angiotensin is directly involved in the motivating and appetite pathways for sodium and for water ingestion. More involved experiments on the extent to which animals will go to find water or salt after such an injection have shown that the motivation is powerful. Rats will run down mazes or learn complicated procedures to get the water and/or salt; therefore, it was suggested that adrenalectomy, while causing the loss of aldosterone, increases the amount of angiotensin in the brain (Fluharty and Epstein, 1983). In animals with adrenal glands intact but suffering salt deprivation, both aldosterone and angiotensin work together to increase salt ingestion. The adrenal gland has a zone surrounding its perimeter (the zona glomerulosa) which has cells with a high density of angiotensin type 1 receptor (AT_1R). Stimulation of these receptors by angiotensin II brings about the immediate release of aldosterone into the blood.

The exact site of action for angiotensin-induced sodium appetite in the brain may be separate from the sites for water-intake induction. The chief focus of attention has for many years been the circumventricular organs. These are organs that do not contain a blood–brain barrier, and it was supposed that angiotensin in the blood could pass through into the brain through these organs (Fig. 2.4). These include the organum vasculosum of the laminae terminalis (OVLT), the subfornical organ (SFO), and other areas in the wall of the anterior ventral third ventricle (AV3V). A simple notion is that the AV3V region is the site for salt appetite induction by angiotensin, and the SFO is the site for water intake induction (this is based on minute injections of angiotensin into either the SFO or the OVLT). Lesioning the AV3V region abolishes the ingestion of sodium. Both areas, the SFO and the AV3V (which includes the OVLT) are dense with angiotensin type 1 receptors. Thus, the difference in location implies a difference in the neural networks that are stimulated by the AT_1R in these two sites.

We have noted above that in adrenalectomized rats that have no mineralocorticoid hormone (aldosterone), the sodium appetite is probably due to angiotensin in the brain; however, in rats, with intact adrenal glands, aldosterone can also elicit an appetite for sodium. So, we have a situation

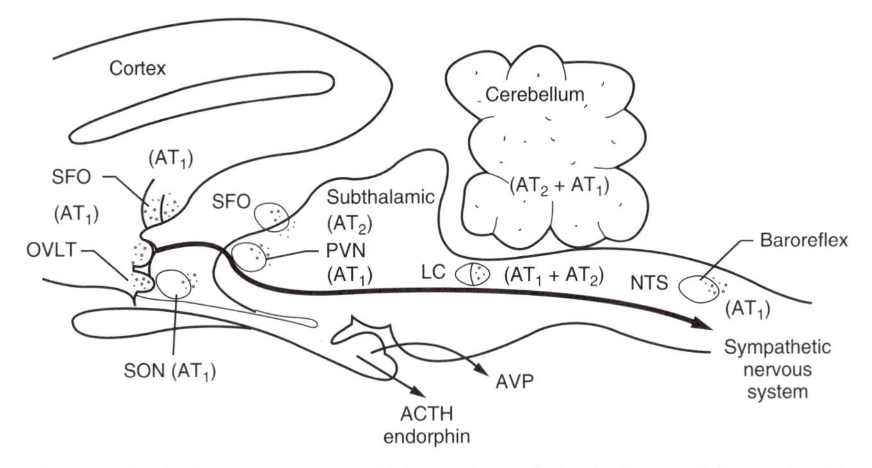

FIGURE 2.4. Angiotensin receptors. Major regions of the brain containing angiotensin receptors. The principle of "one hormone/many actions" is reflected here in multiple receptor sites from basal forebrain through hindbrain. Pictured is a generalized vertebrate brain in sagittal section looking at it from the left side. (From Phillips, M.I. and Sumners, C., *Regul. Pept.*, 78(1–3)1–11, 1998. With permission.)

where the hormone is sufficient but not necessary for inducing a behavior. The reason for this seeming contradiction is that only very high levels of aldosterone induce salt appetite. Aldosterone has many actions related to sodium, in addition to sodium reabsorption in the kidney. Aldosterone redistributes sodium in the salivary glands and the gut and transports sodium from the bone. All of its actions are intended to maintain sodium at physiological levels. Sodium ingestion is a behavioral response to either a need for sodium or a *perceived* need for sodium. Unlike angiotensin, which is a peptide and does not easily cross the blood–brain barrier and may be limited to those circumventricular organs described above, aldosterone is a steroid and therefore lipid soluble. This property of lipid solubility allows it to pass through the blood–brain barrier with ease. Therefore, the mineralocorticoid aldosterone can stimulate multiple sites in the brain wherever there are mineralocorticoid receptors (MRs).

Interestingly, glucocorticoids that interact with glucocorticoid receptors (GRs) in the brain do not elevate sodium ingestion; however, when glucocorticoids are combined with mineralocorticoids, sodium appetite is greatly increased compared to the effect of aldosterone alone (Ma *et al.*, 1993). This is because glucocorticoids (cortisol in humans and corticosterone in rodents) activate both glucocorticoid receptors and mineralocorticoid receptors, whereas aldosterone activates only mineralocorticoid receptors (see also Chapters 3 and 4).

The term for sodium appetite is *natriorexegenia*. This effect clearly depends on the brain. Peripheral blockade of mineralocorticoid receptors does not affect the natriorexegenic action of aldosterone. The hippocampus contains aldosterone receptor sites, and this area is more associated with memory than with appetite; however, it is also known as a target for corticosteroids, particularly elevated steroids during stress. The amygdala is involved in sodium appetite (Schulkin, 2002) and contains both mineralocorticoid receptors and glucocorticoid receptors. A direct injection of a mineralocorticoid antagonist (RU28318) into the medial amygdala abolishes sodium appetite induced by intravenous aldosterone (Sakai *et al.*, 2000).

D. Inhibition of Salt Intake

It is a principle of hormone behavior relationships that if a hormone can induce a behavior, another hormone will inhibit the same behavior. This principle is essential to reach eventual physiological homeostasis, even though there may be a need at one specific time for one behavior or hormone effect. As an illustration of the principle that hormones can both facilitate and repress behavioral responses, centrally administered atrial natriuretic peptide (ANP) decreases sodium appetite in sodium-depleted rats. ANP was originally discovered in the atria of the heart, and its effects were demonstrated by injecting atrial extract into the blood. It induced diuresis (loss of water) and natriuresis (loss of sodium) (deBold, 1985). Since then, related natriuretic peptides have been found in the heart ventricles, and brain (BNP) and a widely distributed CNP. All are endocrine hormones (Fig. 2.6). ANP appears to completely oppose the effects of angiotensin II, both the dipsogenic, vasoconstrictive effects and the natrirexogenic effects. ANP receptors are also found in the circumventricular organs and the AV3V region. ANP reduces angiotensin-induced water drinking; inhibiting ANP with antisense to ANP mRNA (*i.e.*, a specific inhibition of ANP synthesis) increases water drinking in response to angiotensin (Fig. 2.5). This implies that ANP chronically competes with angiotensin II to inhibit the effects of angiotensin. Based on the second messenger action of ANP, it seems likely that ANP opposes the actions of angiotensin receptors at the second messenger level, causing a decrease in cyclic adenosine monophosphate (cAMP) and an increase in cyclic guanosine monophosphate (cGMP).

To understand why these hormones have opposite effects we must refer back to the most basic principle of physiology: homeostasis. The body constantly tries to maintain physiological function within a certain range, whether it is temperature, blood pressure, body weight, or fluid volume. Angiotensin is the primary effector hormone for preventing loss of fluid, and

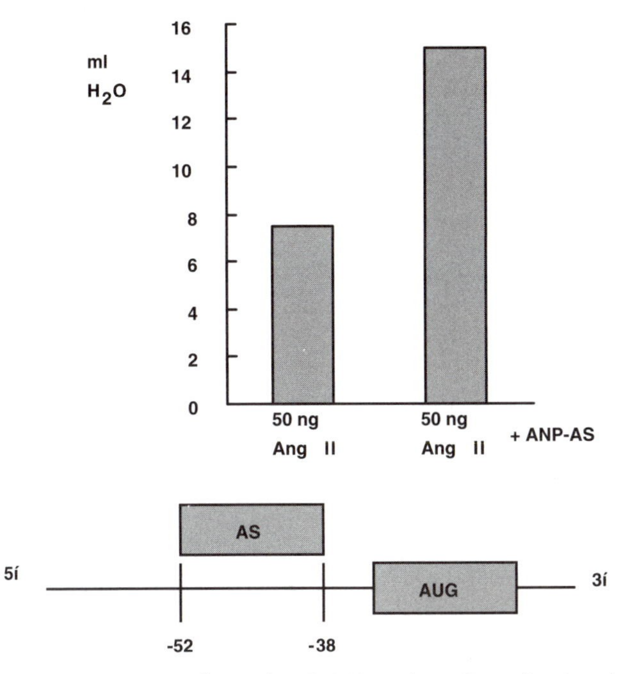

FIGURE 2.5. An antisense DNA oligonucleotide 14 bases long, directed against the mRNA for the ANP peptide just upstream from the translation start site (AUG). Delivered intraventricularly, it significantly increases the drinking of water stimulated by a given dose of angiotensin II. This shows that ANP opposes angiotensin II effects on thirst in the brain. (From Phillips, M.I. and Gyurko, R., *Regul. Pept.*, 59(2): 131–141, 1995. With permission.)

ANP is the primary effector hormone for decreasing fluid volume (Fig. 2.7) (for more on angiotensin, see Chapters 4, 15, and 21).

Oxytocin (OT), operating in a completely different domain, has a number of behavioral effects, in addition to its peripheral reproductive effects. Associated with these reproductive actions is the important OT facilitation of maternal behavior, which in turn provides positive feedback upon central OT release. OT also increases pain thresholds, probably through direct release of oxytocin on the beta-endorphin cells in the arcuate nucleus. In various species, oxytocin is associated with facilitating reproductive and sexual behavior in both males and females and social behaviors in some species (refer to OT discussion in Chapters 1, 3, and 6). For example, in prairie voles, OT is necessary for pair bonding. Infusions of OT increased the preference for social partnering and parental behavior. Showing the full behavioral power of the oxytocin gene required the use of a seminatural environment that would reveal the

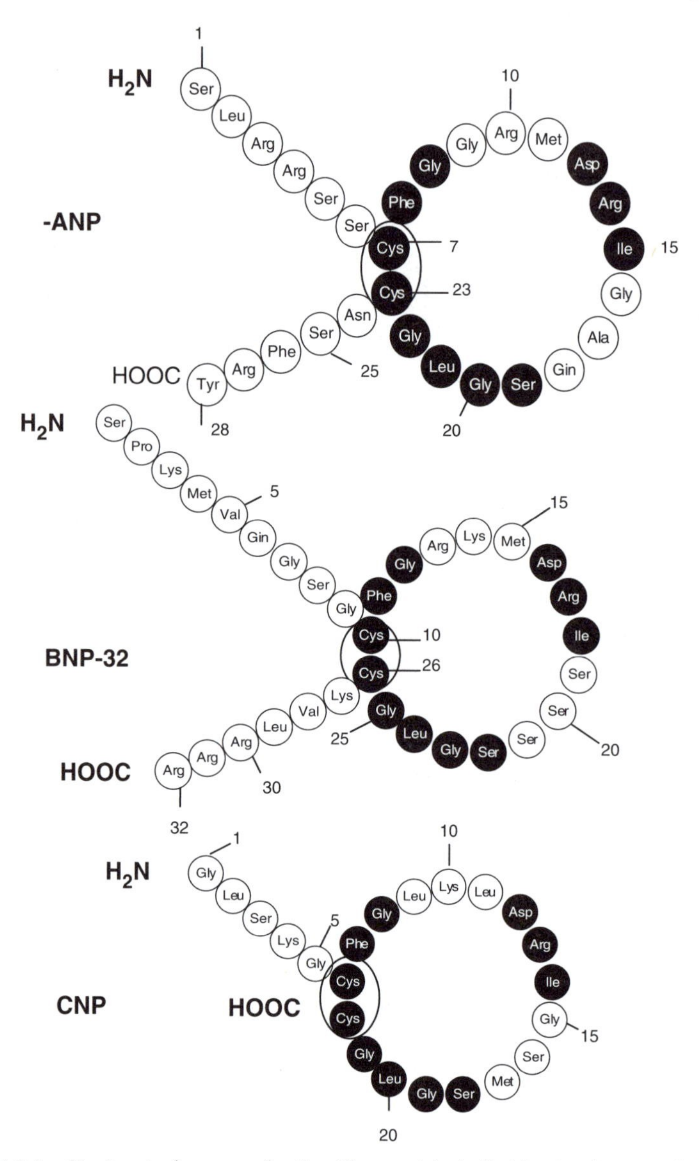

FIGURE 2.6. Natriuretic hormone family. The crucial similarities in the peptides in the natriuretic hormone family are evident. Conserved amino acids are in black. Note the structure imposed by disulfide bonds coordinated by two cysteines. ANP, atrial natriuretic peptide; BNP, brain natriuretic peptide; CNP, originally considered to be a clearance peptide, but now all three are recognized in different tissues and different species. Their functions in fluid balance are similar. (Modified from Sudoh, T. *et al.*, *Bioch. Biophys. Res. Comm.*, 168(2): 863–870, 1990.)

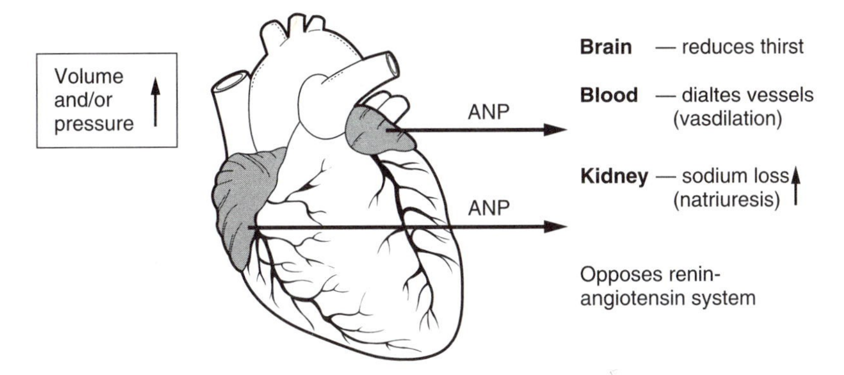

FIGURE 2.7. Atrial natriuretic peptide (ANP) has several actions to decrease fluid volume. It is secreted in response to stretch of atria cells when pressure or volume is increased. (From Christensen, 1995.)

increased aggressiveness of the female mouse with the oxytocin gene knocked out.

The immediate-early gene *fos*-B activates oxytocin (Brown *et al.*, 1996). Knockout mice lacking the *fos*-B gene make poor mothers. Their normal maternal behaviors of nesting and suckling are absent. Oxytocin has also been associated with grooming and with yawning. When OT is injected centrally, the most common response is for animals to yawn (Melis and Argiolas, 1995).

II. CLINICAL EXAMPLES

A. CRH Promotes a Spectrum of Anxiety-Related Behaviors

In Chapter 1, we introduced the neuropeptide, corticotropin-releasing hormone (CRH), which provides the main stimulus for ACTH secretion by the anterior pituitary but also influences many behavioral mechanisms in the brain. Remember that CRH is synthesized by neurons with cell bodies in the amygdala and axons that project to the locus ceruleus in the pons (refer back to Fig. 1.4). The locus ceruleus contains about half the norepinephrine (NE)-secreting neurons in the entire CNS, and CRH can activate these neurons. The projections of these NE-secreting neurons are widespread (to the hypothalamus, thalamus, and limbic systems and the cerebral cortex), and they are involved in orienting and attentional responses.

Any hormone, including CRH, acting on serotonergic neurons or serotonin receptors (for which 14 genes have been discovered) is likely to have multiple, widespread effects on mood. Consider this question: What do depression, anxiety, obsessive–compulsive disorder, posttraumatic stress disorder (PTSD), and premenstrual dysphoric disorder (PMDD; see Chapters 1 and 3) have in common? They all can be treated with serotonin reuptake inhibitors that act on specific serotonin receptors. Long ago, the treatment for depression was monoamine oxidase inhibitors (MAOIs), which interfere with the catecholamine pathways. The catecholamine pathways include norepinephrine, dopamine, and serotonin. The problem with MAOI inhibitors is that they are not specific and they have high levels of toxicity; therefore, their use was limited until the 1980s, when the first of a new series of drugs was introduced: selective serotonin reuptake inhibitors (SSRIs). These had the effect of increasing 5′-hydroxytryptamine (5HT) in the synapses. An increase of 5HT, or serotonin in synapses, was first achieved with fluoxetine (better known as Prozac®), which was the first of a new type of drug for treating depression. Since then, these types of drugs have been used for anxiety, panic disorder, social phobias, bulimia nervosa, anorexia nervosa, obesity, PMDD, and PTSD. Although the effect of the SSRIs are to increase 5HT in synapses, the treatment is slow. It takes about 2 to 6 weeks of taking these drugs daily for them to become fully effective. The reason for this slow effect is that the neurons respond to the drugs through a series of adaptations. The first effect is an increase of serotonin in the synaptic cleft; however, as frequently happens when there is too much ligand, this downregulates the number of receptors on the postsynaptic side. It also reduces the autoreceptors on the presynaptic side that are involved in reducing the amount of serotonin released. Thus, in the first stages of taking the drugs, the increased levels of serotonin are adapted to and nothing changes. Over time, however, the concentration of receptors readjusts and responds to the higher levels of serotonin, and at that point clinical effects are seen.

B. CRH Effect on Norepinephrine Systems

Abnormally increased and prolonged activation of the NE-releasing neurons of the locus ceruleus has been implicated in the production of anxiety states, a component of which may be inappropriately heightened arousal and vigilance. CRH acts in other brain areas as well (see above), so that it is likely that the anxiogenic effect of CRH administration to experimental animals involves multiple sites of action. The relative specificity of CRH in anxiety production can be demonstrated in laboratory rats by examining the anxiolytic effects of CRH receptor antagonists under several conditions:

(1) when exogenous CRH is administered, (2) when rats are subjected to experimental stress, and (3) when rats are selectively bred to show innate heightened anxiety-like behaviors and to have a hyperactive hypothalamo–pituitary–adrenal cortical (HPA) axis.

Because CRH has anxiogenic effects in addition to its hormonal effects, because 30 to 50% of psychiatric patients with major depression have increased HPA axis activity, and because about a third of such patients also have anxiety syndromes, CRH has been proposed as an etiological hormone in the pathogenesis of major depression. Consequently, drug development efforts for the treatment of depressed patients have focused on CRH antagonists. As with most new drugs, the effectiveness of CRH antagonists in the treatment of major depression so far has been modest, and some troubling side effects have led to the discontinuation of trials. Efforts toward the development of safer and more effective CRH antagonists for use in depressed patients are continuing.

C. ACTH Influences Learning

Adrenocorticotropic hormone and related peptides have consistent effects on learning in laboratory rats. ACTH administration prior to testing improves the acquisition of an escape response, and ACTH administration during acquisition of an avoidance response makes the response more resistant to extinction. As well, ACTH-like peptides improve the acquisition and retention of passive avoidance behavior, likely by affecting memory-storage processes. ACTH may trigger a state-dependent condition that enhances memory retrieval. As well, ACTH may directly influence reward mechanisms, as indicated by its enhancing effect on intracranial self-stimulation in rats.

The enhancing effects of ACTH and related peptides appear to be on short-term, trial-to-trial memory. Studies suggest that ACTH improves learning processes, and its related peptide, melanocyte-stimulating hormone (MSH), improves attentional processes. In contrast, as indicated by work with vasopressin and its analogs, posterior pituitary hormones appear to influence long-term memory processes. ACTH-related peptides also influence sexually motivated behavior, increasing the urge to seek contact with the incentive male in ovariectomized, estrogen-primed female rats and enhancing lordosis behavior. The effect of MSH depends on the initial level of lordosis behavior, inhibiting the behavior in female rats showing high lordotic activity and enhancing the behavior in rats showing low activity. Thus, an arousal-inducing peptide may have a stimulatory or inhibitory action, depending on an animal's baseline level of arousal.

Intracranial, but not systemic, administration of ACTH and several ACTH fragments also facilitates stretching, yawning, and grooming in rats. ACTH lengthens grooming episodes but does not change their behavioral characteristics. ACTH-induced grooming behavior does not appear to be stereotypic, and it is not dependent on environmental circumstances.

D. Hormones are Involved in Fasting

Unlike camels, humans have no way of storing water; therefore, dehydration can become a very serious condition after only a few hours. By contrast, fasting is less serious, even for many days, because we do store food in the form of lipids, glycogen, and proteins. Wild animals prepare for fasting by storing large fat reserves. Bears, for example, can accumulate as much as 50% of their body mass as fat before hibernating in the winter. Therefore, in a Darwinian sense, obesity has a survival advantage. Humans have been faced with famine at irregular intervals throughout history, and obese humans can sustain longer periods of fasting than can non-obese individuals. One successful therapeutic fast was reported in an obese patient who starved for a year. His body mass decreased from 207 kg to 82 kg (Stewart and Fleming, 1973). Obviously, a non-obese person could not survive such a draconian fast. Obese Zucker rats survive fasts of 60 to 90 days, while non-obese control rats survive only 10 to 20 days. As the amount of fat is reduced, protein loss begins. The protein loss begins with the obligatory nitrogen loss during fasting and the use of proteins in processes such as glyconeogenesis to form more glucose for the brain.

The hormonal changes that occur during fasting begin with a high glycogen-to-insulin ratio. This ratio signals lipolysis and acceleration of gluconeogenesis. Glucocorticoids stimulate protein utilization and increase with the time of fasting. A drop of glucose in the blood due to fasting raises the levels of growth hormone (GH). Thyroid hormones may also play a role. In hypothyroid states, protein is metabolized. In fasting, thyroid hormones are reduced in the plasma, which may conserve the proteins. Each of these hormones has many other effects—here, fasting is just one part of their spectrum (see Chapter 1).

E. Prolactin is Important in Maternal Care as Well as Having Straight Endocrine Actions

In rodents, prolactin secretion occurs in response to several hormonal and sensory stimuli related to reproduction. Prolactin secretion is enhanced by estrogen, so that a prolactin elevation occurs in conjunction with the preovulatory surge of estrogen during the estrus cycle. During mating, vaginal and cervical stimulation induces prolactin secretion. At parturition, circulating prolactin concentrations increase, secondary to the decrease in progesterone and the increase in estrogen that occur at this time. Following pregnancy, suckling by infants continues to stimulate prolactin over a prolonged period. A likely neurochemical mechanism for all these stimulatory effects on prolactin is decreased dopamine release from dopaminergic tuberoinfundibular neurons into the pituitary portal system, dopamine being inhibitory to, and the major influence on, prolactin secretion.

In addition to promoting lactation in mammals and inhibiting resumption of estrus/menstrual cycling after parturition by inhibiting gonadotropin secretion (prolonged nursing of human infants is an important contraceptive technique in some cultures), prolactin influences maternal behavior in different ways, depending on the species. In a general sense, prolactin is supportive of the effects of estrogen and progesterone in promoting maternal behavior in rodents. Prolactin appears to prime, rather than trigger, maternal care, because prolactin antagonists have no effect once the behavioral sequence is initiated. In rats and mice, prolactin facilitates maternal responsiveness, and in rabbits it facilitates maternal nest building, in particular by loosening maternal hair for use as nesting material. As well, data for wolves and some primates indicate a role for prolactin in facilitating maternal behavior.

It is important to emphasize that, as indicated above, prolactin acts in concert with other hormones in influencing maternal behavior, including expressed feelings of attachment in women with newborn infants. These hormones include not only estrogen and progesterone, but also androgens, oxytocin, and hypothalamo-pituitary-adrenal cortical hormones. While each of these hormones can have certain specific influences on components of maternal behavior, it is the coordinated rise and fall of a number of hormones, including prolactin, during pregnancy and after parturition that evokes the full spectrum of complex behaviors necessary for a mother's successful rearing of her young.

III. SOME OUTSTANDING NEW QUESTIONS

1. What happens when these multiple hormonal effects go wrong? In the domain of metabolism, energy use, and food intake, the problems are multiple and not always in the same direction. On the one hand, the drive to eat beyond caloric needs leads to obesity. On the other hand, especially for young women, the obsession to ignore hormonal demands to meet caloric needs may lead to weight loss to a severe degree

(*i.e.*, anorexia nervosa). These syndromes are difficult to solve, even with 21st-century medicine, precisely because the pleiotropic nature of hormone actions on food intake and energy use make it difficult to know where to look for the problem.

2. When a peptide hormone such as CRH, expressed both in hypothalmic and amygdaloid neurons, has both straight endocrine effects (through the pituitary gland) and a variety of behavioral effects, under what circumstances are these simply in parallel, as opposed to being sequential effects, with an earlier action being causal to a later action?
3. Stress hormones have so many effects—how do possible roles of glucocorticoids and glucocorticoid receptors in depression relate to their very long-term actions (*e.g.*, hippocampal atrophy)?

Hormone Combinations Can be Important for Influencing an Individual Behavior

Already we have encountered at least two sets of results that illustrate the principle highlighted in this chapter. In Chapter 1, several hormones that operate on food intake were covered in some detail (see also the review by Schneider *et al.*, *Hormones, Brain and Behavior*, 2002), and more support for the application of this concept to hunger is offered below. In Chapter 2, combinations of hormones influencing thirst and salt hunger in a physiologically adaptive fashion were mentioned, as was the way in which prolactin works with other hormones to foster maternal behavior. Now we elaborate further examples.

I. BASIC EXPERIMENTAL EXAMPLES

A. Hypoglycemia

In hypoglycemia (glucose levels in serum less than 50 mg/dl), many abnormalities in hormonal signaling can be found. Hypoglycemia initiates counter-regulatory hormone release, including growth hormone, cortisol, glycogen, and epinephrine. These hormonal signals promote the release of amino acids, particularly alanine from muscles, to stimulate gluconeogenesis and release of triglyceride from adipose tissues stores to provide free fatty acids (FFAs).

The *behavioral* signs of hypoglycemia are particularly seen in young children, whose nonverbal, behavioral changes are sometimes the first indication of their becoming hypoglycemic. Symptoms include staring, personality changes (going from happy to crying), headaches, mental confusion, and signs of epinephrine release, which may include perspiration, palpitation, pallor, trembling, anxiety, weakness, nausea, and vomiting.

The equation representing certain physiological aspects of hormone action is that when one side of the equation goes down the opposite hormonal effect goes up. Thus, on the other side of the glucose equation, the adrenal hormone cortisol is needed for gluconeogenesis. Deficient cortisol secretion can be the result of primary adrenal insufficiency. In this case, the adrenal gland is no longer producing aldosterone and the level of adrenocorticotropic hormone (ACTH) released from the pituitary keeps rising because of the lack of negative feedback. Primary adrenal insufficiency is seen in Addison's disease, in which there is hypopigmentation of the skin, a craving for salt, and low levels of sodium and potassium in the blood (see Chapter 5).

B. Protein Energy Malnutrition

The behavioral effects of malnutrition are diverse and depend on both the environment and the degree of malnutrition. Ultimately, a lack of protein will lead to apathy, weakness, hypothermia, and impaired cognitive function. In children, these effects can have long-lasting consequences. Physiologically, the body employs a series of strategies involving hormonal responses to adapt to malnutrition. The first stage is characterized by a behavioral change due to a reduction in voluntary energy expenditure, hence the lethargy and apathy. In the second stage, a reduction in the rate of gain of body mass and, in children, an arrest of growth can be observed. The hormones involved are summarized in Table 3.1.

C. Hunger for Salt

The hunger for salt involves angiotensin and the mineralocorticoid, aldosterone (see Chapter 2). While angiotensin alone, injected into the brain, increases sodium chloride intake (3% NaCl), the intake is greatly increased

TABLE 3.1 Hormones Involved in Protein Energy

Insulin
Growth hormone
IGF's
Epinephrine
Glucorticoids
Aldosterone
Thyroid
Gonadotrophins

when combined with a mineralocorticoid (deoxycorticosterone acetate, or DOCA) (Fluharty and Eptein, 1983). As aldosterone is the naturally occurring mineralocorticoid hormone released from the adrenal gland, the effect of angiotensin in the brain plus aldosterone in the periphery would be an increase in salt intake. DOCA is a precursor of both aldosterone and corticosterone. When DOCA is compared to aldosterone in regard to its effects on salt appetite during intraventricular infusions of angiotensin, the salt appetite of the DOCA group is greater (Sakai, 1996). It has been suggested that, first, these adrenosteroids increase angiotensin receptors in the brain, and, second, they increase angiotensinogen mRNA which would produce more angiotensin in the brain. Thus, the upregulation of either angiotensin receptors or angiotensin synthesis would effectively produce more drinking and increased salt intake. The results confirm that is what happens. At a cellular level in the brain, the steroids increase angiotensin II receptors in neurons associated with circuits linked to drinking behavior. Around neurons are glial cells, and angiotensinogen is predominately synthesized in glial cells. Glial cells also have glucocorticoid receptors (GRs), and corticosteroids activate glucocorticoid receptors (see also Chapter 4). These receptors are intracellular receptors; that is, they are not in the membrane as are the angiotensin receptors. As cystolic receptors, they bind with the steroid and are able to enter the nucleus. There, the complex can bind to a transcription site on the DNA for increased transcription of angiotensinogen. The site of action for this process is in the anterior ventral third ventricle (AV3V) region and also in the central nucleus of the amygdala. The central nucleus of the amygdala has been shown to receive projections from brainstem sites associated with taste (Hamilton and Norgren, 1984). Thus, in this way, the hormones can affect both taste and the behavioral mechanisms for salt ingestion. Salt can be ingested in solution (through a drinking behavior) or by licking or eating foods or materials high in sodium concentration.

Atrial natriuretic peptide (ANP) decreases sodium appetite when injected directly into the brain, but systemically administered ANP in the blood does not have this effect (Fitts *et al.*, 1985). Central administration of ANP inhibits the thirst and sodium appetite induced by angiotensin. If, however, the thirst responses are induced by carbachol, then ANP does not have an effect. We can understand this quite easily by recognizing the difference between intracellular-induced thirst and extracelluar-induced thirst (introduced in Chapter 2). As noted in the previous chapter, intracellular-induced thirst is an osmotic thirst. It can be mimicked by carbachol but not by angiotensin. Extracellular volume depletion, or hypovolemia, stimulates the release of angiotensin. On the other hand, ANP is a hormone with the role of reducing extracellular volume; therefore, it reacts in the opposite way to angiotensin and selectively inhibits sodium appetite and drinking when the stimulus is angiotensin or hypovolemia.

These hormonal changes are all related to fluid balance and blood pressure. The body tries to maintain a physiological range of fluid volume and blood pressure. When blood pressure decreases, cells in the kidney detect this change and release renin, which activates the renin–angiotensin system (see Chapter 2). Therefore, the role of angiotensin is to raise blood pressure. It does this by directly causing vasoconstriction of blood vessels and by releasing more norepinephrine from adrenergic nerve fibers that are wound around blood vessels. These nerves have varicosities full of vesicles to release norepinephrine. Release of norepinephrine to the blood vessels causes vasoconstriction if the receptors are alpha-adrenergic receptors. In hypertension, where blood pressure is constantly above normal, inhibiting angiotensin at the site of its receptors (AT_1 receptors) or inhibiting its synthesis with angiotensin-converting enzyme inhibitors, are effective and very widespread pharmacological methods of controlling blood pressure. If the high blood pressure is due to increased extracellular volume, then the use of diuretics is a powerful way to increase fluid loss through the kidneys.

D. Feeding Behaviors

As we have seen in Chapters 1 and 2, feeding involves multiple combinations of many hormones, some of which stimulate feeding and some inhibit. For example see Fig. 1.2. We are all familiar with the fact that feeding and hunger are not continuous. We break the day up into meals. We eat because we are hungry, and as we eat we become sated and the meal ends long before the food can be absorbed. For over 100 years, there has been an interest in the source of this satiation. In the Civil War, an unfortunate soldier suffered

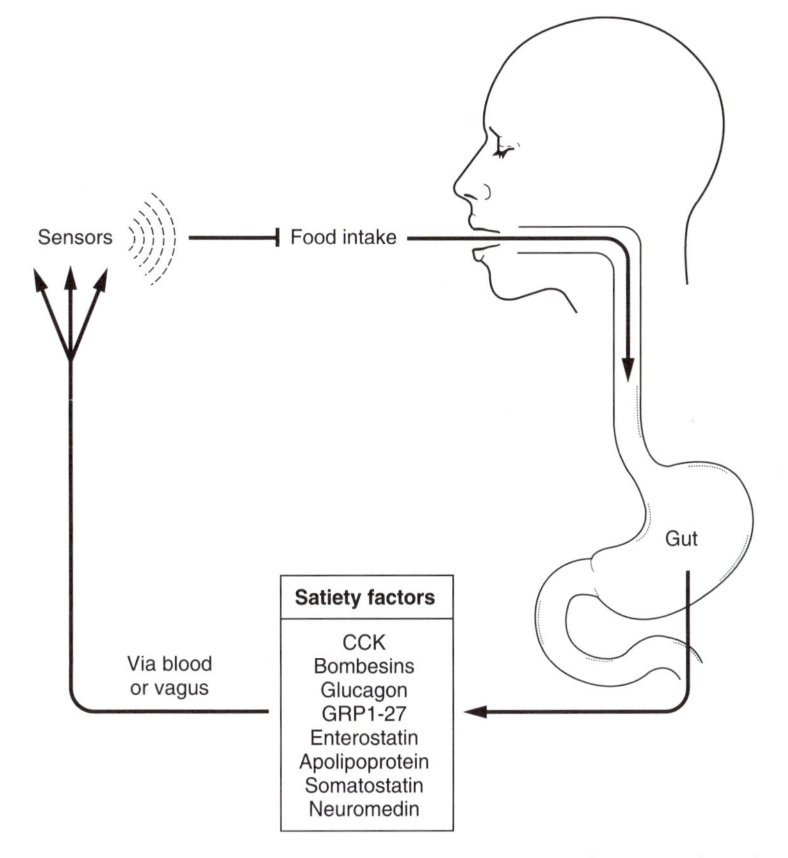

FIGURE 3.1. Several hormones secreted from the gastrointestinal tract, working through different routes to sensors in the brain, can act as satiety factors, thus reducing food intake appropriately.

a wound that opened his stomach to the exterior through a fistula. Because food constantly left the stomach, the patient was never able to satisfy his hunger, which ruled out esophageal and oral origins of hunger. Studies in a laboratory setting indicated that the source of satiety was in the small intestine, and that gastric distension was not a stimulus (for a general picture, see Fig. 3.1).

We have already become acquainted with some of the many hormones and hormone combinations that are necessary for feeding and hunger (see Chapter 1), but following are some more details regarding individual hormones.

1. Cholecystokinin (CCK)

Gibbs *et al.* (1997) discovered that cholecystokinin (CCK), a peptide, was released from the small intestine when ingested food arrived and rapidly inhibited eating. The inhibition of eating was specific. CCK did not inhibit drinking but did inhibit eating, even if the food was provided as a liquid diet so that the motor movements required for eating were identical to those for drinking. This was shown by an experiment where a rat was put on a liquid diet but had a fistula that prevented food from being delivered to the small intestine. With the saline injection, this rat ate continuously. With a CCK injection, the rat stopped eating. In human volunteers with normal weight, CCK also inhibited food intake. The question of how CCK produces satiety, however, is still an open one. As an octapeptide, just as angiotensin II, it cannot penetrate the blood–brain barrier. It can, however, reach circumventricular organs and act on CCK receptors. Injection of CCK directly into the brain inhibits food intake (Figlewicz, 1992). The effects of CCK in the gut can be inhibited by sectioning the abdominal vagus nerve (Bloom and Polak, 1981), because the CCK activates stretch-sensitive receptors in the stomach that are endings of the vagus nerve. Thus, as we have seen with other peptides (angiotensin and oxytocin [OT]), it is possible to have both a neural reflex and a blood-borne reflex. Inhibiting CCK with antagonists increases the size of meals consumed after food deprivation (Moran *et al.*, 1992). This type of evidence points to a physiological role for CCK in normal satiety.

2. Ghrelin

The name of this hormone comes from "growth hormone receptor ligand." Ghrelin is secreted by epithelial cells lining the empty stomach. It works in the opposite direction of leptin (the term *leptin* is derived from a linguistic root meaning "thin") in that it signals hunger and stimulates feeding. During fasting, the blood concentration of ghrelin increases, and its peak levels are just before the expected time of a meal; therefore, the combination of high ghrelin levels with low leptin levels would be ideal for stimulating feeding. (see Chapter 12.)

3. Bombesin

Bombesin is a small protein found throughout the mammalian gastrointestinal tract and also in the nervous system (Brown *et al.*, 1978). It was originally isolated from frog skin (*Bombena bombena*). Injections of bombesin reduced food intake in rats (Gibbs *et al.*, 1979). As with CCK, intracerebral intraventricular infusions of bombesin reduced food intake in rats but intake was not abolished by cutting the vagus.

4. Neuropeptide Y

Neuropeptide Y (NPY) is widely distributed throughout the nervous system and is found in the terminals of several regions of the hypothalamus, including the paraventricular nucleus (PVN) and arcuate nucleus. NPY is closely associated with norepinephrine and is made in the same neurons that synthesize norepinephrine. Norepinephrine and NPY are co-released as transmitters in the brain, and reports of norepinephrine stimulating feeding are probably due to the coincident release of NPY. The role of NPY has become more and more recognized as being an important mediator of feeding in the hypothalamus (Chapter 1). Several NPY receptor subtypes are distributed in the hypothalamus in areas that are connected to neural circuits involved in food-seeking behavior and the motor acts of feeding.

And, now, we move on to a closely related point illustrating the principle of this chapter but with the difference that the hormones in question work in opposite directions to each other on a given molecular or behavioral endpoint. We present first a case related to the previous discussion regarding food intake, then an example related to sex hormones.

E. Corticotropin-Releasing Hormone

When corticotropin-releasing hormone (CRH) is injected into the brain, it reduces food intake. CRH increases sympathetic activity, particularly to brown adipose tissue. The significance of this is that brown adipose tissues contains norepinephrine and plays an important role in thermogenesis. In animals that hibernate in winter, building up brown adipose tissue during the summer months is critical. Injections of NPY decrease sympathetic activation of this fat tissue. Therefore, CRH may be important in the hibernating period when feeding must be inhibited and thermogenesis stimulated.

F. Thyroid and Sex Hormones

Again, while many of the examples of hormone combinations acting on behavior that are illustrated in this chapter deal with synergies—one hormone helping the other to produce a behavior—other examples show how one hormone can oppose another. Thyroid hormone administration can actually reduce the ability of sex hormones to facilitate sex behavior in a variety of female animals. This effect is biologically adaptive in that it brings females

Hormones administered	Estrogenic effects on:	
	Transcription through ER-a and ERE	Lordosis behavior
None	O	O
Estradiol (E)	↑↑↑↑	↑↑↑↑
Thyroxine (T)	O	O
E + T	↑↑	↑↑

FIGURE 3.2. Co-administration of thyroid hormones can reduce the magnitude of an estrogen effect at the molecular level (transcription assays in neuroblastoma cells) and at the behavioral level (the female reproductive behavior lordosis). Considering combinations of hormones acting on behavioral mechanisms, therefore, the interactions can be subtractive as well as additive. ER-a = Estrogen Receptor-alpha. ERE = Estrogen Response Element on DNA.

out of the breeding season at a time when a new pregnancy would have babies being born at the wrong time of year. The opposition of estrogenic effects by thyroid hormones at the behavioral level finds its equivalent in molecular mechanisms. Vasudevan at Rockefeller has found that, by using transient transfection assays, liganded thyroid hormone receptors can interfere with estrogen-facilitated transcription (Fig. 3.2).

In the same vein, while leptin can reduce food intake primarily by reducing meal size, neuropeptide Y can increase food intake. Again, mechanistic studies are in line with observed behavioral effects. That is, Schwartz and Moran, at Cornell Medical School, reported opposite effects of the two hormones on neurons in the nucleus of the solitary tract in the brainstem. Leptin increased NTS responses to gastric loading, whereas NPY reduced such responses. In summary, hormone combinations can achieve a sensible physiological result in some cases by synergizing and in other cases by opposing each other's effects.

II. FOR HORMONE COMBINATIONS, ORDER OF ADMINISTRATION CAN BE CRUCIAL

A. Estrogen Plus Progesterone Acting on Reproductive and Maternal Behaviors

Even for the same pair of hormones, different temporal patterns of combinatorial action will be important for different behaviors. For example, in

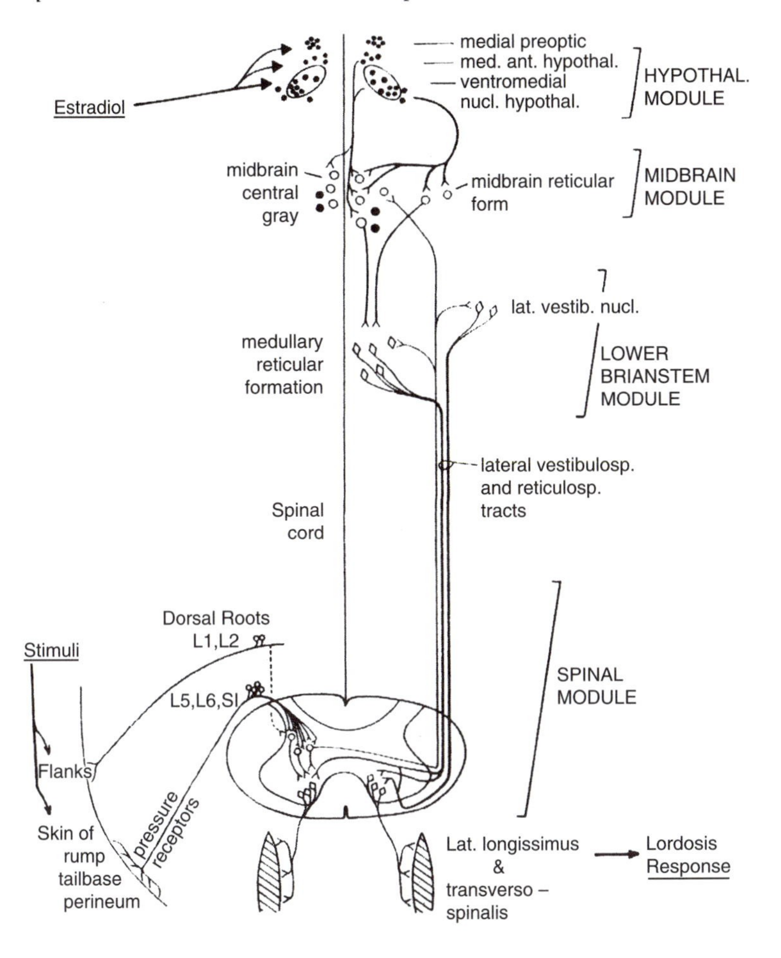

FIGURE 3.3. The neural circuit for lordosis behavior is bilaterally symmetric and is plotted here on just one side for convenience of illustration. Briefly, there are opportunities for ascending sensory mechanisms to influence the relevant motor control pathways at spinal, lower brainstem, and midbrain levels; however, genomic actions of estrogens (amplified by progesterone actions) at the hypothalamic level are required for the rest of the circuit to operate and produce the back muscle contractions typical of lordosis. (From Pfaff, D.W., *Drive: Neurobiological and Molecular Mechanisms of Sexual Motivation*, MIT Press, Cambridge, MA, 1999. With permission.)

many female laboratory animals, a long priming treatment of estrogens (48 hours or more) followed by a brief progesterone exposure will promote female-typical sexual behaviors. The neural circuit and some of the genes

	Ovulatory release of Leuteinizing hormone (LH)	Expression of Lordosis behavior
No hormones	O	O
Estrogens (E) only	↑	↑
Estrogens (E) followed by progesterone (P) (tested P + 2-5 hrs)	↑↑↑	↑↑↑
E followed by P administration longer (tested P + 20-25 hrs)	O	O
P (long administration) concomitant to E	↑	↑
P (long administration) concomitant to E, followed by high E and declining P	O	O*

* This hormonal pattern is optimal for parental behavior

FIGURE 3.4. There are many parallels between the hormonal conditions sufficient for ovulation in the female laboratory animal and those sufficient for the primary sex behavior lordosis.

supporting the mechanisms for this type of behavior have been worked out (Pfaff, 1999) (Fig. 3.3). In a biologically adaptive fashion, the requirement of estrogen (E) followed by progesterone (P) for female sex behavior synchronizes the behavior with ovulation, which shows the same (E + P) combinatorial effects. On the other hand, following the estrogen priming, if the progesterone is allowed to stay around for a long time, the opposite effect on sexual behavior is seen. Finally, in a pregnant animal, estrogen levels are high and remain high during parturition. In contrast, progesterone levels start high but decline around the time of giving birth. High estrogens and a decline in progestins make the optimal combination for maternal behavior (Fig. 3.4).

A different kind of combination of hormone actions underlies social recognition, motivation, and memory in female laboratory animals. Insel *et al.* have shown the powerful effects of oxytocin and, in some species,

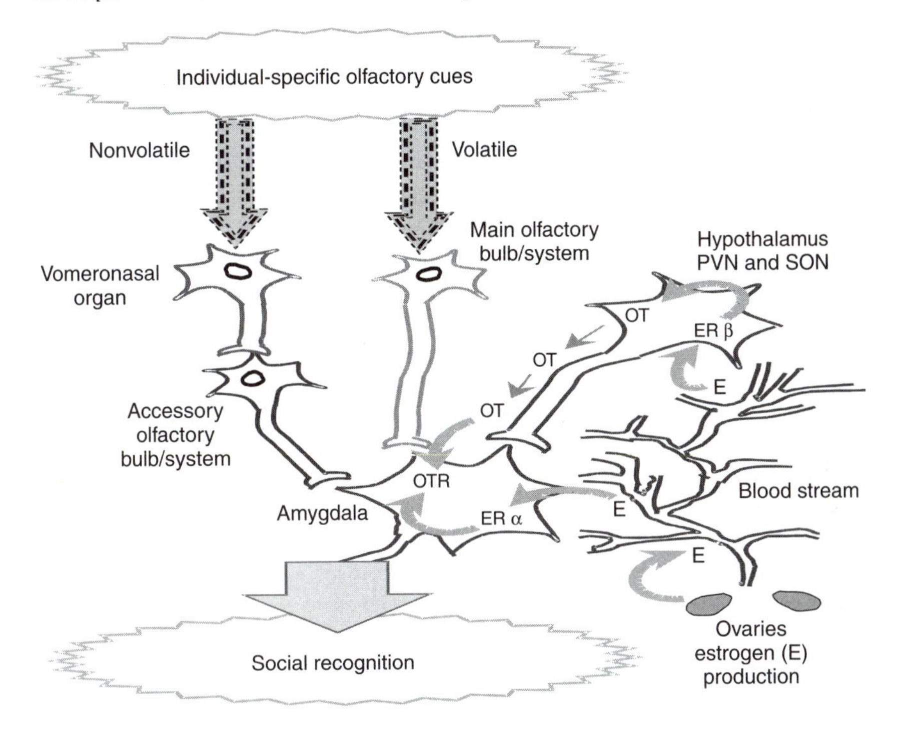

FIGURE 3.5. Genomic and neuronal mechanisms for hormonal influences on social recognition. Estrogens produced in the ovaries circulate to the paraventricular nucleus of the hypothalamus (PVN) where, following binding to ER-β, they stimulate oxytocin (OT) transcription. OT is carried to the amygdala. Meanwhile, estrogens circulate to the amygdala where, following binding to ER-α, they stimulate OT-receptor transcription. OT, operating through OT-R in the amygdala, fosters social recognition. In mice, highly olfactory animals, olfactory signaling through both the main and the accessory olfactory systems, provides the stimuli to the amygdala, to which the mice react using the estrogen-influenced OT/OTR system. (From Choleris, E. *et al.*, *Proc. Natl. Acad. Sci. USA*, 100(10): 6192–6197, 2003. With permission.)

vasopressin on this cluster of behaviors. In addition, however, estrogens have an overriding effect in two ways (Fig. 3.5). Working through estrogen receptor-β, estrogens turn on the oxytocin gene; working through estrogen receptor-α, they turn on the oxytocin receptor gene. Therefore, in this combination of hormones (*i.e.*, estradiol and oxytocin), one has a superordinate relation to the other. Estrogens have to come first, to turn on the gene for the oxytocin receptor, for instance, and oxytocin later.

III. CLINICAL EXAMPLES

A. Hyponatremia

A number of hormones contribute to hyponatremia, which is defined as a serum sodium level of less than 135 mEq/L (refer to introductory material in Chapter 2). In normal people, the range is 135 to 145 mEq/L, even with fluctuations in drinking fluids. Hyponatremia is caused by an inability to excrete sodium-free water by the kidney. It occurs in hypoglycemia, hypoproteinemia, and hyperlipidemia. The cause appears to be an increase in osmolality, which brings water out of cells and dilutes sodium in the serum. Hyponatremia leads to headache, nausea, vomiting, and weakness and can, in a severe state, cause seizures and decorticate posturing. Decorticate posturing occurs when the brain is decerebrated by lesion or widespread damage to the cortex. It results in the arms and the forelegs extending rigidly and the hands turning inwards. It is a cardinal sign of cerebral impairment. In hyponatremia, intracellular water expansion causes edema, and the brain swells inside the skull. This swelling leads to an encephalopathy with the neurological symptoms described above. Hyponatremia is severe in the syndrome of inappropriate antidiuretic hormone (SIADH).

B. Syndrome of Inappropriate Antidiuretic Hormone

The syndrome of inappropriate antidiuretic hormone (SIADH) secretion is characterized by severe hyponatremia in which sodium serum levels are 120 mEq/L. This condition is brought on by increased levels of antidiuretic hormone (ADH) in the absence of an osmotic or hypovolemic stimulus. The increase in ADH causes the kidney to retain water, thereby diluting the total amount of sodium. The causes of SIADH are multiple and include CNS infection, neoplasms, hydrocephalus, pulmonary asthma, carcinomas, and overuse of medicines such as serotonin reuptake inhibitors used for depression.

C. Antidiuretic Hormone

Antidiuretic hormone, which in the human is arginine vasopressin (AVP), plays a major role in the defense against hypernatremia. ADH is exquisitely sensitive to changes in osmolality. A change as small as 3 mOsm/kg is enough to stimulate the release of vasopressin from the hypothalamus.

Vasopressin is released into the blood directly through the neurohypophysis and circulates in the plasma to reach the kidney. In the kidney, ADH acts on collecting duct cells to concentrate urine. The effect of AVP on the cell is to stimulate cyclic adenosine monophosphate (cAMP), which activates aquaporin molecules (see Chapter 15). Aquaporin molecules are proteins which, when inserted into the membrane of the cell, become channels through which only water molecules can pass. The effect of AVP is to cause aquaporin (AQP-2) to migrate to the lumenal membrane of the cell. As hundreds of aquaporin molecules are embedded into the membrane, water can now enter the cell and exit through different aquaporins (aquaporins 4 and 5) on the opposite end of the cell to be reabsorbed into the plasma. This increases plasma volume and concentrates the urine. In this way, ADH (vasopressin) concentrates urine.

Vasopressin is very effective and can reabsorb as much as 99.7% of filtered water. Antidiuresis is a vitally important function, and vasopressin is an ancient peptide that has been conserved for millions of years (see Chapter 20). Arginine vasopressin is synthesized in the paraventricular nuclei and the supraoptic nuclei of the hypothalamus and is released by hyperosmotic changes of 3 mOsm/kg and by angiotensin II. Angiotensin II also defends against hypernatremia by releasing vasopressin. Although angiotensin II is mainly responsive to hypovolemia, brain angiotensin II responds to small changes in Na^+. Angiotensin II acts on the angiotensin type 1 receptor (AT_1R) in the PVN. When pNa^+ is increased, angiotensin II stimulates AT_1R to release AVP from the PVN (Hogarty *et al.*) (see also diabetes insipidus discussion in Chapter 5).

IV. AGAIN, IN CLINICAL EXAMPLES, THE SEQUENCE OF HORMONE TREATMENT CAN BE IMPORTANT

A. Postmenopausal Hormone Replacement Therapy

Menopause denotes the cessation of menstrual cycles and occurs in women around the age of 50 (see also Chapter 11). It is due to a rapid decrease in the number of ovarian follicles. Circulating estradiol, produced during the follicular phase of the menstrual cycle, and progesterone, produced during the luteal phase, fall to very low values, and luteinizing hormone (LH) and follicle-stimulating hormone (FSH) production by the pituitary rises substantially. Many women experience a wide array of symptoms at menopause, including hot flashes, sweating, insomnia, headaches, dizziness, lack of energy, palpitations, digestive disturbances, difficulty concentrating, nervous tension, and decreased sex drive. Also, bone density decreases, thus increasing the risk of fractures, and the occurrence of cardiovascular disease increases with increasing age.

Estrogen replacement therapy can ameliorate many of these symptoms, but it carries certain risks. Initially, estrogen alone was used, but it became apparent that a major side effect was an approximately sevenfold increase in the occurrence of cancer of the uterine lining (endometrium). Thus, estrogen alone can be used in women after total hysterectomy (removal of both the uterus and the ovaries), but alternative hormone replacement strategies are necessary in anatomically intact postmenopausal women to guard against the development of endometrial carcinoma.

The next strategy shown to be effective in relieving menopausal symptoms was sequential hormone replacement, with estrogen followed by a progestin, which mimicked the normal monthly hormonal sequence and resulted in induced monthly shedding of the endometrium. This strategy reduced the risk of endometrial carcinoma while still ameliorating the symptoms of menopause and protecting against osteoporosis, but it did lead to monthly bleeding, an inconvenience for menopausal women. More recently, continuous administration of a combination of estrogen and a progestogen has been used, which confers a similar therapeutic effect and maintains an atrophic endometrium, thereby eliminating induced menses (although breakthrough bleeding can occur in some women).

A recently confirmed complication of combined estrogen and progestogen hormone replacement therapy is increased risk of breast cancer. A number of epidemiological studies have concluded that a low daily dose of estrogen does *not* increase breast cancer incidence. An ongoing, large prospective study, however, in which women were randomized to estrogen alone, estrogen and progestogen, and placebo, discontinued the estrogen and progestogen arm after 5 years of follow-up, because the adverse effects (significantly increased risk of breast cancer, pulmonary embolism, and stroke) outweighed the beneficial effects (significantly lowered risk of colorectal cancer and hip fracture). This finding has severely impacted the clinical practice of combined hormone replacement therapy for postmenopausal women, and many women are now reexperiencing distressing menopausal symptoms because they are concerned with the risks of hormone treatment. Of note, the estrogen arm of the ongoing study is still continuing, because a significantly increased risk of adverse events has not emerged. Alternative treatments directed at specific components of the menopausal state, such as decreased bone density, fortunately are under active development.

B. Assisted Reproduction

Some women of reproductive age do not ovulate regularly during their menstrual cycles and therefore are candidates for assisted reproduction techniques. The first goal of such techniques is to promote mature follicles and, if possible, ovulation. Follicle growth and induction of ovulation can be accomplished hormonally by increasing circulating pituitary gonadotropins. Clomiphene citrate, a nonsteroidal estrogen that behaves as a competitive estrogen receptor antagonist, enhances release of pituitary LH and FSH, resulting in follicular growth, dominance of one follicle, and ovulation. For clomiphene to be effective, the patient must have an intact hypothalamic–pituitary–gonadal (HPG) axis. A similar result can be achieved by the direct administration of gonadotropins and is indicated when there is an inherent deficiency of LH or FSH or when clomiphene treatment has not been successful.

The goal of hormonally induced ovulation without assisted reproduction techniques is to produce one to two mature follicles, thereby minimizing multiple births; however, with assisted reproduction such as *in vitro* fertilization, the goal is to produce multiple mature follicles so that several eggs can be harvested, inseminated, and transferred into the patient's uterus, because the success rate with this technique is not high. With assisted reproduction, a sequential hormonal regimen is followed. Withdrawal bleeding may be initiated with progesterone or medroxyprogesterone administration for a week to cause shedding of the endometrial lining. A long-acting, gonadotropin-releasing hormone agonist such as leuprolide is given to downregulate the menstrual cycle, following which follicular induction is accomplished with gonadotropin administration. Eggs are harvested from several mature follicles for insemination and insertion into the uterus, and progesterone is then administered to support the endometrium and promote successful implantation. Thus, when assisted reproduction techniques are used, sequential administration of several hormones is necessary to suppress endogenous cycling, to develop mature follicles, and to prepare the endometrium for successful implantation of the inserted fertilized egg.

C. Multiple Endocrine Deficiency Syndromes

There are many endocrine deficiency syndromes, some of which involve several hormones. In partial hypopituitarism, deficiencies of gonadotropins and growth hormone often occur, and in panhypopituitarism virtually all the anterior pituitary hormones are involved. Causes can be genetic or traumatic and can include tumors and vascular insults to the pituitary. Other diseases involve the target endocrine glands (*e.g.*, Schmidt's syndrome, which is an autoimmune disease of the adrenal and thyroid glands that can result in complete exhaustion of basal glucocorticoid secretion as well as hypothyroidism). Replacement of the deficient hormones must be done sequentially in illnesses such as panhypopituitarism and Schmidt's syndrome, when the adrenals and thyroid gland are both hypoactive. If thyroid hormone replacement is begun before the adrenal insufficiency is treated, the increase in overall metabolism produced by exogenous thyroid hormone can precipitate an Addisonian crisis with severe consequences. Therefore, the adrenal insufficiency is treated first, and then thyroid hormone administration is begun.

D. Premenstrual Dysphoric Disorder

A troubling set of symptoms reported by some adult women is characterized by falling estrogen levels in concert with falling progesterone levels (Table 3.2). Negative (dysphoric) emotions and negative behaviors and social interactions can occur for 7 to 10 days prior to menses and during menstruation itself. There also can be painful physical changes, including breast swelling, uterine cramping, and fluid retention. If this occurs with most menstrual cycles for at least a year and involves a variety of symptoms that include depression, anxiety, mood swings, irritability, fatigability, appetite and sleep changes, and physical changes, the condition can be formally diagnosed as premenstrual dysphoric disorder (PMDD), and treatment is recommended. The assumption has been that, because the ovary undergoes profound changes in its production of estrogen and progesterone across the menstrual cycle, these hormones must be intimately involved in the etiology of this syndrome.

Many studies have examined basal and stimulated activity levels of the HPG axis in affected women. Studies of circulating estrogen and progesterone have been quite variable, in sum indicating that there is no excessive or deficient secretion of either hormone or testosterone (converted in women in small amounts from adrenal androgens) in patients compared to normal women. There may, however, be a positive correlation between severity of premenstrual symptoms and circulating female sex steroids, although they are still within the normal range.

More recent studies have considered other possibilities. The rate of change of gonadal steroids in the late luteal phase (*e.g.*, progesterone decline) may be important—depressive symptoms become more frequent in girls than in boys at the time of puberty, when major sex hormone changes occur, and

TABLE 3.2 DSM-IV Research Criteria for Premenstrual Dysphoric Disorder

In most menstrual cycles during the past year, five or more of the following symptoms were present most of the time during the last week of the luteal phase (between ovulation and menses), began to remit within a few days after onset of the follicular phase (onset of menses), and were absent during the week postmenses, and at least one of the symptoms was 1 through 4:

1 Markedly depressed mood, hopelessness, self-deprecating thoughts

2 Marked anxiety, tension, feeling "keyed up" or "on edge"

3 Marked affective lability (*e.g.*, feeling suddenly sad)

4 Persistent and marked anger or irritability

5 Decreased interest in usual activities

6 Difficulty concentrating

7 Lethargy, fatigability, or markedly decreased energy

8 Marked change in appetite, overeating, or specific food cravings

9 Hypersomnia or insomnia

10 Sense of being overwhelmed or out of control

11 Other physical symptoms (*e.g.*, breast swelling or tenderness, headaches, joint or muscle pain, bloating, weight gain)

Source: Adapted from the *Diagnostic and Statistical Manual of Mental Disorders*, 4th ed., American Psychiatric Association, Washington, D.C., 1994. With permission.

there is a predominance of cyclic mood disorders in women throughout adult life. Studies in which similar hormone changes have been experimentally produced in women with and without premenstrual dysphoria indicate that those without antecedent premenstrual complaints do not develop such complaints when their sex steroids are experimentally altered, whereas women who do have antecedent premenstrual dysphoria develop similar complaints with experimental alteration of their hormones. These findings suggest a heightened sensitivity of the CNS to changes in circulating hormones in women with premenstrual dysphoria.

Another consideration, and fitting with the differential CNS sensitivity hypothesis, is the role of neuroactive steroids, given that they have a predominantly inhibitory effect on CNS excitability, via GABA-A receptor modulation, as discussed above. Studies of circulating concentrations of, for example, allopregnanolone, however, have been conflicting, with both higher and lower concentrations reported in premenstrually dysphoric versus normal women. Again, the rate of change may be an important factor. And, a potentially important influence on CNS sensitivity to gonadal hormone changes may be CNS-produced neurosteroids, which, as indicated above, may be concentrated differently in local CNS areas important in the mediation of affect and mood.

Finally, the role of sociocultural factors must be considered in premenstrual dysphoria (see Chapter 20). Some symptoms, such as sleep and appetite disturbances, may be relatively environmentally independent, whereas others, such as depression and irritability, can be provoked by interpersonal and other environmental stressors. Given that treatment of this major cause of discomfort in women remains empirical, attention to social as well as to biological factors during treatment is quite important.

V. OUTSTANDING NEW BASIC OR CLINICAL QUESTIONS

1. Why do premenstrual physical and mental symptoms differ so markedly among women? There is very little information available to answer this question, except that, as noted above, the rate of change of hormones appears to be a relevant factor, suggesting that differing receptor sensitivities among women predisposed to premenstrual dysphoria versus those without such symptoms play a role. The systems involved, however, remain obscure. Do gonadal steroid receptors within the CNS play a role? Do changes in neuroactive steroids, secondary to late luteal phase changes in gonadal steroids, play a role (*e.g.*, falling allopregnanolone concentrations secondary to falling concentrations of its precursor, progesterone)? Do these hormone changes affect CNS amine neurotransmitter systems, such as serotonin?
2. Why do women with similar premenstrual mental symptoms respond so differently to similar hormone treatments? This question is a corollary of question 1, in that the vulnerabilities in hormonal and neurotransmitter systems leading to premenstrual dysphoria are the targets of hormonal and other replacement therapies. Because the hormonal and neuroendocrine links are not well understood, it is difficult to speculate on why symptomatic women respond differently to hormone treatment. Factors include the blood and tissue levels achieved with a specific dose of hormone, which may differ among women depending on differences in hormone pharmacokinetics across individuals, as well as differences in the innate sensitivity of hormone receptors and the metabolic cascades following receptor activation.
3. On the principle that the effects of hormones can be due to too much or too little (see Chapter 5), we wonder if the lack of counter-regulatory hormones can be the principal cause of food intake and metabolic disorders.

Hormone Metabolites Can be the Behaviorally Active Compounds

This chapter provides examples of integrating a tremendous amount of experimental data to gain an understanding of clinical findings, as well as examples of basic scientific investigations that were inspired by clinical findings; therefore, these two domains have not been separated in the following discussion.

I. TESTOSTERONE IS A PROHORMONE FOR BOTH MALE AND FEMALE SEX STEROIDS

Testosterone, produced mainly by the testes in men and the adrenals in women, is a potent androgen (male sex hormone) that influences many tissues throughout the body, including the central nervous system (CNS), in which testosterone receptors are present. Metabolites of testosterone, however, also are important hormones and have specific effects that are different from the effects of testosterone itself. Two important testosterone metabolites are dihydrotestosterone (DHT) and estradiol (E2), which are male and female sex steroids, respectively.

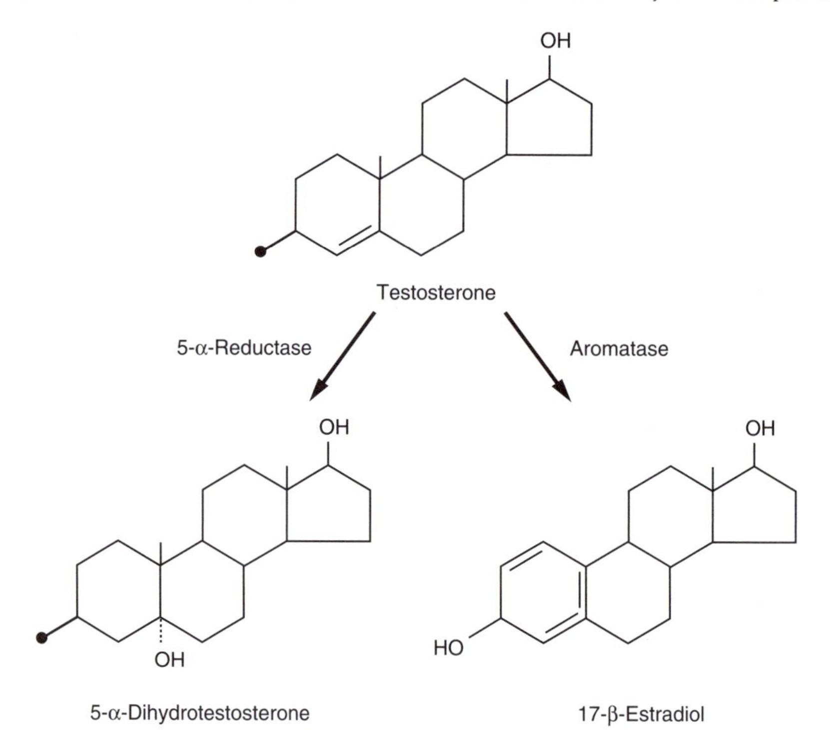

FIGURE 4.1. Testosterone is a prohormone for dihydrotestosterone (DHT) and estradiol (E2). The conversions to DHT by 5-alpha reductase and to E2 by aromatase appear minor in this two-dimensional representation, but there are major conformational changes that confer receptor specificity to these steroid compounds.(From Gorski, R.A., in *Principles of Neural Science*, 4th ed., Kandel, E.R. *et al.*, Eds., McGraw-Hill, New York, 2000, p. 1138. With permission.)

Figure 4.1 illustrates the conversion of testosterone to these two active hormonal metabolites. There are specific receptors for each; for DHT these are in peripheral tissues related to the development of secondary sex characteristics, and for E2 these are in both peripheral tissues and the CNS. Figure 4.1, because it is two dimensional, does not convey the major structural changes inherent in the enzymatic conversions of testosterone to DHT and E2 that confer receptor specificity to these steroid compounds.

The conversion of testosterone to E2, a sequence of three enzymatic steps that go under the name *aromatase* within the CNS, is considered to be important for the early development of the "masculine brain." The aromatase enzyme is concentrated in areas of the brain related to sexual differentiation of the CNS, such as the hypothalamus. In the male, exposure of the developing brain to high concentrations of testosterone (and therefore to high concentrations of E2 converted from testosterone in specific regions)

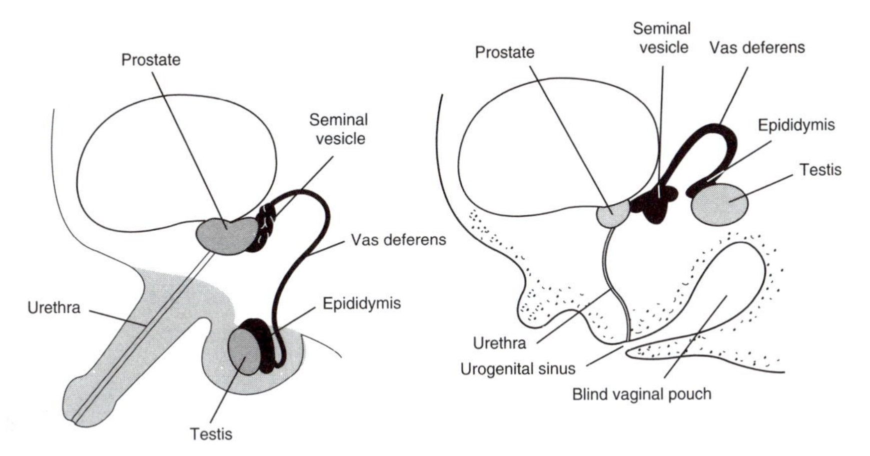

FIGURE 4.2. Schematic representation of the genitalia of a normal man (left) and a patient with 5-alpha reductase deficiency (right). Organs dependent on testosterone for normal development are in black, and those dependent on dihydrotestosterone (DHT) are stippled. (From Forest, M.G., in *Endocrinology*, 4th ed., DeGroot, L.J. and Jameson, J.L., Eds., WB Saunders, Philadelphia, PA, 2001, p 1993. With permission.)

leads to, for example, the tonic secretion of gonadotropins in the adult male versus cyclic secretion of these hormones in the adult female. Many differences in behavior, especially in aggressive behavior, also result from the differential exposure of the developing male and female CNS to testosterone and E2.

While testosterone has many androgenic effects throughout the body and is responsible for virilization of internal structures (*e.g.*, testicular development), DHT is required for virilization of the external genitalia (*i.e.*, penile growth and scrotal development). If 5-alpha reductase, the enzyme that converts testosterone to DHT, is genetically deficient, the external genitalia of the newborn infant appear female or at least ambiguous (male pseudohermaphroditism). Figure 4.2 illustrates the genitalia of a normal man (left) and the genitalia of a 5-alpha-reductase-deficient prepubertal boy. The female-appearing external genitalia belie the presence of an XY sex chromosome complement, functioning testes (although undescended), and a masculinized brain secondary to fetal testosterone exposure. The behavioral consequences of this anatomical alteration are obvious, and they also illustrate how hormone metabolites can influence behaviors through indirect routes.

Before it was recognized that this syndrome is inherited and therefore concentrated in certain families, affected infants were raised as girls. Upon reaching puberty, however, the increased secretion of testosterone from the pubertal testes led to some DHT being produced, so that many prepubertal "girls" developed a male phallus with erections, scrotal testes, male hair distribution, deepened voice, and male body habitus and psychological characteristics, to the initial consternation of parents and family members. However, the syndrome was quickly recognized to occur in affected families in several parts of the world (*e.g.*, in the Dominican Republic), so that newborns with ambiguous genitalia in these families were not forced to grow up as girls. Indeed, some adult individuals have entered into heterosexual relationships and have been able to function physically and emotionally as men. Others have led isolated lives or retained some female gender identity.

The outward switch of sex and gender identity from female to male at puberty in 5-alpha-reductase-deficient individuals was initially interpreted as the primacy of nature over nurture; that is, sex hormones are more influential than psychosocial factors imposed since infancy. However, despite the early recognition that newborns from affected families and with ambiguous genitalia might have masculine pubertal development, these infants have been raised either as boys or ambiguously as girls. The nature-versus-nurture dichotomy, therefore, is not as straightforward as some would believe.

It is important to indicate that not all discrepancies between physical sex and psychological gender are hormonally based. Table 4.1 summarizes the possible relationships among genital sex, sex of rearing, and gender identity in genetic (XY) men. In experimental animals, the neurochemical and molecular bases of sexual differentiation of brain and behavior have been explored. The major findings add mechanistic detail to the clinical picture given above. Testosterone, having entered the brain of the neonatal rat or

TABLE 4.1 Sex/Gender Identity Relationships in XY Men

Clinical Description	Pathology	Genital Sex	Sex of Rearing	Gender Identity
Normal male	None	Male	Male	Male
Transsexual	None	Male	Male	Female
Androgen insensitivity (complete)	Lack functional androgen receptors	Female	Female	Female
5a-reductase deficiency	Lack dihydrotestosterone	Female (±)	Female (±)	Female to male

Source: Adapted from Gorski, R.A., Sexual differentiation of the nervous system, in: Kandel, E.R., Schwartz, J.H., and Jessel, T.M., Eds., *Principles of Neural Science*, 4th ed., McGraw-Hill, New York, 2000, pp. 1131–1148; Wallen, K. and Baum, M.J., 2002. Masculinization and defeminization in altricial and precocial mammals: comparative aspects of steroid hormone action, in: Pfaff, D.W., Arnold, A.P. *et al.*, Eds., *Hormones, Brain and Behavior*, Vol. 4, Academic Press, San Diego, CA, 2002, pp. 385–423.

mouse, is converted to estradiol, and it is largely in the chemical form of estradiol that the hormone exerts defeminizing activity. Three lines of evidence have been well established:

1. McCarthy and colleagues temporarily interrupted the function of the gene for the classical estrogen receptor, ER-alpha, in the hypothalamus of the neonatal rat and were able to protect against the defeminizing actions of systemically administered testosterone.
2. Blocking the action of the enzyme responsible for converting testosterone to estradiol reduces the masculinizing effects of the androgen.
3. An estrogen receptor blocker can reduce the amount of neuroendocrine masculinization during the neonatal period. The physiological import of this androgen-metabolism step is manifold; the defeminized brain can not manage to produce an ovulatory surge of luteinizing hormone (LH) nor can it initiate feminine courtship and sexual behaviors.

II. SIX ADDITIONAL EXAMPLES

A. Proopiomelanocortin Gives Rise to Several Behaviorally Active Hormones

An important mechanism in endocrinology is the cleavage of protein hormones, yielding smaller peptides with different neuroendocrine activities. A prominent example is proopiomelanocortin (POMC), a prohormone for adrenocorticotropic hormone (ACTH) and several other hormones. In the human, POMC is encoded by a single gene on the short arm of chromosome 2 and contains three exons and two introns. The structure is well conserved across species. In healthy humans, expression of the POMC gene occurs in abundance in anterior pituitary corticotrophs, to a limited extent in the hypothalamus, and to a minimal degree in some peripheral tissues. It also may be expressed in relatively high levels in certain tumors (*e.g.*, in ACTH-secreting pulmonary carcinomas). Corticotropin-releasing hormone (CRH), arginine vasopressin (AVP), and the proinflammatory cytokine leukemia inhibitory factor (LIF) stimulate POMC gene expression, and glucocorticoids (cortisol, corticosterone) inhibit it. POMC is first cleaved into two subunits, pro-ACTH and beta-lipotropin (β-LPH). Pro-ACTH is further cleaved into ACTH1–39 (the active, circulating corticotropin) and a second fragment that in turn is cleaved into N-proopiocortin (N-POC) and joining peptide (JP).

ACTH1–39 itself is cleaved into alpha-melanocyte-stimulating hormone (α-MSH or ACTH1–13) and corticotropin-like intermediate lobe peptide

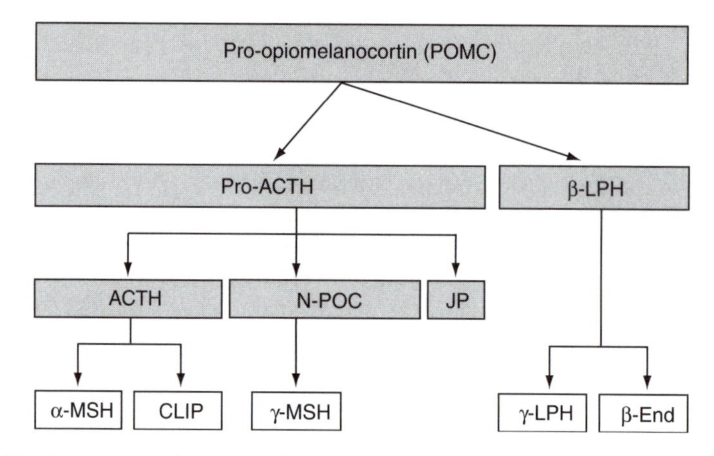

FIGURE 4.3. Proopiomelanocortin (POMC) serves as a prohormone for smaller peptide hormones that have a variety of metabolic actions. ACTH, adrenocorticotropic hormone; N-POC, N-proopiocortin; JP, joining peptide; MSH, melanocyte stimulating hormone; CLIP, corticotropin-like intermediate lobe peptide; LPH, lipotropin; End, endorphin. (Modified from White, A. and Ray, D., in *Endocrinology*, 4th ed., DeGroot, L.J. and Jameson, J.L., Eds., W.B. Saunders, Philadelphia, PA, 2001, p. 225. With permission.)

(CLIP or ACTH18–39). Cleavage of N-POC results in the production of γ-MSH. Finally, β-LPH, a product of the original prohormone, POMC, is cleaved into γ-LPH and β-endorphin. POMC, Pro-ACTH, ACTH, N-POC, JP, and β-LPH all can be detected in the human circulation. The other products do not circulate in detectable amounts but can have important hormonal effects (*e.g.*, the opiate-like analgesic property of β-endorphin in the CNS. Some of these products are more abundant in non-primate species—for example, α-MSH, which is produced in abundance in the intermediate lobe of the pituitary in species such as rat and mouse. Figure 4.3 illustrates the peptide hormones and fragments produced from POMC.

Proopiomelanocortin processing varies, depending on the tissue and species. In the human anterior pituitary, cleavage products are pro-ACTH (which is then cleaved to ACTH), N-POC, JP, and beta-LPH. There is minimal production of γ-MSH from N-POC or β-endorphin from β-LPH. In the rodent intermediate lobe, POMC present in melanotrophs is converted to smaller fragments including α-, β-, and γ-MSH, CLIP; and β-endorphin. α-MSH stimulates melanocytes (pigment cells) in skin and influences, for example, coat coloration in rodents and skin pigmentation in frogs. In the hypothalamus, ACTH is processed to CLIP and desacetyl α-MSH, and β-LPH is processed to β-endorphin. Thus, POMC serves as a prohormone for a number of smaller peptide hormones that have a variety of metabolic actions throughout the body.

B. Dehydroepiandrosterone and DHEA-Sulfate Are Interconvertible and Have Different Potencies

The unconjugated steroid dehydroepiandrosterone (DHEA) and its sulfate ester, DHEA-S, are present in the CNS and are both neurotrophic. They are neurosteroids; that is, they are synthesized *de novo* in the CNS (see Chapter 1) and are interconvertible. DHEA stimulates axonal growth, an effect that can be blocked by the glutamatergic *N*-methyl-D-aspartate (NMDA) receptor antagonist, dizocilpine (MK801). DHEA-S stimulates dendritic growth, but this effect is not blocked by NMDA receptor antagonism. DHEA and DHEA-S also are excitatory neurosteroids that increase the neuronal firing rate. In this regard, DHEA-S is an allosteric γ-aminobutyric acid (GABA) type A receptor antagonist, decreasing the activity of inhibitory GABA systems in the brain. DHEA mimics this effect of DHEA-S, but with only about one-third its potency.

C. Angiotensin Metabolites

In Chapter 2 we saw how angiotensin II, the octapeptide, is involved in thirst and drinking. Angiotensin II has been intensively studied, and its synthesis and receptors are now well known. It now appears that many of its actions are actually due to its metabolites. Circulating plasma angiotensin is formed from renin in the kidneys in response to lower volume or low pressure in the renal artery. Renin is secreted from the juxtaglomerular apparatus (JGA) and is the specific enzyme for hydrolyzing angiotensinogen (AGT), found abundantly in the liver. The renin enzymatic action cleaves angiotensinogen to form angiotensin I. Angiotensin I (10 amino acids) is then converted by angiotensin converting enzyme (ACE) to form angiotensin II (8 amino acids). ACE is predominately found in the lungs and blood vessels. In this way, angiotensin II is formed peripherally and circulates in the blood. Thus, in response to hypovolemia (see Chapter 3), angiotensin is the primary hormone secreted. It reduces blood vessel size by direct vasoconstriction, it induces thirst to increase water intake, it releases vasopressin to concentrate urine, and it releases aldosterone from the adrenal glands to increase sodium reabsorption so that the serum sodium is increased. However, angiotensin II itself may be metabolized into a number of metabolites by other enzymes (Fig. 4.4). Aminopeptidase A (AspAP) cleaves the aspartyl amino acid in position 1 from angiotensin II to form angiotensin III (or angiotensin 2–8). Angiotensin III has been shown to be potent in the brain in inducing thirst and is released at the same time as angiotensin II. This has given rise to the proposition that the effects of angiotensin II are

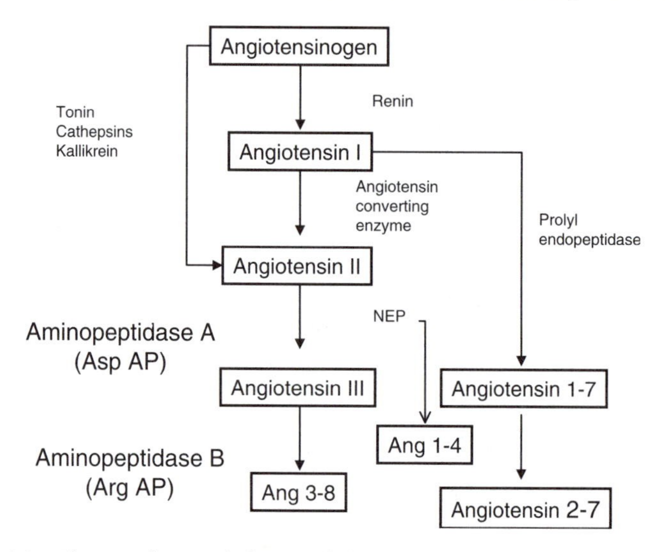

FIGURE 4.4. The complexity of the metabolic pattern starting with angiotensinogen is apparent. Metabolites of angiotensinogen are produced by a number of different enzymes. The activity of specific enzymes will generate metabolites that have different effects. NEP, neutral endopeptidase. The metabolites act on specific receptors.

in fact mediated by angiotensin III (Wright *et al.*, 2003). Angiotensin III is itself metabolized by aminopeptidase B (ArgAP), which removes the arginine in position 2 to form the hexapeptide angiotensin 3–8. This metabolite may have specific receptors. Angiotensin III has physiological effects on aldosterone release and vasoconstriction, and the hexapeptide angiotensin 3–8 has been shown to release vasopressin. Neutral endopeptidase (NEP) can produce angiotensin 1–4 by cleaving off the position 5–8 of angiotensin II. Angiotensin 1–4 is an active metabolite. An entirely different route is through prolylendopeptidase, which metabolizes angiotensin I by removing the phenylalanine position at the carboxyl terminal to produce angiotensin 1–7. This metabolite appears to have the opposite effects of angiotensin II in that it causes vasodilation. To do this, there have to be specific receptors for these metabolites because they do not activate the angiotensin II receptors (AT_1R or AT_2R). High-affinity binding sites for angiotensin 3–8 are widespread in many tissues, including the kidney, heart, brain, and blood vessels. Angiotensin 3–8 stimulates the release of plasminogen-activating factor inhibitor (PA-I), which inhibits plasminogen activator, increasing thrombosis in the blood (Kerins *et al.*, 1995). When ACE inhibitors are given, the conversion of angiotensin I to angiotensin II is inhibited. This increases angiotensin I as a substrate, and the levels of angiotensin 1–7 (formed from angiotensin I) are significantly increased.

Because ACE inhibitors are used as antihypertensive drugs, one possibility is that their effects not only reduce angiotensin II and its vasoconstrictive effects but also permit increases in the metabolite angiotensin 1–7, which has vasodilatory effects.

D. Adrenal Steroids

Among steroid hormones produced in the adrenal cortex, one fascinating metabolic conversion is between two sets of steroids each with important but different spectra of behavioral actions. Corticosterone is a classic glucocorticoid stress hormone, while aldosterone, a mineralocorticoid, protects blood fluid volume by, among other actions, stimulating salt appetite. The enzyme steroid-18-hydroxylase accounts for the metabolic conversion from corticosterone to aldosterone. In molecular terms, this steroid conversion becomes even more interesting because of the different affinities of two nuclear receptors, both transcription factors. Mineralocorticoid receptors (MRs) are high-affinity receptors with high selectivity for their ligands and a restricted neuroanatomical distribution, notably in the hippocampus. Glucocorticoid receptors (GRs) have lower affinity, readily bind both glucocorticoids and mineralocorticoids, and show a much broader neuroanatomical distribution. Most important, GRs mediate effects that are often (as shown by the work of deKloet and colleagues in Leiden) opposite those of MRs. Within the field of stress-related behaviors themselves, liganded MRs seem to work through the rapidly responding CRH-1 receptor system. In contrast, GRs are more connected with the slower CRH-2 receptor system, through which they help to restore physiological balance following stress.

At the intersection between stress physiology and learning, as well, adrenal steroids acting primarily through MRs or GRs have opposing roles. In the hippocampus, the electrophysiological phenomenon called *long-term potentiation* (LTP) is widely used as a tractable experimental model for some forms of learning. Pavlides and co-workers at Rockefeller University used LTP in adrenalectomized rats to study the effects of adrenal hormones on the amplitude of the electrical response of neurons in the dentate gyrus of the hippocampus to electrical stimulation of hippocampal inputs. (Fig. 4.5). Strikingly, while the mineralocorticoid aldosterone produced a significant enhancement of LTP, a pure GR agonist produced the opposite result, a significant reduction of LTP.

Another enzyme for metabolic conversion among adrenal steroids, 11-beta-hydroxysteroid dehydrogenase-1 (11-β-HSD-1), reveals its medical importance in an entirely different way. Glucocorticoids can be produced locally from inactive metabolites through the action of this enzyme. Because

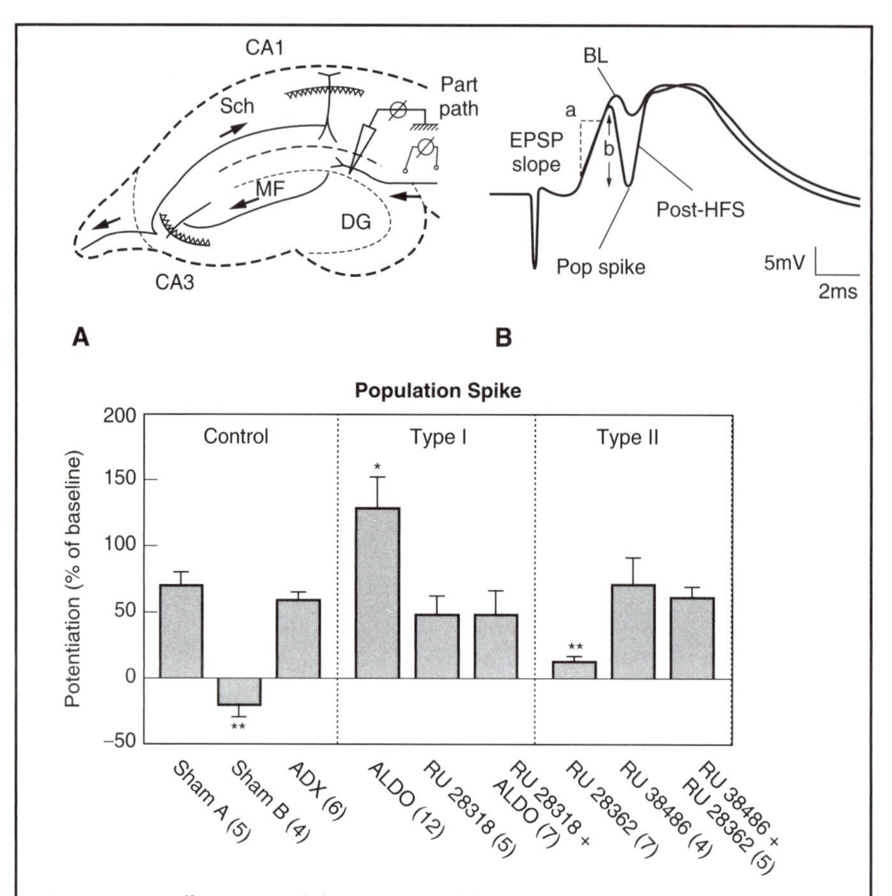

FIGURE 4.5. Different steroid hormone metabolic patterns that yield agonists either for mineralocorticoid receptors (MRs) or glucocorticoid receptors (GRs) have opposite effects on long-term potentiation (LTP) during recordings from hippocampal neurons. *Top panel:* The electrophysiological setup; a stimulating electrode was inserted in the perforant pathway, an input to the hippocampus. A recording electrode was inserted into the dentate gyrus granule cell layer. BL, baseline; DG, dentate gyrus; HFS, high-frequency stimulation; MF, mossy fibers; Sch, Schaffer collaterals. Figure shows examples of electrical recording before potentiation (BL) and after potentiation (Post-HFS). *Bottom panel:* In the left box, the sham-operated animals were split into two groups. Sham A animals had low corticosterone levels, and showed normal LTP, while Sham B animals had high corticosterone levels and significantly suppressed LTP. The middle box illustrates MR receptor manipulations (also called type I receptors). A natural agonist, aldosterone (ALDO) stimulated increased LTP, whereas the receptor antagonist RU28318 did not. The antagonist blocked the ALDO effect. The righthand box illustrates GR receptor manipulations (also called type II receptors). The GR agonist RU28362 had the opposite effect of ALDO in that it reduced LTP, whereas the antagonist did not. Note that the GR antagonist blocked the agonist effect. (Modified from Pavlides, C. and McEwen, B.S., *Brain Res.*, 851(1–2): 204–214, 1999; Pavlides, C. *et al.*, *Neuroscience*, 68(2): 379–385, 1995; Pavlides, C. *et al.*, *Neuroscience*, 68(2): 387–394, 1995.)

high levels of glucocorticoids are associated with visceral fat, diabetes, and consequent mortality, the molecular controls over this enzyme and its consequence are of interest. When the gene for 11-β-HSD-1 is knocked out by homologous recombination, the resultant mice have reduced activation of GR-sensitive liver enzymes and, notably, have a diabetes-resistant phenotype. Flier and colleagues at Harvard Medical School then created a mouse in which a transgene for 11-β-HSD-1 was overexpressed under the control of the strong aP2 promoter. These mice had pronounced diabetes, hyperlipidemia, and, despite high leptin levels, hyperphagia (increased food consumption). The relative importance of a variety of behavioral changes to the complete health picture of these mice remains to be determined.

E. Thyroid Hormones

The thyroid hormone predominantly secreted from the thyroid gland and circulating in the blood is thyroxine (T4, to reflect its four iodine atoms); however, the thyroid hormone predominantly found in the nuclei of cells in the brain and pituitary gland is T3. One iodine has been lost. The enzyme deiodinase type 2, studied intensively by Larsen *et al.*, at Harvard Medical School, is in fact present and active in the CNS and in the pituitary and accounts for the severance of one iodine atom. On nuclear thyroid hormone receptors, T3 is a more effective ligand than T4. As a consequence, the regulation of deiodinase-2 synthesis and enzymatic activity powerfully controls the concentration of active thyroid hormone in nerve and glial cell nuclei and can buffer the brain's T3 defense against problems with thyroid gland production. For rapid actions of thyroid hormones, (for example, on Na^+/H^+ exchanger activity), the metabolite T3 is more effective than the prohormone T4. All of these details of thyroxine metabolism must bear on the ability of thyroid hormones to support actions requiring physical and mental energy and to affect mood.

F. Progestins

Progesterone (P) acts in the brain in the chemical form of progesterone itself and also acts as a prohormone. While P binding to the nuclear progesterone receptor (PR) in hypothalamic neurons is proven to be a behaviorally important mechanism, P is also metabolized into reduced progestins such as 5-alpha-pregnane-3alpha-ol-20-one. The behaviorally important responses to these reduced metabolites, called *neurosteroids*, typically are not due to binding to nuclear PRs but instead are due to a potentiation of the action of the inhibitory neurotransmitter GABA through its GABA-A receptor. The

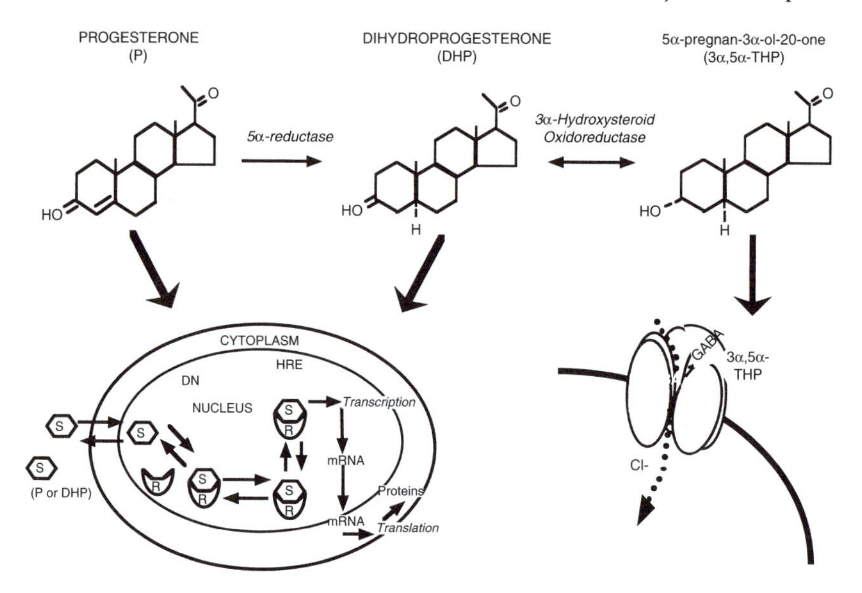

FIGURE 4.6. When progesterone is administered to an animal or a human, some of its behavioral effects are consequent to progesterone's metabolic products. Figure depicts CNS mechanisms of action of progesterone (P) and its metabolites, dihydroprogesterone (DHP) and 5α-pregnan-3α-ol-20-one (3α,5α-THP). P and DHP have a high binding affinity for intracellular progestin receptors (PRs), whereas 3α,5α-THP is devoid of affinity for PRs. 3α,5α-THP acts as a positive modulator for GABA-A receptors to increase GABA's binding to its site on the receptor and thereby increase chloride ion influx. In the ventral tegmental area, progestins mediate sexual responses of rodents in part through actions of 3α,5α-THP at GABA-A receptors (Frye, 2001a,b).

effects of neurosteroids that are progesterone metabolites on behavior are rapid, not allowing time for new gene expression and protein synthesis (Fig. 4.6). Behavioral effects include not only the rapid facilitation of lordosis, a female sex behavior, but also the reduction of anxiety. The clinical importance of these developments is illustrated by the ability of high doses of progesterone metabolites, given to human patients, to act as anesthetics.

III. SOME OUTSTANDING NEW QUESTIONS

1. How might genetic conditions such as 5-α-reductase deficiency be detected early enough in gestation so that gene therapy or hormone therapy might be instituted?
2. What sorts of intrauterine therapies might be developed for genetically determined hormone deficiencies?
3. Should angiotensin II be the target for drug treatment of high blood pressure or should its metabolites?

CHAPTER 5

There are Optimal Hormone Concentrations: Too Much or Too Little Can be Damaging

Because the primary importance of this principle is clinical, we have not included a Basic Science section in this chapter. Most endocrine problems are the result of hormones being too much, too little, too early, or too late. It should be noted that the mechanisms of hormone action involve not simply the temporal envelope of its concentration, but also the sensitivity of the cellular response to the hormone. This is mediated by specific receptors that detect and bind to the hormone and then signal to the cell over a variety of pathways, thus shaping the response of the cell. Therefore, determining optimal levels for a given hormone must be done keeping all of these mechanistic steps in mind.

I. TOO MUCH

To begin with a subtle example, it is possible to have the appearance of too much hormone but no cellular response because the receptors of the hormone are downregulated. An example of this was thought to occur in some cases of diabetes mellitus type 2 (T2DM), where the insulin concentration in plasma can be elevated, yet there is no effect (*i.e.*, glucose is not taken up by the cells). If the number of insulin receptors on the surface of the cell is reduced (downregulated) or the signaling component of the receptor is defective, the cell would be resistant to insulin. Insulin resistance is the hallmark of T2DM; however, other factors such as fatty acids, adiponectin, and peroxisome-proliferator-activated receptor (PPAR) alpha and gamma are also involved. PPAR gamma agonists, such as rosiglitazone, increase insulin sensitivity and are used to treat T2DM. Arizona's Prima Indians have a very high incidence of type 2 diabetes which is associated with obesity and low levels of adiponectin, suggesting adiponectin protects against insulin resistance.

A. Testosterone

When the steroid testosterone is abused by athletes, supraphysiological levels are taken to increase muscle mass and strength. Too much hormone switches off releasing hormones from the pituitary (luteinizing hormone [LH] and follicle-stimulating hormone [FSH]) and also downregulates the testosterone receptors on the germinal cells necessary for sperm formation. Low sperm count and infertility are associated with steroid abuse in males. Importantly, excess testosterone has profound behavioral effects. This is because testosterone, being a steroid, can easily cross the blood–brain barrier to enter the brain. In the case of abuse, testosterone acts on receptors directly or is metabolized by a steroid reductase to dihydrotestosterone (DHT) and activates receptors in the amygdala, hypothalamus, and limbic system. The receptors are inside the cell (cytosolic steroid receptors) and bind DHT. This forms a receptor complex that migrates into the nucleus and binds to DNA and activates it. This process, *transactivation*, leads to a cellular response: the production of proteins, including neurotransmitters such as norepinephrine. The release of these transmitters affects behavior. Norepinephrine serves the sympathetic nervous system for a "fight or flight" response. Unfortunately, the steroid addict too often chooses to fight. With excess testosterone, there is an increased amount of aggression, sometimes including a lowered threshold for violent behavior. There is also increased libido or sex drive, despite the lowered sperm count.

II. TOO LITTLE

A. Insulin

Obviously, a paucity of hormone is not going to be as effective as normal levels. Right? Well, just as too much can cause receptors to downregulate so can too little cause receptors to upregulate; however, there is a limit to how effective this compensatory action can be. In the case of diabetes mellitus type 1, where insulin is not produced due to death of the insulin-synthesizing cells in the pancreas (the beta cells of the islets), there is too little insulin no matter how much upregulation of receptors occurs. The first signs of insulin-dependent diabetes are polyuria and thirst (polydipsia), because the glucose in the blood (hyperglycemia) that is not entering cells acts as an osmotic stimulus to draw water out of cells. Also, patients are weak and literally starving to death without the insulin required to activate glucose uptake. (See also Chapter 1.)

B. Vasopressin

We are 75% water. There is a constant need to maintain fluid balance by drinking water. The amount of water ingested has to be balanced by the quantity and concentration of urine produced in order to regulate fluid volume and osmolarity within a normal range. The principal hormonal mechanism for controlling urine output is vasopressin. Vasopressin is released from the posterior pituitary gland (the neurohypophysis), which is directly connected to the magnocellular cells of the paraventricular nucleus (PVN) and supraoptic nucleus (SON) of the hypothalamus where vasopressin is synthesized. Vasopressin acts on the renal collecting duct to enhance water reabsorption. It does this by activating aquaporins, which are proteins with water channels that run from inside the cell to the apical membrane of the renal collecting duct cells. The aquaporins allow water to be reabsorbed across the membrane from the collecting duct back into cells and then into the blood. In diabetes insipidus, there is excessive production of diluted urine. This constant loss of fluid leads to increased thirst and drinking. (See also Chapter 15.)

C. Androgens

Another example is too little testosterone. This can range from hypogonadism, where the testes are not producing enough hormone, to complete loss of testes by castration (orchiectomy). History has many cases of the latter—for example, in China and Arabia where eunuchs were used in households and administration on the assumption that they lacked sexual desire and physical strength so they would not be a threat to their owners. In Italy in the 17th and 18th centuries, prepubertal boys were castrated so they could continue to sing with a high soprano voice (*castrati*). Without testosterone, their voices remained soprano, but as they grew into adulthood the increased lung capacity gave a physical strength to the voice that was unique. The custom lasted at least until the late 1800s; the last of the *castrati* recorded his voice in 1902.

III. TOO EARLY, TOO LATE

The correct timing of the actions of hormones in development is essential for normal growth (see also Chapter 9). The wrong amount of hormone too early can have profound behavioral consequences. An enzymatic defect in 21-hydroxylase in girls with congenital adrenal hyperplasia (CAH) causes a masculinization of their genitals, and they develop as children with pronounced boy-typical behavior. Likewise, animal experiments have shown that the exposure of a fetus to testosterone determines future sexual characteristics; if the exposure is late, it can lead to masculinization of female fetuses, both physically and mentally. Finally, stress in the prenatal period can alter offspring behavior, as stress results in release of cortisol and testosterone. Studies of maternal stress in women 32 weeks pregnant, when corticoid receptors are developing, showed a high risk for behavioral problems in their children compared to those offspring of mothers whose stress came at 18 weeks (see also Chapter 9).

IV. CLINICAL CONDITIONS, ORGAN BY ORGAN

Clinical conditions in which target endocrine glands are underactive or overactive can be caused by pathology at several levels of the endocrine axis. The site of pathology can be in the target endocrine gland itself, the pituitary gland, the hypothalamus, or higher brain centers. Changes in gland size often result from under- or over-stimulation; for example, the adrenal cortex atrophies (shrinks in size) and becomes hypoplastic (loses cells) without adrenocorticotropic hormone (ACTH) stimulation, and it hypertrophies (becomes overly large in size) and hyperplastic (increases in cell number) with excess ACTH stimulation. Different behavioral pathologies also result, depending on the hormone affected. Diagnosis is confirmed by measurement of circulating hormone concentrations under baseline conditions and following stimulation and suppression tests. Following are examples of clinical conditions related to adrenal, thyroid, and growth hormone function that illustrate these circumstances.

A. Adrenal Gland Hypofunction

Adrenal insufficiency is classified into two types, depending on the site of pathology producing it. Primary adrenal insufficiency, or Addison's disease, results from hypofunction of the adrenal gland itself. In this condition, therefore, both of the main adrenal cortical hormones, cortisol and aldosterone, are deficient. Because feedback of cortisol to the hypothalamus and pituitary is reduced, circulating ACTH levels are correspondingly elevated. Secondary adrenal insufficiency, in contrast, results from pathology in the hypothalamus and/or pituitary gland; therefore, ACTH secretion is deficient, which then leads to reduced secretion of cortisol by the adrenal cortex. Because aldosterone secretion is regulated mainly by the renin–angiotensin system, its secretion is not impaired in secondary adrenal insufficiency.

The actions of cortisol that are most affected in adrenal insufficiency are reductions in negative feedback regulation of ACTH secretion, modulation of the vasoconstrictor response to β-adrenergic agonists, maintenance of cardiac muscle inotropy, and antagonism of insulin secretion. In Addison's disease, the actions of aldosterone that are most affected are reductions in the retention of sodium ions and excretion of potassium and hydrogen ions by the kidney. Patients with adrenal glucocorticoid (cortisol) insufficiency show hypotension secondary to decreased peripheral vascular resistance and reduced cardiac output, increased heart rate (tachycardia), and, in some cases, hypoglycemia. Patients with primary adrenal insufficiency also can have isosmotic dehydration secondary to reduced mineralocorticoid (aldosterone) secretion and hyperpigmentation secondary to increased ACTH secretion.

Behavioral changes are secondary to the physiological abnormalities and include weakness, fatigue, and loss of appetite. Additional symptoms can include weight loss, dizziness, nausea, diarrhea, and abdominal, muscle and joint pain. Adrenal insufficiency can be fatal; both glucocorticoid and, when indicated, mineralocorticoid hormone replacement therapy are mandatory. Figure 5.1 illustrates some clinical features of a patient with primary adrenal insufficiency.

B. Adrenal Gland Hyperfunction

Adrenal hyperfunction, or Cushing's syndrome, results from excessive tissue exposure to cortisol. The excess cortisol production may be ACTH-dependent or ACTH-independent. The former condition results from increased

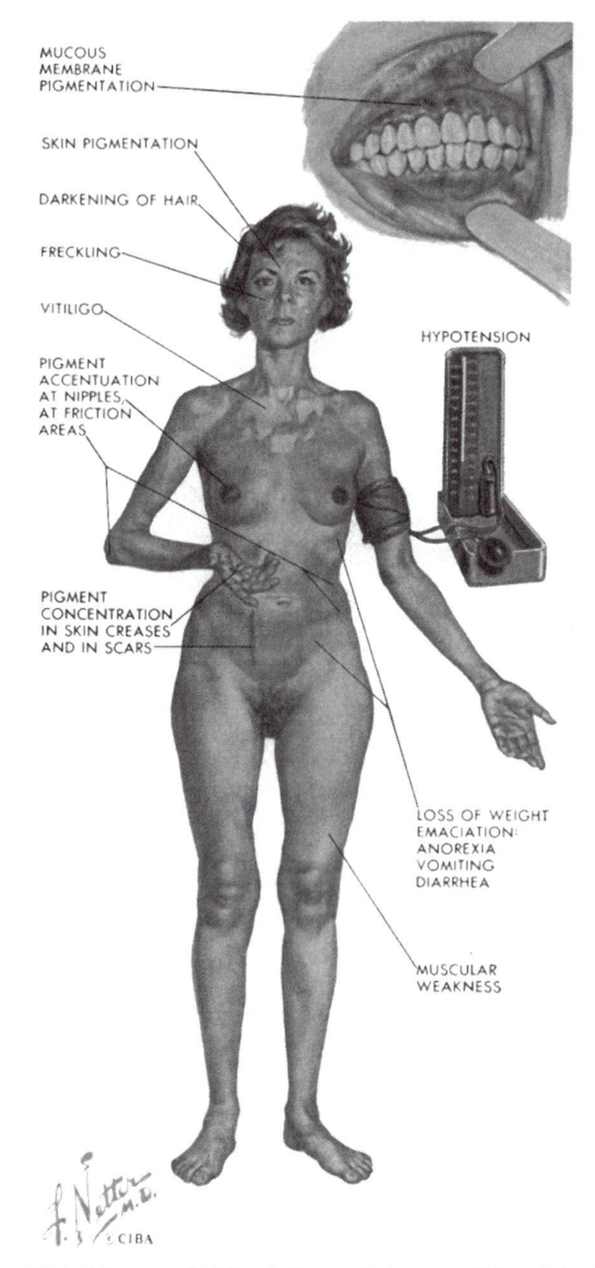

FIGURE 5.1. Clinical features of Addison's disease. Patients would have behavioral symptoms of inadequate glucocorticoid levels. (From Netter, F.H., Icon Learning Systems, LLC, 1965. With permission.)

ACTH secretion from the pituitary, ectopic sources of ACTH (*e.g.*, malignancies of the lung), and ectopic production of corticotropin-releasing hormone (CRH). Because many of the first Cushing's patients identified had basophilic adenomas of the anterior pituitary (corticotroph tumors) which produced excessive ACTH, patients with this particular etiology are considered to have Cushing's disease. ACTH-independent Cushing's syndrome results from adrenal gland tumors and other pathologies and from exogenous glucocorticoid administration for therapeutic purposes; in these instances, negative feedback of cortisol or administered glucocorticoid results in very low circulating ACTH concentrations.

The secretion of ACTH and cortisol normally has a prominent circadian (24-hour) rhythm that is related to the sleep/wake cycle. Little or no hormone secretion occurs in the early morning hours (midnight to 3:00 or 4:00 a.m.). Both hormones are then episodically secreted, with ACTH driving cortisol, which reaches peak blood concentrations between 7:00 and 8:00 a.m., about the time of awakening. The amplitude of the circadian rhythm of circulating cortisol is about threefold, so that over a 24-hour period tissues are exposed to widely varying cortisol levels. In contrast, in Cushing's syndrome, cortisol is secreted not only excessively but also consistently around the clock; that is, its circadian rhythm is blunted or lost. The high cortisol levels may fluctuate very little, or they may show episodic spikes. For example, tumors of the adrenal gland usually produce a high, steady output of cortisol, whereas driving of the adrenal gland by excess ACTH from a pituitary tumor often produces a high, fluctuating cortisol output.

Irrespective of the cause of excess cortisol production, the clinical features of Cushing's syndrome are similar. One of the earliest signs, increased fat deposition, occurs in the vast majority of patients, and difficulty in maintaining weight is a common first complaint. The fat deposition is primarily central (face and trunk) and leads to the characteristic "moon facies," "buffalo hump" on the upper back, and a collar of fat above the clavicles. Concomitant loss of subcutaneous tissue produces several signs, including thinning of the skin, easy bruising, facial flushing, and reddish-purple striae over the lower trunk. Thinning of bones can result in fractures, typically of the feet, ribs, and vertebrae, and weakness of proximal muscles can lead to difficulties in mobility. As well, biochemical changes include retention of sodium ions, loss of potassium ions, glucose intolerance or frank diabetes, reduced lymphocyte count, and suppressed thyroid and gonadal hormone output. In ACTH-dependent Cushing's syndrome, increased circulating ACTH also can cause hyperpigmentation, similar to primary adrenal insufficiency.

Prominent changes in several behavioral domains are characteristic of Cushing's syndrome. Alterations in mood and affect occur in many patients. Irritability occurs in most patients and is often the first behavioral symptom, beginning with the earliest manifestations of fat deposition and weight gain. Irritability varies in intensity, with some patients simply being more sensitive to minor irritations and others feeling close to exploding emotionally. Depressed mood also occurs in most patients, varying in intensity from short spells of sadness to, rarely, feelings of helplessness and hopelessness. Crying spells may accompany these feelings. In contrast to primary major depression, the depressed mood of Cushing's syndrome is usually episodic rather than sustained, often occurring suddenly and lasting 1 to 2 days, interspersed with nondepressed intervals. Autonomic activation, such as shaking, sweating, and palpitations, may be an accompanying feature. Also in contrast to patients with major depression, Cushing's patients do not often experience social withdrawal, excessive guilt, or marked severity of depressive symptoms; they usually feel worse in the evening, not in the morning; and they often are cognitively impaired (see below).

A few patients may experience episodes of elation or hyperactivity early in their illness, feeling more ambitious than usual and having increased activities and pressured speech. This phase often disappears as the illness progresses and other behavioral characteristics emerge.

Alterations in cognition are prominent. Memory impairment, especially for new information, is very common. Other symptoms include difficulty concentrating, shortened attention span, distractibility, slowed and scattered thinking, and, in more severe cases, thought blocking. Difficulty with tests such as mental subtraction and recall of common information occurs in many patients. As with affect and mood symptoms, cognitive impairments can range from minimal to severe. Of note, they are not accompanied by any disorientation or clouding of consciousness (delirium).

Alterations in biological functions comprise a third domain of disturbance. Fatigue occurs in all patients, and reduced sex drive occurs in most. Appetite disturbances also occur in most patients, increased appetite being more common than decreased appetite. Sleep disturbances occur in most patients as well and usually consist of insomnia in the middle and late parts of the sleep period, along with frequent, intense, bizarre, and vivid dreams.

In contrast to most patients with endogenous Cushing's syndrome, some patients who are receiving exogenous glucocorticoid treatment for medical conditions develop mental disturbances of a psychotic or confusional nature. These patients often receive high doses of potent synthetic steroids over a short period of time, in contrast to endogenous Cushing's patients in whom steroid increases can occur over months to years.

Treatment of Cushing's syndrome is, of course, directed toward the locus of pathology and often involves surgery for removal of a pituitary adenoma, an adrenal tumor, or an ectopic source of ACTH or CRH. If pituitary surgery fails to remove the entire tumor, irradiation of the remaining tissue can be undertaken. Drugs that interrupt the enzyme pathways for the synthesis of cortisol are effective in some cases. Fortunately, impairments in all behavioral domains improve with reduction of excess hormone secretion. In some patients with lingering depression, antidepressant medication can be helpful. Figure 5.2 provides some clinical features of a patient with Cushing's syndrome.

As well, failure to turn off stress responses associated with high glucocorticoid levels can lead to a variety of problems; among them, depression and signs of hippocampal impairment are prominent possibilities. McEwen, in his book *The End of Stress as We Know It*, has summarized pathologies linked to adrenal gland stress hormones (Table 5.1).

C. Thyroid Gland Hypofunction

As with several other endocrine deficiencies, hypothyroidism is classified into two types, depending on the site of pathology. Primary hypothyroidism, the most common type, is caused by hypofunction of the thyroid gland itself. Because feedback of thyroid hormones to the pituitary and hypothalamus is decreased, circulating thyroid-stimulating hormone (TSH) concentrations are correspondingly elevated. Secondary, or central, hypothyroidism results from pathology in the pituitary gland or hypothalamus; TSH secretion is deficient, and the thyroid gland is understimulated.

The thyroid gland secretes two active hormones, triiodothyronine (T3) and thyroxine (T4). These are interconverted by enzymes in the thyroid and peripheral tissues. T4 is the major hormone secreted and serves primarily as a prohormone for T3, which is biologically more potent. T3 and T4 are rich in iodine, a necessary substrate for normal thyroid function that is supplied in the diet (*e.g.*, iodized salt). Thyroid hormones have important effects in two broad areas: cellular differentiation during development and maintenance of metabolic pathway activity in adulthood. The most characteristic finding of hypothyroidism is a general slowing of physical and mental activity involving many organs. Aberrant metabolic pathways lead to a build-up of hyaluronic acid and related compounds in interstitial tissues. Because these substances are hydrophilic, their presence leads to a mucinous edema of the skin and internal organs (myxedema).

During fetal and postnatal development, thyroid hormones in the central nervous system (CNS) promote neuronal and glial cell proliferation and maturation, myelination, and synthesis of enzymes critical in neurotransmitter and neuromodulator pathways. Thyroid hypofunction during these early periods therefore can result in neurological abnormalities and mental retardation, in addition to delayed body growth (cretinism).

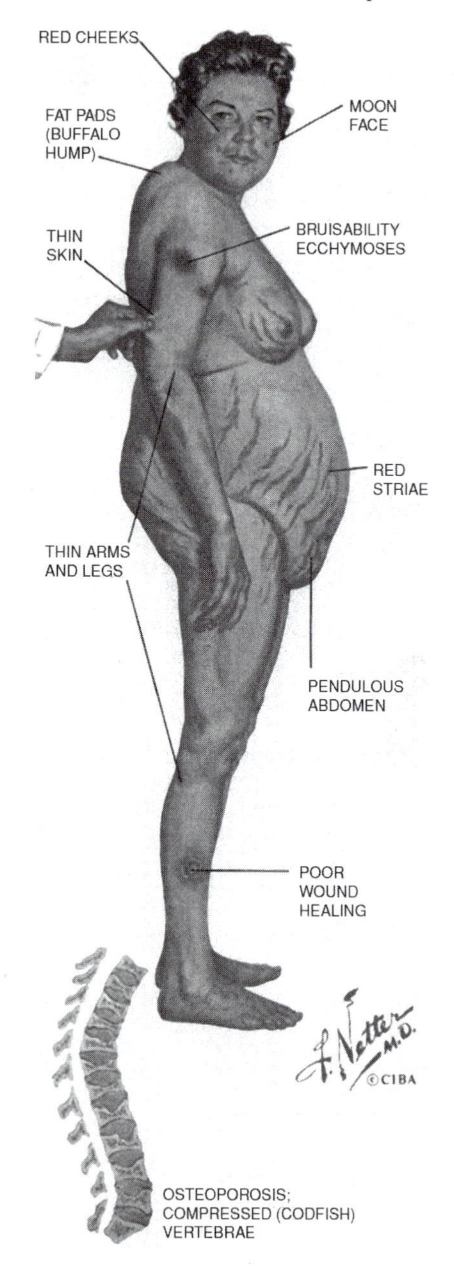

FIGURE 5.2. Clinical features of Cushing's syndrome. Patients would have behavioral symptoms of a chronic excess of glucocorticoid hormones. (From Netter, F.H., Icon Learning Systems, LLC, 1965. With permission.)

TABLE 5.1 Disorders Linked to Over- and Underproduction of Cortisol

Overproduction	Underproduction
Cushing's syndrome	Atypical/seasonal depression
Melancholic depression	Chronic fatigue syndrome
Diabetes	Fibromyalgia
Sleep deprivation	Hypothyroidism
Anorexia nervosa	Nicotine withdrawal
Excessive exercise	Rheumatoid arthritis
Malnutrition	Allergies
Panic disorder	Asthma
Chronic active alcoholism	
Childhood physical and sexual abuse	
Functional gastrointestinal disease	
Hyperthyroidism	

Adapted from McEwen (2002).

In adulthood, hypothyroidism also results in metabolic slowing of all tissues, including the CNS. Patients complain of being tired, cold, and unable to think clearly. They gain weight, due to increased body fat and sodium and water retention. Their skin becomes cold, dry, rough, and scaly, because of epidermal thickening and myxedema; their hair becomes dull, coarse, brittle, and thinned; and the tongue becomes edematous. Heart rate and other organ functions are slowed, and complaints include constipation and menstrual irregularities. Such patients also are slow in movement and thought, and they have impaired concentration and memory. Excessive sleepiness is common, as are depressive symptoms. Left untreated, patients may lapse into coma. Rarely, in addition to their other, severe mental impairments, patients may become extremely anxious and agitated, a condition termed *myxedema madness*. In the elderly, hypothyroidism may present first as depression or cognitive decline, with few or no physical symptoms. Because of the shared symptoms between hypothyroidism and these behavioral disorders, many psychiatric facilities routinely screen patients for thyroid status.

Treatment of hypothyroidism at all stages of life is straightforward: thyroid hormone replacement to achieve the euthyroid state in all tissues. This is most often done with T4 and usually slowly, so as not to overtax the heart or produce additional mental disturbances such as mania. Fortunately, if patients are diagnosed and treated early in the course of their illness, most physical and mental symptoms are completely reversible. Figure 5.3 provides some clinical features of an adult patient with myxedema.

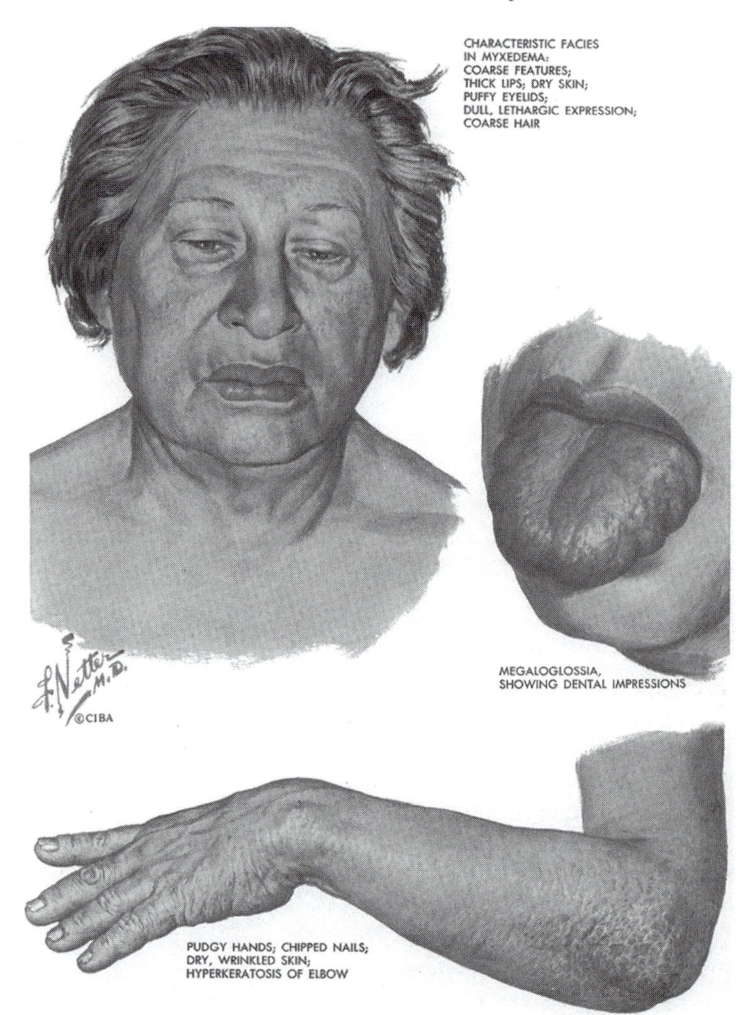

FIGURE 5.3. Clinical features of myxedema. Patients would have behavioral symptoms of inadequate thyroid hormone levels. (From Netter, F.H., Icon Learning Systems, LLC, 1965. With permission.)

D. Thyroid Gland Hyperfunction

As might be expected, hyperthyroidism results in metabolic changes opposite those in hypothyroidism. Primary hyperthyroidism, the most common form, is caused by hyperactivity of the gland itself, frequently on the basis of an organ-specific autoimmune disease (Graves' disease) resulting from circulating autoantibodies to the TSH receptor that mimic the effect of TSH on the gland. Psychological stress has been implicated as one precipitating factor, perhaps as an immune-response rebound following immunosuppression by stress-related hypothalamo–pituitary–adrenal cortical (HPA) axis activation (see also Chapter 6). Prominent pathology includes diffuse enlargement of the thyroid gland and protruding eyes, secondary to an inflammatory increase in retro-orbital tissue volume (Graves' ophthalmopathy). Hypermetabolic symptoms include heat intolerance, warm skin, sweating, increased heart rate and palpitations, fatigue, weight loss, and muscle weakness and wasting.

Common mental and behavioral complaints include nervousness, jitteriness, irritability, anger, anxiety, panic-like episodes, and emotional lability; rapid and disjointed speech; and insomnia and fatigue. Changes in cognition and memory are apparent on neuropsychological testing. About 10% of patients may have severe psychiatric illness including mania, psychotic states, and delirium; these conditions occur mainly when patients have a sudden, severe increase in thyroid hormone production (*thyrotoxic storm*). As with hypothyroidism, because of shared symptoms between hyperthyroidism and several behavioral disorders, psychiatric patients presenting with severe anxiety, panic attacks, and unexplained psychosis or delirium are often tested for thyroid status.

The primary treatment of hyperthyroidism is identifying the cause of excessive thyroid hormone production and eliminating it. Pharmacotherapy is often the immediate approach to controlling the hypermetabolic state, with treatments such as surgery to remove a hormone-secreting tumor and radioiodine therapy being more permanent approaches. Because the thyroid gland has a strong avidity for iodine, radioactive iodine will concentrate in and destroy glandular tissue. Following radioiodine therapy, the thyroid is usually underactive, so that T4 replacement must be given, generally for a lifetime. Fortunately, when patients are adequately treated and tissue metabolic rate returns to normal, most physical and mental symptoms are completely reversible. Figure 5.4 provides some clinical features of a patient with thyroid hyperfunction.

E. Growth Hormone Excess

Growth hormone (GH) is secreted by the anterior pituitary in response to stimulation by hypothalamic GH-releasing hormone and other secretagogues such as ghrelin and to inhibition by GH-inhibiting hormone (somatostatin), which are themselves regulated by higher CNS centers. GH acts directly

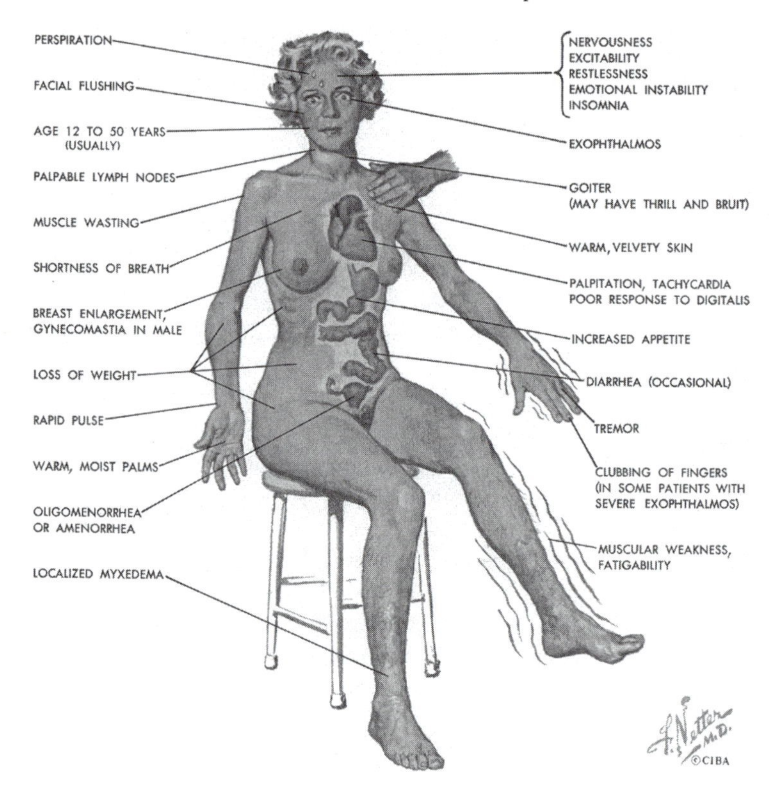

FIGURE 5.4. Clinical features of Graves' disease. Patients would have behavioral symptoms of a chronic excess of thyroid hormones. (From Netter, F.H., Icon Learning Systems, LLC, 1965. With permission.)

to stimulate target stem cells, including neural stem cells. As these cells differentiate, they develop receptors for (and produce) insulin-like growth factor 1 (IGF-1; somatomedin-C). In response to GH stimulation, IGF-1 and its binding protein are produced in the liver and circulate in blood; IGF-1 also is locally produced in kidney, gastrointestinal tract, muscle, cartilage, and pituitary. Cells that express IGF-1 receptors are responsive to the growth-promoting effects of both circulating (endocrine) and locally produced (paracrine) IGF-1. GH acts through IGF-1 to enhance DNA, RNA, and protein synthesis and the growth of many tissues, including bone, muscle, cartilage, and CNS tissue (neural maturation and glial cell formation). Metabolic effects of GH/IGF-1 include antagonism of the action of insulin, stimulation of the breakdown of adipose tissue (lipolysis), and retention of nitrogen. IGF-1 also feeds back to the pituitary and hypothalamus to stimulate somatostatin secretion, thereby inhibiting GH release.

Growth hormone hypersecretion is usually caused by hyperplasia or a tumor of anterior pituitary somatotrophs. GH hypersecretion in early childhood results in overall excessive somatic growth (gigantism). Such individuals are tall, with a large trunk, long limbs, and proportionally large hands and feet. GH hypersecretion that begins in adulthood occurs after the epiphyses of long bones are fused and growth is complete, so that overgrowth of only acral (peripheral) and soft tissue occurs (acromegaly). Prominent acromegalic features are enlargement of the jaw, nose, and frontal bones of the skull, hands, and feet. The process is generally so slow that features other than cosmetic lead patients to seek treatment; these often are joint and back pains secondary to arthritis and joint degeneration.

Skin thickening, increased sweating and hair growth, voice deepening, and obstructive sleep apnea (due to tongue, laryngeal, and pharyngeal soft tissue enlargement), and entrapment neuropathies such as carpal tunnel syndrome (due to tissue overgrowth) also occur secondary to GH hypersecretion. Cardiac enlargement and diabetes are important causes of disability and death, and the occurrence of gastrointestinal cancer is increased. Excessive GH, therefore, shortens life expectancy, whether the onset is in childhood or adulthood. Behavioral effects of GH hypersecretion appear to be related primarily to the somatic difficulties patients develop. Gigantism produces an additional set of psychosocial problems related to peer-group relationships in school and other venues.

The treatment of GH excess includes surgery to remove a hormone-secreting tumor and pharmacotherapy with octreotide, a synthetic eight-amino-acid analog of somatostatin. Octreotide inhibits GH/IGF-1, glucagon, and insulin secretion. Figures 5.5 and 5.6 provide clinical features of adults with GH hypersecretion beginning in childhood (*i.e.*, gigantism). Clearly, attitudes of patients toward various kinds of social behaviors, including both romantic and aggressive behaviors, will be influenced secondarily by this kind of medical condition.

F. Growth Hormone Deficiency

Growth hormone hyposecretion usually becomes apparent in early childhood, based on a failure to conform to established growth curves. The wide range of causes includes genetic deficiencies, birth trauma, CNS tumors, congenital abnormalities of the hypothalamus or pituitary, and psychosocial stress. Lack of normal amounts of GH and IGF-1 can result in mental retardation to varying degrees. In general, the longer the GH/IGF-1 deficiency persists, the more profound are the CNS deficits, such that some individuals, even

FIGURE 5.5. Gigantism, due to an oversupply of growth hormone since childhood. Shown here is George Auger, who worked with Ringling Brothers Circus. and is shown with a midget and normal-sized persons. Imagine the large number of indirect causes of behavioral changes in such an individual. (Courtesy of Circus World Museum, Baraboo, Wisconsin.)

though treated with adequate amounts of GH or IGF-1, do not develop normal intelligence and may remain significantly mentally retarded. This has occurred in conditions such as Laron syndrome (hereditary IGF-1 deficiency) when hormone treatment was not begun until adulthood. Most often, however, persons with GH deficiency are normal behaviorally and can be very accomplished during their lives. Figure 5.7 shows a group of small stature men and women ("midgets"), likely on the basis of genetically determined GH hypofunction, who were successful entertainers.

Psychosocial dwarfism, or emotional deprivation dwarfism, has been reported in children who have suffered emotional and/or physical deprivation and abuse, including sexual abuse. Such children have reduced GH secretion, and they display failure to thrive and short stature. They may be hyperphagic as well, but remain underdeveloped. Fortunately, most of these children improve their neuroendocrine and physical status when they are removed from the stressful environment (*e.g.*, placed in a hospital for a period of weeks or in a foster home with a nurturing family). The syndrome

FIGURE 5.6. Shown on the left is a 22-year-old man with pituitary-gland-caused gigantism, standing with his identical twin brother who is normal size. Such a comparison allows hormone-dependent size differences and differences in rearing to be analyzed in regard to their influences on behavioral change, separate from genetic differences. (From Gagel, R.F. and McCutcheon, I.E., *N. Engl. J. Med.*, 340: 524, 1999. With permission.)

of psychosocial dwarfism provides another example of the important interplay between stress-responsive, higher CNS centers and neuroendocrine function.

G. Implications for Behavior

It goes without saying that all of the above endocrine maladies have consequences for the body type of the patients; thus, they can affect these patients' behavior in a number of ways, such as depriving them of the ability to enjoy some normal human behaviors. Somatic changes also raise issues of self-image and influence social perceptions of these people by their families and associates. Therefore, these endocrine pathologies not only illustrate

FIGURE 5.7. Abnormally small stature, due to a deficiency of growth hormone since childhood. Shown are The Eagle Midgets, who worked with the Ringling Brothers Circus. (Courtesy of Circus World Museum, Baraboo, Wisconsin.)

the "optimal level of hormone" principle, which is the main point of this chapter, but also remind us of the *indirect* routes by which human behavior can be altered.

V. OUTSTANDING NEW BASIC OR CLINICAL QUESTIONS

As indicated by the above examples, hormone secretion outside the normal range can have far-reaching pathological consequences and can even be life-threatening in some circumstances. Prompt diagnosis and treatment often returns the individual to a normal hormonal state, and the behavioral changes, as well as the physical changes, become normal as well. Nevertheless, there remain some intriguing questions about the antecedents of these abnormal hormonal states and their severity in different individuals, the answers to which will allow more precise diagnosis and treatment:

1. What are the interplays between genetic and environmental factors in the genesis of endocrine disorders? The answer, of course, depends on the disorder; some are primarily constitutional, in that they have a prominent genetic component, and others may develop after an unusual stress situation. What is not understood is how environmental stress may trigger a pathological endocrine response in some individuals (characterized, after the fact, as "susceptible" because they developed the illness).
2. What determines individual human CNS sensitivity to hormone changes; that is, why do some individuals develop severe mental symptoms with a certain hormone change and others do not? For example, why can the same degree of hypercortisolemia in Cushing's disease patients produce profound depressive symptoms in one patient but minimal symptoms in another? Why can the same degree of hyperthyroidism in Graves' disease produce severe symptoms of anxiety in one patient but little or no anxiety in another? What factors determine the degree and type of disruption of neuronal function by abnormal hormone levels?
3. Regarding the neuroendocrine disorders mentioned here, what is the metabolic significance of episodic versus constant secretion of hormones? Do disorders of timing thus contribute to behavioral disorders?
4. For endocrine pathology, what is the metabolic significance of the circadian rhythms of hormones? When do rhythm disturbances lead to behavioral pathologies?
5. Obviously, there are other examples of "too much, too little, too early, too late." Can you think of some?
6. This chapter refers to optimal levels of any given hormone at any given time. Are there some hormone actions that are opposite, according to whether the initial condition, endocrinologically, is to the low side versus the high side of optimal concentration?

CHAPTER 6

Hormones Do Not "Cause" Behavior; They Alter Probabilities of Responses to Given Stimuli

"His hormones caused him to do it?" Wrong. It is not the case that hormones, *in vacuo*, simply spew out behavioral responses without regard to the stimuli coming in. Instead, to understand what hormones really do, we must undertake systematic thinking, beginning with the stimuli themselves. In a specific set of circumstances, well-defined stimuli will have a quantitatively determined chance of triggering a measurable response. In the absence of the hormone in question, the number of occurrences, their latency, and their amplitude are recorded. Then the hormone in question is added, using a dose, route of administration, and time course deemed likely to be behaviorally effective. Again, the same well-defined stimuli are applied and the response measures taken. If the hormone facilitates the behavior of interest, the probability of occurrence goes up, latency goes down, and/or amplitude goes up. If the hormone represses that behavior, the opposite changes are seen. Several examples are given below.

I. BASIC SCIENTIFIC EXAMPLES

Laboratory animals popular for biomedical research include rodents because of their relatively low financial and space requirements, because of animal welfare concerns, and because some aspects of the central nervous system (CNS) are remarkably similar to humans. Most rodents live lives that are heavily dependent on olfactory stimuli. As documented for female rodents by Moffatt in his *Brain Research Reviews* paper, hormones fluctuating during the estrus cycle significantly influence sensitivity to weak olfactory stimuli. The same is true for some women as they pass through the menstrual cycle.

Male rodents also have behavioral repertoires that depend heavily on olfactory stimuli. Consider territorial responses. A male under the influence of testicular androgenic hormones might well do a certain amount of territorial "flank marking" behavior in the absence of any stimulation; however, he certainly will respond to pheromones from a foreign male's scent mark by increasing his own territorial responses, including even covering up the other male's mark with his own flank gland odor (Fig. 6.1). A similar story obtains with sex. A castrated male rodent might casually investigate sources of sexual pheromones from the estrus female of his species; however, the duration and avidity with which he approaches the estrus odor will be greatly increased if adequate levels of testosterone are circulating in his bloodstream.

Many neuroanatomists and neurophysiologists, such as Sarah Newman at Cornell and Barry Keverne at Cambridge, would contend that both olfactory and vomeronasal receptors, signaling through the main and the accessory olfactory bulbs, respectively, impact the amygdala, deep in the primitive basal forebrain, to govern these responses. Certainly, the roles played by estrogens, involved with both estrogen receptor-α and estrogen receptor-β for the support of social recognition are conceived as working this way (see Chapter 3, Fig. 3.5).

The principle behind the discussion in this chapter is especially easy to illustrate with social behaviors because of the obvious participation of stimuli coming from the social partner (or contender) for governing subsequent behaviors by an animal or human being. Therefore, testing animals under circumstances where a full range of social behaviors can be displayed normally is becoming especially important in this field (Fig. 6.1). A well-founded example is the role steroid sex hormones play in raising levels of aggression in humans as well as in animals (see Chapter 1, Fig. 1.1; Chapter 10, Fig. 10.1).

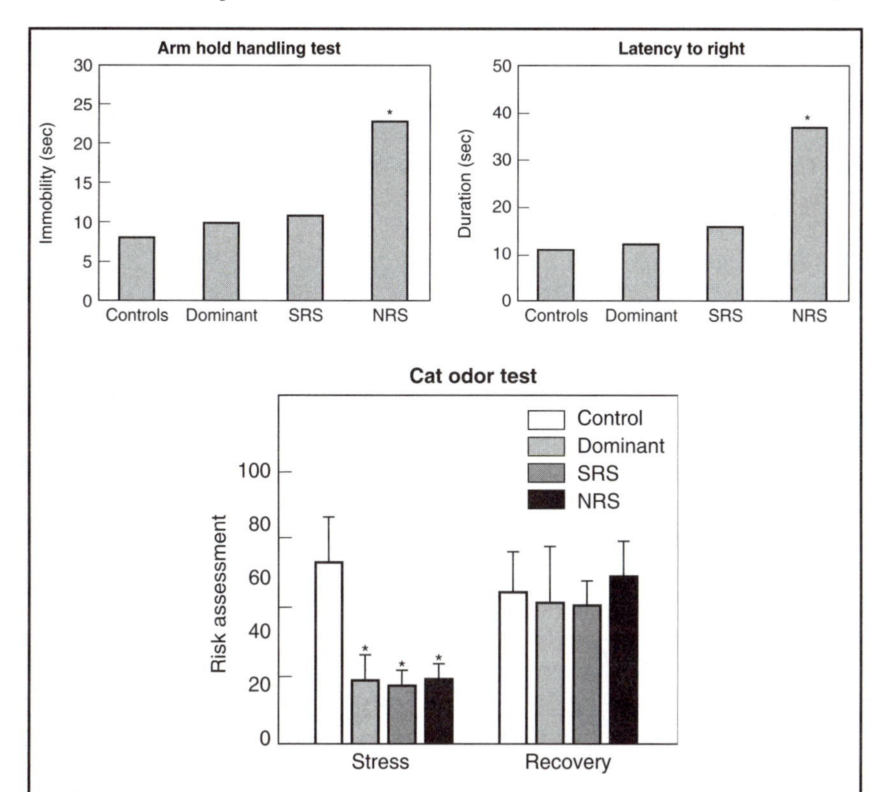

FIGURE 6.1. Hormones by themselves do not simply drive isolated behavioral responses in a vacuum but instead change the likelihood of certain behaviors in response to various stimuli. One of the experimental setups richest in its wide variety of natural social stimuli has been invented by the Blanchards in their lab at the University of Hawaii. In their controlled environment, aggressive and other emotional responses to social stresses can be studied in a manner that takes into account the natural structure of behavior. The Visible Burrow System (VBS) is a habitat providing burrows and an open area for mixed-sex rat groups, typically 4 to 5 males and 2 females. This attractive habitat, plus the presence of females, induces a high level of dominance motivation in males, and their fighting results in rapid formation of a very stable dominance hierarchy that may persist over disbandment and reformation of the colony group. VBS subordinate males have higher basal levels of plasma corticosterone (CORT) and lower levels of corticosterone-binding globulin (CBG) and testosterone (T) than do dominants or controls. They show regional changes in a number of neurotransmitter systems, including serotonin, dopamine, enkephalin, galanin, and mineralocorticoid and glucocorticoid receptors. A subset of stress nonresponsive subordinates (30–45% over a dozen studies) failed to mount a normal CORT response to a novel restraint stress. This group is particularly deficient in corticotropin-releasing hormone (CRH) mRNA in the paraventricular nucleus (PVN) of the hypothalamus and shows elevated D2 density in the nucleus accumbens. All VBS males, compared to controls, show higher levels of potential reorganizational changes such as reduced numbers of dendritic spines and dendritic length in the CA3 region of the hippocampus. Some of these changes are rapid in onset, while others may require several periods of VBS housing, each 1 to 2 weeks in length, to develop.

Males are more aggressive than females, in a variety of species and under a variety of circumstances. Both androgenic hormones (*e.g.*, testosterone) and estrogenic hormones play a role in facilitating aggression. Among men, bodybuilders taking androgens of high potency to build muscles are storied, anecdotally, for occasional hostile acts known as "roid rage" (see Chapter 5). In the normal case, however, aggression in animals and humans does not occur in the absence of provocative stimuli from a conspecific. Among humans, this could be irritating communication, for example, from a person nearby. Another form of aggression, predation, would involve the appropriate set of stimuli from the prey. Picture the classic encounter between two male animals, where both are impacted by testicular androgens (and estrogens, both directly and by metabolism of the androgens). The males notice each other. If and only if influenced by androgenic hormones, one goes into a defensive posture—in rodents, a "boxing" stance. As a result, the second

---

Figure 6.1. (*Continued.*)

VBS subordinate males show dramatically reduced fighting; in fact, they do not fight among themselves at all after dominance is established, and all male–male fights are between the dominant and a subordinate. Subordinates avoid the VBS open area, which is often patrolled by the dominant, and do not enter it without a prolonged period of scanning from the protection of a tunnel. In VBS studies with food and water available in burrows as well as on the surface, subordinates show tunnel-guarding behaviors, remaining in (and blocking) tunnels to the open area for hours, presenting their snouts—and the possibility of a defensive bite—to the dominant as it attempts to enter the tunnel. Subordinates also show reduced social and sexual behaviors with reference to colony females, disturbances in sleep cycles, reduced levels of general activity, and reduced response in the dexamethosone suppression test, all consistent with important symptoms of depression. Tests outside the VBS suggest that subordinates are more anxious and that the stress-nonresponsive subordinates, in particular, utilize passive rather than active defenses; they show reduced locomotion, more freezing, and heightened latencies to right themselves when flipped over.

Although dominants attack female colony members more often than male subordinates, the females show few signs of stress. They move freely through the VBS and actually spend more time in the vicinity of the dominant over days of colony formation. This paradox apparently reflects the fact that females are seldom wounded and that the initial site of male contact with a female is the female's anogenital area, suggesting male attack on females often reflects a response to the rejection of the male's sexual advances. This rejection involves female defenses that effectively protect nonestrus females from mounting and copulation, plus an inhibited (nonwounding) attack by the female on the male. This inhibited female attack tends to repulse the male without inducing a fight, which the much smaller female would inevitably lose.

These findings illustrate some complex interactions between gender, individual differences, and changes in social and environmental conditions in the effects of social stress on behavior. While intraspecific social stress is largely mediated by fighting, animals that are attacked the most frequently (females) or that are involved in the largest number of serious (*i.e.*, wounding) fights (dominant males) are relatively little stressed. For subordinate males who consistently lose serious fights, hormonal, neurochemical, and behavioral changes are profound and long lasting. These animals are a particularly relevant group for detailed investigation of the consequences of natural social stress. (Courtesy of R.J. and D.C. Blanchard, University of Hawaii.)

---

does, too. Then, one might make a rapid head movement toward the other. As a result, the other one will defend against the attack and make a biting movement back. Quickly, facilitated by androgens, the vicious, biting attack of one will lead to a response by the other that could lead to a rolling, biting, jumping fight. Blood will be drawn, and one might eventually be killed. What we are witnessing is a series of transitional probabilities, many raised by adrogenic and estrogenic steroid hormones, from low-intensity, preliminary aggressive responses to vicious, high-intensity fighting. No response is independent of the social stimuli that preceded it. Androgens and estrogens do not cause the aggression; rather, they elevate aggressive response probabilities.

The same is true for the case of sexual behavior When adequate schedules of estradiol and progesterone are given to female rats or mice to mimic the normal sex hormone conditions that lead to ovulation, they do not simply spring into the swaybacked posture (lordosis) that allows the male to fertilize the newly released eggs, nor does the naturally estrus female that is in heat do lordosis "in a vacuum." Instead, the behavior is triggered by tactile stimuli on the flanks and rump consequent to mounting by the male. In the absence of estrogen and progesterone, lordosis will not occur. Consistently, estrogens followed by progesterone raise the probability that such tactile stimuli will cause lordosis to occur (Fig. 6.2).

Oxytocin (OT) priming for affiliative behavior makes another fine example. The very nature of social behavior dictates that the hormone (in this case, oxytocin) cannot cause friendliness all by itself. It can only influence the probability of positive and friendly responses to social stimuli. For mice, the nonapeptide oxytocin, acting as both a hormone and a neuropeptide/transmitter, relies on the olfactory stimulus coming in from the potential mate or the potential opponent for its behavioral effect, which is revealed as an increased response probability, appropriate to the particular conspecific.

II. CLINICAL EXAMPLES

A. Hypercortisolism as a Factor in Major Depression

Major depression is a psychiatric syndrome defined by a number of signs and symptoms, as indicated in Table 6.1. Crucial is the point, according to this chapter's principle, that a terrible mood, in and of itself, is not sufficient to diagnosis clinical depression. It is the altered probability of negative emotional responses in the face of environmental stimuli (that would

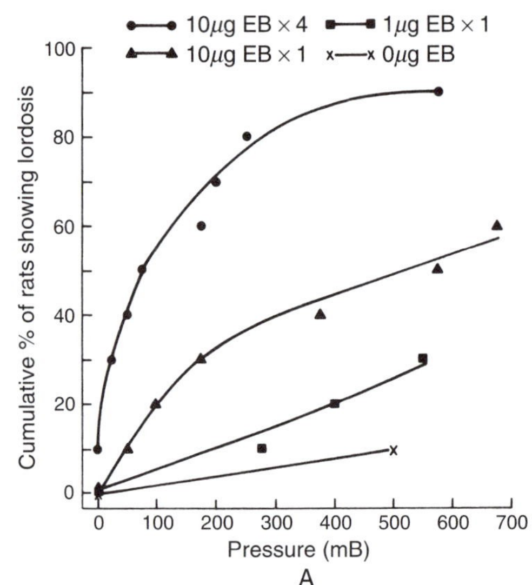

FIGURE 6.2. (A) Estrogenic hormones act to increase behavioral responses (due to electrophysiological responses) to specific stimuli. Increasing doses of estradiol benzoate (EB) amplify the ramps of behavioral responses to pressure on the skin of ovariectomized female rats sufficient for eliciting lordosis. (B) and (C) Electrophysiological responses of single primary sensory neurons in the dorsal root gangion of an anesthetized female rat to pressure stimuli on the skin (pressure intensity shown quantitatively on bottom trace of each figure). Sustained pressure led to action potentials in a neuron which had no spontaneous discharge. A sudden peak of pressure, as would be caused by stimulation from a male rat during mounting, causes a high rate of firing. (From Kow, L.-M. *et al.*, *J. Neurophysiol.*, 42: 195–202, 1979; Kow, L.-M. and Pfaff, D.W., *J. Neurophysiol.*, 42: 203–213, 1979. With permission.)

not evoke such gloom in the normal person) that makes a major depressive state. With this in mind, consider the role of glucocorticoid hormones.

Many studies have shown that 30 to 50% of patients with major depression have increased hypothalamo–pituitary–adrenal cortical (HPA) activity, as indicated by increased circulating adrenocorticotropic hormone (ACTH) and cortisol concentrations, as well as resistance of the HPA axis to suppression by the synthetic glucocorticoid dexamethasone. Ancillary indicators of HPA axis hyperactivity include elevated cerebrospinal fluid corticotropin-releasing hormone (CRH) content, downregulation of pituitary corticotroph CRH receptors, and adrenal gland enlargement. The more severe the depression, the more likely there will be increased activity of this endocrine axis. Successful treatment results in return of HPA activity to normal levels, as well as amelioration of depressive symptoms. If clinical improvement occurs with treatment but HPA axis activity remains elevated, there is a greater likelihood that relapse will occur when treatment is discontinued.

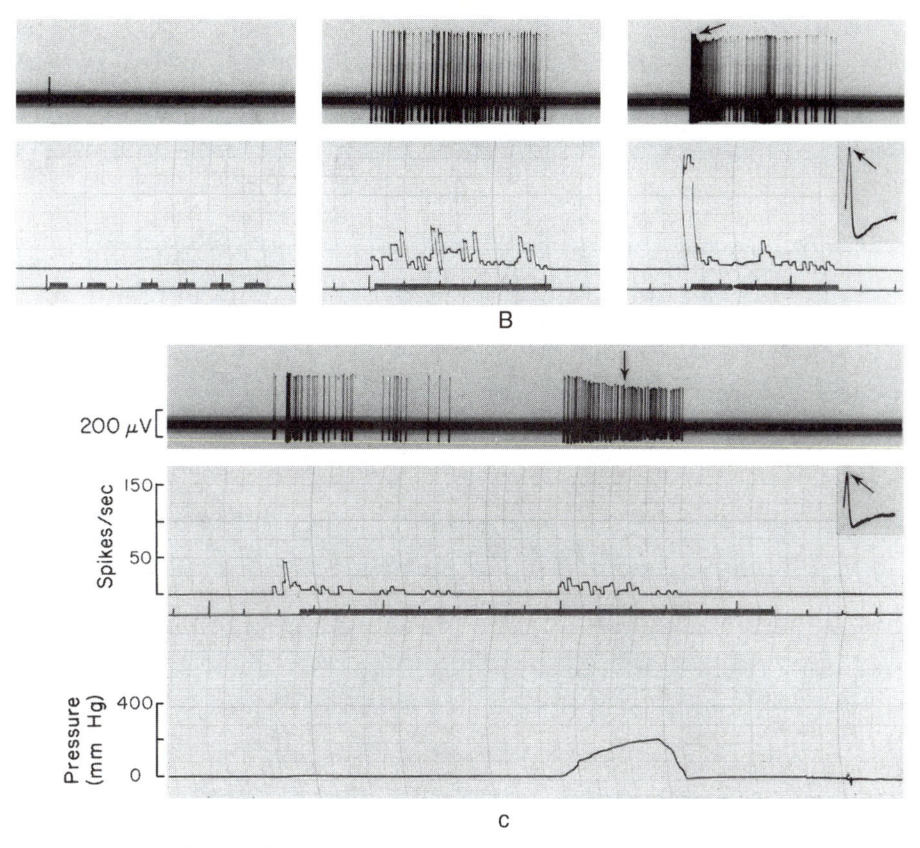

FIGURE 6.2. *(Continued.)*

There has been some debate as to whether the increased HPA axis activity is an epiphenomenon resulting from altered CNS neurotransmission and has no influence on the depressive illness or whether it compounds the severity of, or even is causally related to, the depression. Some investigators believe CNS glucocorticoid receptors are subsensitive in depressed patients, leading to reduced hippocampal negative feedback on the HPA axis and toxic influences of the increased cortisol on hippocampal neurons, resulting clinically in the memory disturbance seen in some depressed patients (depressive pseudodementia). A proposed treatment for major depression, therefore, has been reduction of circulating cortisol with glucocorticoid synthesis inhibitors such as ketoconazole or blockade of glucocorticoid receptors with receptor antagonists such as mifepristone (used as an "abortion pill" because it also blocks progesterone receptors). These drugs, however, have not proven to be effective stand-alone treatments for depression; at best,

TABLE 6.1 DSM-IV Diagnostic Criteria for a Major Depressive Episode

A. Five or more of the following nine symptoms (one of which is either depressed mood or loss of interest or pleasure) have been present for at least two weeks and represent a change from previous functioning:

1. Depressed mood (*e.g.*, feels sad) or affect (*e.g.*, appears tearful) most of the day nearly every day (in children and adolescents, can be irritable mood)
2. Markedly diminished interest in all or almost all activities most of the day nearly every day
3. Significant weight loss or gain or decrease or increase in appetite nearly every day (in children, can be failure to make expected weight gains)
4. Insomnia or hypersomnia nearly every day
5. Psychomotor agitation or retardation nearly every day as observed by others
6. Fatigue or loss of energy nearly every day
7. Feelings of worthlessness or excessive or inappropriate guilt nearly every day
8. Diminished ability to think or concentrate or indecisiveness nearly every day
9. Recurrent thoughts of death, suicidal ideation, a specific plan for committing suicide, or a suicide attempt

B. The symptoms do not meet criteria for a mixed episode.

C. The symptoms cause clinically significant distress or impairment in social, occupational, or other important areas of functioning.

D. The symptoms are not due to the direct physiological effects of a substance (drug or alcohol) or a medical condition.

E. The symptoms are not better accounted for by bereavement (*i.e.*, the recent loss of a loved one).

Source: Adapted from the *Diagnostic and Statistical Manual of Mental Disorders*, 4th ed., American Psychiatric Association, Washington, D.C., 1994. With permission.

they may provide some adjunctive help to primary antidepressant therapy. Perhaps the most important observation in the relationship between the HPA axis and major depression, as noted above, is that one-half to two-thirds of patients meeting diagnostic criteria for major depression have normal HPA axis function.

B. Hyperthyroidism as Predisposing to Anxiety and Irritability

In Chapter 5, we introduced the subject of thyroid gland hyperfunction; here, we expand upon that treatment. It is well understood that abnormally high thyroid hormone levels are associated with irritability, but this is not a temperamental change divorced from the patient's environment. Rather, the syndrome refers to a tendency to express annoyance (or worse) under circumstances that might be irritating, but which a euthyroid person would suffer gracefully. Clearly, there are many situations in which no people, including hyperthyroid patients, are angry. Across a range of increasingly difficult problems, more and more people become fed up, until circumstances are reached in which *everyone* has had enough. The irritability curve for the hyperthyroid person is shifted significantly to the left.

C. Psychosis

What about patients for whom the principle discussed in this chapter is, in fact, not true? In such a case, certain hormone levels (be they sex hormones, stress hormones, or thyroid hormones) would allow a behavior to be set off *in vacuo*. The objective external situation would not affect the behavioral outcome. Such a patient is, in fact, psychotic. He is operating independent of reality, and the hormone is simply allowing the psychotic behavior to be expressed.

III. SOME OUTSTANDING QUESTIONS

1. In each case, basic or clinical, what are the loci for hormone effects? Do hormone actions at several levels of a neural circuit interact with each other? Multiplicatively? When are the hormone actions directly on the sensory apparatus? Are hormone-caused changes in responses to stimuli sometimes physiologically on the motor side, with a strong alteration in response readiness?
2. Among human subjects, why do some individuals' thresholds for responsiveness to sensory stimulation, secondary to hormonal changes, vary so widely?
3. Multiple chemical sensitivities—abnormal responses to a wide variety of odors including headache, dizziness, and nausea—are seen much more frequently in women than in men. Why?

SECTION II

History: Hormone Effects Can Depend on Family, Gender, and Development

Familial/Genetic Dispositions to Hormone Responsiveness Can Influence Behavior

A person's family background can predispose him or her in several ways toward more or less sensitivity to a given hormone in a given situation. Most obvious is the person's genetic inheritance, pure and simple. In addition, however, genomic changes beyond DNA nucleotide sequences—epigenetic changes—can play a role. These can be manifest as differential DNA methylation or transcriptional barriers composed of DNA binding proteins and other nuclear proteins, among other mechanisms. Also, the common environment shared by the family can impose a familial disposition on an individual's behavior quite separate from any DNA-linked mechanisms. Of course, the most important feature of this environment is the treatment of the individual by family members. This would include the distinct possibility that the individual will copy behavioral tendencies exhibited by his or her parents. Thus, during this individual's adulthood, if a given hormone-dependent behavior is displayed and is exactly like that shown, for example, by his father, then this individual might be considered to be more sensitive to this hormone as it affects this particular behavior. If the reaction is the opposite, then he is correspondingly less sensitive. The examples of experiments in this chapter are so obviously relevant to questions of human behavior—and the effects of family background and early environment are so important for later human behavioral pathology—that this chapter is not separated into separate experimental and clinical discussions.

A huge phenomenon in this field is the long-lasting effects of maternal deprivation and neonatal handling on hypothalamo–pituitary–adrenal cortical (HPA) axis function. As a stress-responsive endocrine system, the HPA axis is important in regulating metabolic activity when animals are confronted with environmental stressors. The widespread effects of glucocorticoids have been discussed elsewhere in this volume. In the fetal and neonatal rat, components of the HPA axis (as well as other endocrine axes) mature at different rates and reach peak activity at different times. For example, plasma adrenocorticotropic hormone (ACTH) and corticosterone reach their peaks during late embryogenesis, whereas glucocorticoid receptor binding does not peak until more than 3 weeks after birth. Some components of the HPA axis peak perinatally and then decline to adult levels in the neonatal period.

Early experimental work applying stressors to neonatal rats suggested a stress-nonresponsive period in the first 2 weeks of life, but subsequent data (*e.g.*, the fact that corticosterone secretion is blunted but ACTH secretion is not significantly altered following a variety of stressors) have led to the newer concept of a stress-hyporesponsive period, during which adrenal cortical responses to stressors are blunted but not absent. During the hyporesponsive period, however, a specific stressor (separation of pups from their mother) leads to enhanced adrenal cortical sensitivity to ACTH and, consequently, exaggerated corticosterone responses compared to non-separated pups. Several aspects of the maternal separation paradigm must be considered in this regard. First is body temperature regulation of the pups, which has been controlled in the experiments. Second is nutrition of the pups; separation paradigms that cause adrenal hypersensitivity to ACTH (*e.g.*, for 24 hours) result in nursing and food deprivation. Third is maternal grooming of the pups, which appears to be a powerful component of the paradigm in that HPA axis changes were prevented by stroking the anogenital region of separated pups with a wet brush to mimic the maternal licking of the area that stimulates micturition.

A particularly important aspect of maternally separated neonatal rats is that HPA axis hyperresponsiveness to stressors persists into adulthood. While basal ACTH and corticosterone secretion appear to be normal, the ACTH/corticosterone stress response is exaggerated and resistant to dexamethasone suppression. Corticotropin-releasing hormone (CRH) content of the hypothalamic paraventricular nucleus (PVN); where neurosecretory cells for CRH are located, the median eminence, and the pituitary stalk is increased. Of importance, there also appears to be reduced glucocorticoid receptor density in several central nervous system (CNS) areas, including the hippocampus, that are important for negative feedback regulation of the HPA axis and in the prefrontal cortex, a cortical area implicated in several psychiatric syndromes.

In contrast to the effects of maternal deprivation, neonatal handling of rats appears to have the opposite (*i.e.*, "calming") effects on the HPA axis: lessening the ACTH/corticosterone stress response; reducing CRH and arginine vasopressin (AVP) content of the PVN, median eminence, and pituitary stalk; and increasing glucocorticoid receptor density in hippocampus and prefrontal cortex.

There are clear, long-term neuroanatomic and behavioral consequences of the interrupted mothering of neonatal rats, among them the loss of CA3 neurons, altered mossy fiber morphology, and upregulation of CRH expression in the hippocampus, along with impairment of long-term memory in adult rats. This may be a direct effect of CRH, in that elevated corticosterone levels are not necessary for these effects to occur, and they can be mimicked by CRH injections during the neonatal period. As mentioned, impaired negative feedback regulation of the HPA axis secondary to reduced glucocorticoid receptors in the hippocampus may be the key element in the excessive and prolonged HPA axis responses in stressed adult rats exposed to neonatal maternal deprivation. This, however, is likely not the only mechanism, as maternal separation also causes decreased CNS expression of neurotrophins, particularly brain-derived neurotrophic factor (BDNF), and decreased GABAergic neurotransmission, γ-aminobutyric acid (GABA) being a widespread CNS inhibitory neurotransmitter that inhibits the HPA axis.

Increased CRH levels secondary to maternal deprivation also may affect CNS circuits underlying the anxiety response. Several pathways are putatively affected by decreased BDNF and GABAergic input, a major one being from the central nucleus of the amygdala to the locus ceruleus. CRH produced in the amygdala stimulates the locus ceruleus to release norepinephrine, a neurotransmitter that has activating and anxiogenic effects. The locus ceruleus has 50 to 70% of the norepinephrine content of the CNS.

Genetic influences modulate the familial (*i.e.*, parenting) aspect of neonatal stress, in that there are strain differences in the resiliency of exposed pups. For example, BALB/cByJ mice exhibit more stress-related HPA axis activity and behavioral disturbances as neonates than do C57BL/6ByJ mice. Early-life handling of BALB/cByJ mice and cross-fostering of BALB/cByJ neonates to C57BL/6ByJ dams reduce the HPA-axis stress response and prevent associated behavioral disturbances. In contrast, cross-fostering C57BL/6ByJ neonates to BALB/cByJ dams does not induce stress hyperreactivity in the C57BL/6ByJ pups. These findings indicate that qualitatively different familial interactions can have overriding effects in certain animal strains that are genetically predisposed to high stress reactivity but not in other strains that are more stress resistant.

As well, there are major species differences in physiological responses to neonatal handling. In contrast to rats and mice, neonatally handled boars had greater corticosteroid binding globulin and lower circulating free and total cortisol concentrations at age 7 months than did non-handled control boars. Also, in contrast to rodents, 7-month-old handled and control boars had similar ACTH and cortisol stress responses and similar glucocorticoid receptor densities in frontal cortex, hippocampus, hypothalamus, and pituitary gland (Fig. 7.1).

With reference to non-human primates, there is a considerable literature on maternal separation studies in monkeys. In some studies, after the infant monkeys and their mothers have lived together for a short time, the infants are separated from their mothers for a certain length of time and then are returned to their mothers. Upon separation, the infants go through a stage of protest and then a stage of despair and withdrawal. Upon reunion with their mothers, the infants show considerable mother-directed behavior, with increased clinging and contact. Figure 7.2 shows social withdrawal in young monkeys that had been separated from their mothers.

If young monkeys are fully socially isolated for an extended period (*e.g.*, 6 months), for months thereafter they continue to show disturbed behavior including self-mouthing, self-clasping, and huddling, with almost no locomotive or exploratory behavior or social behavior when group-housed.

	CRH	Hyper-response to Stress? Depression, Anxiety?
Separation from mother	↑	YES
Neonatal handling & contact	→	NO

FIGURE 7.1. Treatment of baby rats by mother or experimenter can have long-lasting consequences for the animal's temperament, especially its responses to stress. Separation from the mother—which surely deprives the baby of the licking and handling it would receive in the nest and may also allow body temperature to fall—is associated with permanent changes in the release of corticotrophic releasing hormone levels (CRH; see neural circuitry in Fig. 1.4) and with a disposition toward hyperresponsivity to stressful situations, depression, and anxiety. In contrast, excess handling and contact can actually render the adult response to stress more adaptive than the normal case.

How this relates to human depressive behavior, however, is a subject of continuing debate, having to do with how strictly adherent one requires animal behavior to be to the human condition. At the very least, such disturbed-rearing models in rodents and primates may serve as test beds for pharmaceutical interventions that may prove useful in the treatment of human depressive disorders. For example, clinically useful antidepressants have been shown to ameliorate some of the behavioral disturbances in maternally deprived animals.

Thus, there are established connections among early adverse experiences, life-long neurohormonal and other CNS disruptions, and behavioral disturbances. The nature of the long-term physiological and behavioral changes depends on both the genetic makeup of the animal (species and strain differences) and the type of stressor. In some situations, the long-term changes can be modified by corrective parenting after the stressful period.

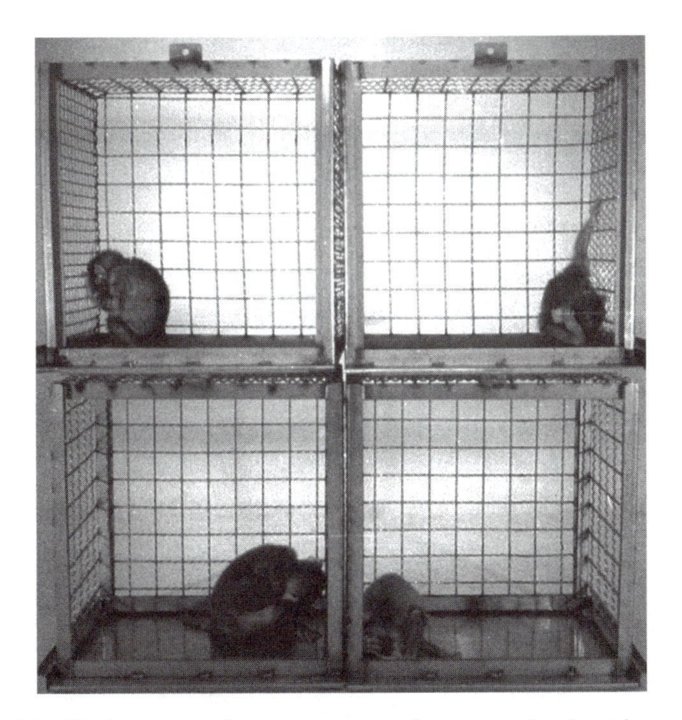

FIGURE 7.2. Monkeys separated at very young ages from maternal and social supports, even while adequately nourished, developed abnormal emotional and social behaviors. Seen here, they appeared sad and withdrawn. (From McKinney, W.T. *et al.*, *Am. J. Psychiatry*, 127: 1313–1320, 1971. With permission.)

These basic experimental findings have clear implications for human behavior and psychopathology, even though some investigators have drawn tenuous extrapolations between the neonatal maternal separation studies in rats to specific psychiatric syndromes such as major depression. The conservative implications are the following:

1. Maternal stresses during pregnancy may have important consequences in the fetus, because of the relatively lengthy period of intrauterine development in primates versus rodents.
2. As well, early childhood experience can shape one's perception of the world throughout adult life, particularly one's physiological and psychological responses to stressful life situations.
3. Both physical and psychological childhood abuse, depending on its severity and extent, can predispose an individual to, or even directly result in, pathological physical conditions as well as psychiatric disturbances.
4. The specific type of behavioral/psychiatric disturbance depends on many factors, including one's genetic predisposition to developing depression, panic disorder, chronic anxiety, schizophrenia, etc.
5. Some hormonal interventions (*e.g.*, reducing excessive glucocorticoid activity with synthesis blockers or receptor antagonists in those patients with major depression who exhibit excessive HPA axis activity) may aid in treatment, but evidence to date indicates such hormonal interventions are clearly not useful as the sole therapeutic agents.

Complete androgen insensitivity in XY males is relevant to the topics of several chapters in this text. An example in XY boys of the disconnect between genotypic and phenotypic sex is complete androgen insensitivity, a genetically determined failure of androgen receptor function. All patients are XY, their testes are capable of normal testosterone secretion, and they undergo normal fetal regression of Müllerian derivatives. However, because of the absence of androgen receptor function, their external genitalia remain female. These genetic males come to medical attention in infancy because of undescended, inguinal testes or at puberty because, as phenotypic girls, they do not develop menstrual cycles. As adults they have a paucity of pubic hair, normal female breast development secondary to unopposed estrogen action, and generally tall stature. They are always raised as girls and can have satisfactory vaginal intercourse without surgery, but, lacking ovaries, they are sterile. Figure 7.3 illustrates an XY male with complete androgen insensitivity The implications for sexual and social behaviors in such individuals are clear.

Repeating earlier material in a different context, a related example comes from 5-alpha reductase deficiency in XY males. While testosterone has many androgenic effects throughout the body and is responsible for virilization of

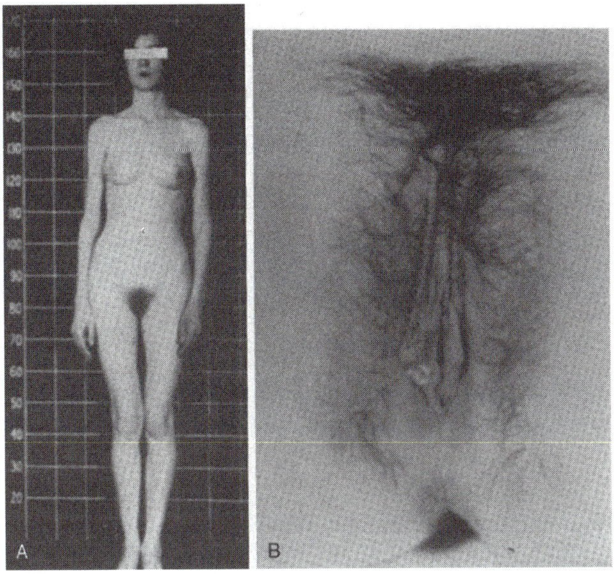

FIGURE 7.3. XY male with complete androgen insensitivity. The subject is tall, with normal breast development, a paucity of pubic hair, and female external genitalia. (From Forest, M.G., in *Endocrinology*, 4th ed., DeGroot, L.J. and Jameson, J.L., Eds., Philadelphia, PA: W.B. Saunders, 2001, pp. 1974–2010. With permission.)

internal structures (*e.g.*, testicular development), dihydrotestosterone (DHT) is required for virilization of the external genitalia (*i.e.*, penile growth and scrotal development). If the enzyme converting testosterone to DHT, 5-alpha reductase, is genetically deficient, the external genitalia of the newborn infant appear female or at least ambiguous (male pseudohermaphroditism). Figure 4.2 (Chapter 4) schematically illustrates the genitalia of a normal man (left) and the genitalia of a 5-alpha-reductase-deficient prepubertal boy. The female-appearing external genitalia belie the presence of an XY sex chromosome complement, functioning testes (although undescended), and a masculinized brain secondary to fetal testosterone exposure.

Before it was recognized that this syndrome is inherited and therefore is concentrated in certain families, affected infants were raised as girls. Upon reaching puberty, however, the increased secretion of testosterone from the pubertal testes led to some DHT being produced, so that many prepubertal "girls" developed a male phallus with erections, scrotal testes, male hair distribution, deepened voice, and male body habitus and psychological characteristics, to the initial consternation of parents and family members. However, the syndrome was quickly recognized to occur in affected families in several parts of the world (*e.g.*, in the Dominican Republic), so that newborns with ambiguous genitalia in these families were not forced to grow up as girls. Indeed, some individuals have entered into heterosexual relationships and have been able to function physically and emotionally as men. Others have led isolated lives or retained some female gender identity.

The outward switch of sex and gender identity from female to male at puberty in 5-alpha-reductase-deficient individuals was initially interpreted as the primacy of nature over nurture; that is, sex hormones are more influential than psychosocial factors imposed since infancy. However, despite the early recognition that newborns from affected families and with ambiguous genitalia might have masculine pubertal development, these infants have been raised either as boys or ambiguously as girls. The nature-versus-nurture dichotomy, therefore, is not as straightforward as some would believe.

I. THE PHENOMENON OF PRECOCIOUS PUBERTY

Puberty, the process of sexual maturation, extends over several years and, in the United States, usually begins at 10.6 years of age in Caucasian girls, 8.9 years in African-American girls, and 11 years in boys of both races. Precocious, or abnormally early, puberty occurs more often in girls and usually results in sexual maturation between ages 6 and 8, although it can occur earlier. Such children exhibit physical sexual maturity, but their emotional and erotosexual behavior more closely matches their chronological age. Onset of sexual activity may be earlier than average, but it usually occurs around the normal age for such activity. Because of the disparity between physical appearance and chronological and emotional age, such children still need age-appropriate interactions with parents and older adults, although they may seek peer friendships with older, more mature children. As their age cohort normally matures into adolescence, children with precocious puberty become less of an anomaly, and their social adjustment becomes less difficult. They may, however, remain of short stature, because their long-bone growth has been prematurely halted by their accelerated hormonal development.

II. RESISTANCE TO THYROID HORMONE

An important inherited hormonal syndrome is that of resistance to thyroid hormone (RTH). The hallmark of RTH is a significant decrease in tissue responsiveness to thyroxine. Refetoff and colleagues at the University of Chicago (Refetoff, 2000) have pioneered both the elucidation of this

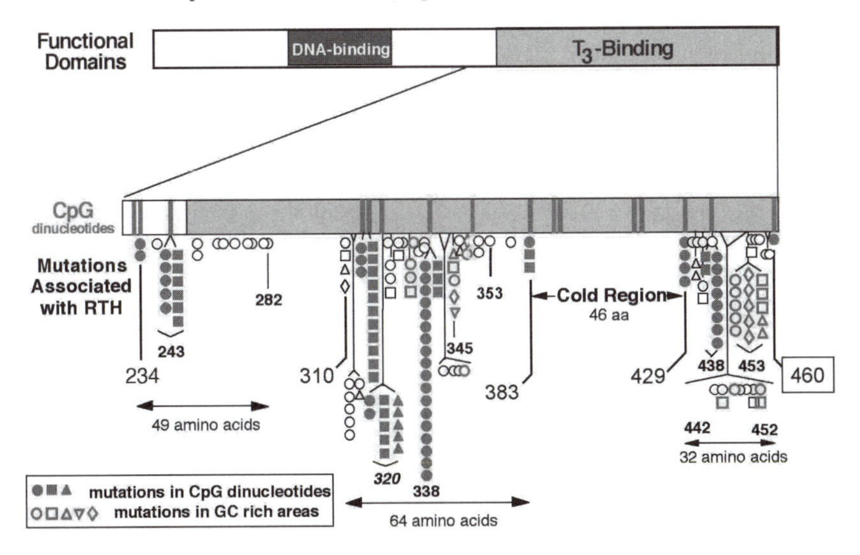

FIGURE 7.4. Familial lack of response to thyroid hormone would have obvious behavioral implications, and it has been associated by Refetoff and colleagues at the University of Chicago with particular mutations. Shown here is a rendition of the TR-β gene and certain genetic alterations. (Top) Location of natural mutations in the TR-β molecule associated with RTH; schematic representation of the TR-β gene and its functional domains for interaction with TREs (DNA-binding) and with hormone (T3-binding). (Bottom) The T3-binding domain and distal end of the hinge region which contain the three mutation clusters are expanded and show the positions of CpG dinucleotide mutational hot spots in the corresponding TR-β gene. The locations of the 99 different mutations detected in 158 unrelated families are indicated by various symbols. Identical mutations in members of unrelated families are represented by the same color and pattern of vertically placed symbols. Cold regions are areas devoid of mutations associated with RTH. Amino acids are numbered consecutively starting at the amino terminus of the TR-β1 molecule according to the consensus statement of the First International Workshop on RTH. TR-β2 has 15 additional residues at the aminoterminus. (From Refetoff, S., in *Werner & Ingbar's The Thyroid: A Fundamental and Clinical Text*, 8th ed., Braverman, L.E. and Utiger, R.E., Eds., Philadelphia, PA: Lippincott, Williams & Wilkins, 2000, pp. 1028–1043. With permission.)

syndrome and the discovery of its genetic mechanisms. The first two mutations they reported for the explanation of RTH were in the ligand-binding domain of the thyroid hormone receptor-beta (TR-β) gene. Now, a large number of mutations has been identified, including some in the so-called "hinge" region of TR-β adjacent to the ligand-binding domain (Fig. 7.4). On the one hand, John Baxter *et al.* in San Francisco have used X-ray crystallography to determine the three-dimensional structure of this region of TR-β and then to explore the consequences of the clinically important mutations. The structural consequences of these ligand-binding-domain mutations allow increased flexibility and disorganization of this region of TR-β. Importantly, thyroid hormone binding is reduced, and release of corepressors is impaired. On the other hand, Refetoff has taken the lead in identifying patients with RTH who do *not* have mutations in either TR-α or TR-β, the only two known thyroid hormone receptors. The search for the mechanisms of disease in these patients devolves upon the analysis of other, related nuclear proteins (see Chapter 18). All of these familial resistance to thyroid hormone syndromes have obvious implications for patients' hormonal controls over mental energy and mood, likely to affect both their intellectual and social capacities.

III. PHENYLKETONURIA, DOWN'S SYNDROME, HYPOTHYROIDISM, AND MENTAL RETARDATION

Phenylketonuria (PKU) is an autosomal recessive disease in which phenylalanine hydroxylase in the liver is absent, which leads to a lack of production of tyrosine and to hyperphenylalaninemia. This is caused by mutations of the genes coding phenylalanine hydroxylase, an enzyme that catalyzes the conversation of L-phenylalanine to L-tyrosine. Until recently, it was supposed that a lack of tyrosine production was the cause of mental retardation; however, a new theory proposes that hyperphenylalaninemia specifically impairs the *N*-methyl-D-aspartate (NMDA) receptor (of glutamate) function. The NMDA receptor appears to play a role in the formation of neural networks during development. Pharmacological blockade of NMDA receptors significantly disrupts memory and learning in animals. In PKU, where L-phenylalanine levels in the blood reach 1200 μM versus 55–60 μM in normal subjects, mental retardation and postnatal brain damage are characteristic. Electrophysiological studies on the hippocampus of the brain showed that L-phenylalanine selectively depresses glutamate receptor activation, indicating that it is the excess phenylalanine and not the lack of tyrosine that leads to mental retardation (Glushakov *et al.*, 2002). Children are normal at birth but steadily develop severe mental retardation. The effects of PKU are primarily on the brain. The lack of glutamate action is believed to reduce the excitation of cells needed for normal synaptic development. Another example is Down's syndrome, which is a chromosome 21 trisomy; that is, there is an extra chromosome 21 in addition to the two normally present. Children with Down's syndrome develop mental retardation, and hypothyroidism is very prevalent. Thyroid hormone acts powerfully on brain development, particularly to increase cell metabolism and the growth of neurite processes; presumably, the lack of adequate neuronal growth and metabolism contributes to the mental retardation.

IV. OBSESSIVE–COMPULSIVE DISORDER (OCD)

In the movie *As Good As It Gets*, Jack Nicholson memorably portrayed an obsessive–compulsive man who unwrapped new bars of soap while washing his hands. This is one example of obsessive–compulsive disorder (OCD). OCD includes hoarding behavior, Tourette's syndrome, compulsive grooming, and other behaviors that appear to have a hormonal basis and a genetic basis. In many cases, too little serotonin (5HT) is a factor. So, too, is dopamine, through its D1 receptor in the brain. Human brains are difficult to study; however, people displaying excessive handwashing and those diagnosed with trichotillomania (pulling out their hair) have parallels in mouse behavior. Recently, a study of mice with a mutation in the homeobox (or *Hox*) gene, showed that the mice groomed themselves (and their cage-mates) with such compulsive aggressiveness that they had bald patches and skin lesions. There seemed to be no secondary cause for this behavior such as pain or irritation of the skin (Greer and Capecchi, 2002). Although the proteins and possible hormones associated with this ancient gene are not known, the extraordinary behavior is a potential example of how a gene mutation can illuminate a behavior of importance in psychiatry.

V. THE THRIFTY PHENOTYPE HYPOTHESIS

Genetic contributions to hormone/behavior relations can also be considered in light of the theoretical approach of the thrifty phenotype hypothesis, which addresses glucocorticoid effects on brain. The thrifty phenotype hypothesis was proposed to account for the finding by Barker that the early environment in which a fetus develops, can have a lasting effect on the individual's life. Fetal malnutrition, which has several different causes (*e.g.*, starvation of the mother or a dysfunctional placenta), causes the growing fetus to be malnourished. To survive, the fetus develops several strategies to maintain growth in the brain at the expense of growth in other organs and tissues (*e.g.*, liver, pancreas, muscle). However, when the fetus is born and experiences abundant nutrition, these survival strategies continue and eventually cause obesity, overeating, and metabolic syndromes, such as diabetes mellitus type 2, ischemic heart attack, and hypertension. In conditions where both prenatal and postnatal nutrition are poor, these survival tactics persist without serious consequence. An example of this is the low occurrence of diabetes among sub-Saharan Africans compared to the high incidence of diabetes in western populations.

In famine, the fetus is likely to be undernourished. The many examples of famines include the famous "Dutch Hunger Winter" in 1944 in Holland.

A study was conducted that included people born immediately before, during, and 21 months after the famine. In adulthood, both men and women who were starved *in utero* during the famine had significantly higher plasma glucose concentrations than those not exposed to the famine. Increased plasma glucose was highest in those individuals who experienced famine in the last trimester. Increased plasma glucose is an indicator of insulin resistance, which occurs in diabetes mellitus type 2. All of these children had lower birth weights and were smaller in size than the controls. Only the head size was larger, indicating the fetal survival strategy of protecting brain growth. Insulin levels were also high in the late pregnancy. Low-birth-weight subjects were at the greatest risk of developing diabetes.

The thrifty phenotype hypothesis has been studied to test if protein is the most important factor for fetal development. Pregnant rats fed a low-protein (8%) diet had smaller pups than those from mothers fed a diet of 20% protein (Snoeck, 1990). As these rats grew older, they all became diabetic and obese. Although the brain growth is spared during malnutrition at a cost to the other organs, the brain is not well vascularized, suggesting that behavioral functions could be impaired (Bennis-Taleb *et al.*, 1999).

How is all this related to hormones, brain, and behavior? The thrifty phenotype hypothesis illustrates that the genes and environment interact initially to protect the genotype in a particular situation, but ultimately this may backfire and reduce lifespan. This "grow now, pay later" adaptation has other consequences. Individuals who have low birth weights tend to have elevated plasma glucocorticoid concentrations in adulthood. This reflects an overresponsiveness to stress and can have consequences in health, income, and quality of life. An excess of fetal glucocorticoids leads to a decreases in the number of nephrons in the kidney, thereby increasing fluid retention and expanding volume, which puts a strain on cardiac output. Fetal glucocorticoid excess also alters brain development and gene expression of brain angiotensinogen and angiotensin receptors. This has been shown by injecting dexamethasone in early gestation. Dexamethasone increases gene expression in the brain for the angiotensin system. Studies in rats have demonstrated that dexamethasone in the prenatal stage also sets up a permanent increase in plasma corticosteroid levels in adults (Wellberg *et al.*, 2001). The early natal experience reduces corticoid receptors in the hippocampus which reduces feedback control on the hippocampal–hypothalamic–pituitary–adrenal corticoid (HHPA) system. By interfering with this hormonal feedback system, memory and coping with stress are put at risk in adulthood.

Experiments closely related to the maternal separation phenomena discussed above are those in which hormones have been related to maternal care. Rats that receive a lot of mothering (which, for a rat pup, is licking and grooming by the mother) grow up to be less fearful and more maternal than pups that did not receive a high amount of mothering. Oxytocin and vasopressin receptors in the brain have been implicated in this difference. Males receiving the attention of their mothers had higher levels of vasopressin V1a receptors in the amygdala than the controls, which were licked less. Females that had received the benefit of more maternal attention grew up to have more oxytocin receptors in the amygdala and in the bed nucleus of the stria terminalis. Thus, maternal care may involve receptors to oxytocin and vasopressin, but in a gender-specific way. (Francis *et al.*, 2002)

VI. MECHANISMS

Some of the fastest progress analyzing detailed mechanisms of familial/genetic contributions to hormone responsiveness is coming from experiments delving into genetic determinations of simple animal behaviors. This is a massive field, extending well beyond the boundaries of laboratory science. For generations, dogs have been bred to achieve desirable behavioral tendencies and to avoid undesirable ones. Likewise, breeding of horses achieves behavioral changes as well as improved strength and speed. Among rat studies carried out in a traditional neurobiological setting, the high attack latency (HAL) and low attack latency (LAL) rats produced by the laboratory of deKloet at the medical school at Leiden are of special interest for further neuroendocrine experimentation. Of course, in light of 21st-century functional genomics, studies with genetically altered mice swamp all of these examples. For example, the ability of Nelson to produce hyperaggressive mice by altering genes affecting nitric oxide provides the opportunity to obtain chemical and cellular detail with respect to this behavior pattern, which is so sensitive to steroid sex hormones (see previous discussion). Likewise, as we carried out at Rockefeller University, knocking out the classical estrogen receptor-α, produces a female mouse that will not respond to estrogen plus progesterone treatment with increased courtship and locomotor behaviors or with normal female sex behavior. In male mice with estrogen receptor-α knockouts, Ogawa and colleagues tried to elicit aggressive behaviors and were unable to, in spite of the fact that normal wild-type littermate controls were very aggressive in response to androgenic hormones. In these and many other examples with gene knockout and transgenic mice, a detailed, functional genomic understanding of familial influences on hormone/behavior relations is being realized.

Applying the material of this chapter to an important medical and ethical question: what will be the technically feasible and ethically acceptable use of pharmacogenomics to predict the metabolism of hormones, the activation of particular hormone receptors, etc., for the purpose of ameliorating behavioral disorders? The traditional approach to both endocrine and behavioral studies is a linear one, tracing the causal relationship of a change in a single hormone or neurotransmitter to its effects on metabolism and behavior. The current genomic/proteomic era, in which massive amounts of data on gene activation and suppression, transcriptional changes, protein formation and conformation, etc. are being produced, requires a major shift in conceptual approach, toward the consideration of interlocking and interactive endocrine, neurotransmitter, and receptor networks that influence multifaceted behavioral changes. A beginning clinical application of genomic informatics, via large datasets collected with gene-chip technologies, is the prediction of exogenous hormone and drug effects in individuals. Gene activation and suppression profiles following hormone and drug administration may someday be of predictive value as to who will respond therapeutically, who will have unacceptable side effects, and what dosage schedule will be required, based on each person's pharmacokinetic and pharmacodynamic handling of the hormone or drug in question. Pharmacogenomics will aid in dissecting the multiple molecular/metabolic steps that result in the final, physiological outcome.

A series of methodological issues must be confronted before this clinical application can be accepted with confidence. These include achieving sufficient sensitivity, reliability, and reproducibility of large-scale genomic screens in individuals; ensuring consistency of methodologies across laboratories; developing mathematical algorithms for handling extremely large multivariate datasets, including setting thresholds for gene activation and suppression compared to reference standards; amalgamating these large datasets across laboratories into megasets that can be accessed and analyzed by investigators in many laboratories; and, of course, securing financial support for all these expensive activities.

VII. SOME OUTSTANDING QUESTIONS

1. To avoid the development of mental retardation in a newborn, what should you look out for?
2. Low birth weight has a number of causes. What is the difference between low birth weight and low gestational weight? What do you think the consequences might be of one versus the other?
3. How do the hormonal consequences of differential early environments translate into the neural differences that will affect adult behavior?

The Sex of the Recipient Can Influence the Behavioral Response

In both animal experimentation and clinical medicine, the impacts of sex differences on hormone/behavior relations are so obvious and pervasive that only a small fraction of the best examples can be recounted here. From this type of scientific work, the social and environmental factors as well as the chromosomal determinants of sex differences will inform the growing medical field of women's health.

I. BASIC EXPERIMENTAL EXAMPLES

In animals and in human beings, the most obvious neuroendocrine sex difference is the ability of the female to demonstrate an ovulatory surge of luteinizing hormone (LH), while the male cannot. Correspondingly, in mammalian animals, the same schedule of estrogen priming followed by progesterone amplification leads to female-typical courtship behaviors followed by lordosis behavior in the female but not in the male (if the animal is a quadruped). In contrast, long-term testosterone treatment of laboratory animals leads to intromission and ejaculation in the male but, clearly, not in the female, even though simple mounting behavior might be exhibited by both sexes.

A. Parental Behaviors

Across a wide range of studies of mammalian behavior an implicit assumption is that females will display effective parental behaviors and males will not. In a large number of human societies, as well, for social and economic reasons documented by Chodorow in *The Reproduction of Mothering* (1999), the larger burden of parental care has been assumed by the mother. Regarding mechanisms, at the purely biological level things are complicated (see review by Gonzalez-Mariscal, *Hormones, Brain and Behavior*, 2002). First, among rodent species, true paternal care can be observed in several of them. Males as well as females lick gently, retrieve, huddle over, and manipulate the young. In some cases, these behaviors are shown by males, but with longer latencies. One factor that determines whether paternal care will be evidenced is the degree to which the female allows the male near the young. In some species, the male is allowed to approach his young only when he has proven that he will not harm the pups. Thus, males who have been absent are less likely to be accepted when they return. In species where paternal care is not shown, its lack is likely due to neonatal actions of sex hormones on the brain. Castration of the male rodent on the day after birth increases the likelihood of maternal-like behavior. Conversely, exposure of the genetic neonatal female to androgens reduces it and instead favors infanticide. Later, in adulthood, high levels of testosterone weigh against short-latency, maternal-like behaviors. Estrogens, combined with progesterone and prolactin (as shown by the results of Robert Bridges), favor these behaviors.

B. Sex Differences in Stress and Fear Responses

Sex differences in stress and fear responses comprise an important part of animal behavior studies (Fig. 8.1). McCormick's laboratory at Bates College has endeavored to compare the effects of sex hormones on neuroendocrine stress hormone responses between gonadectomized females and males in the very same experiments. Consistent with the results discussed above, peripheral administration of estradiol raised stress reactivity, as measured by the circulating stress hormone corticosterone, in females much more than in males. Most interesting was the fact that females with estradiol implanted directly into the medial preoptic area had the highest circulating corticosterone levels after stress, of all the experimental groups compared.

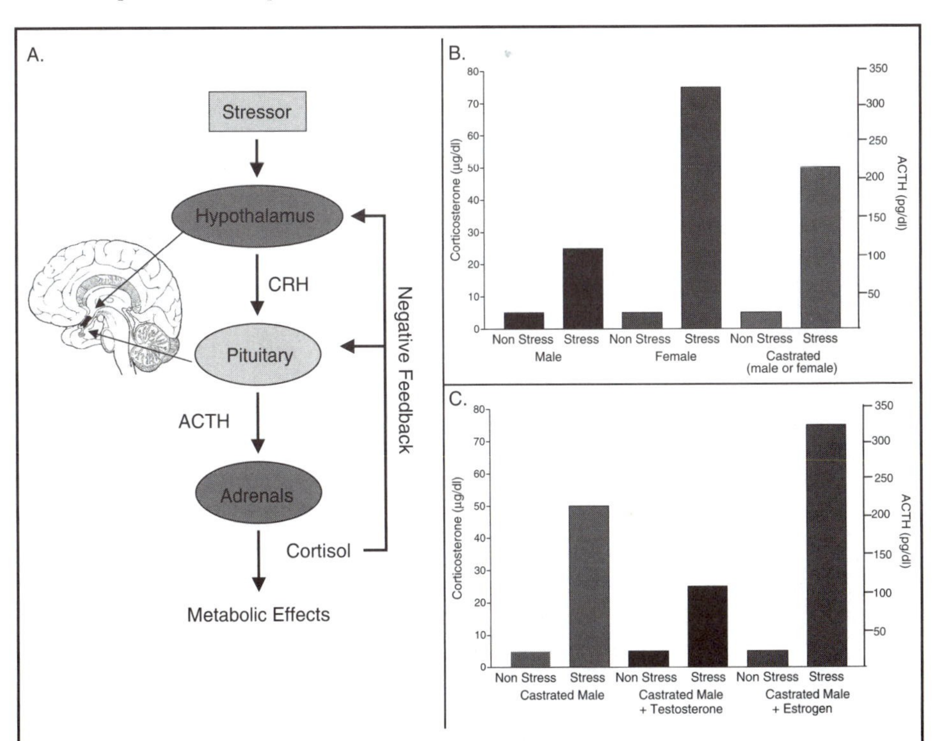

FIGURE 8.1. The results of a variety of studies have reported the existence of a sex difference in the severity of affective changes as well as specific symptoms following a depressed mood state. These studies have shown that in some instances women seem to respond more robustly to stressful events than do men and exhibit more severe symptoms of depressive experience. It has also been reported that the incidence of depressive episodes is two- to three-fold greater in women and that men tend to adapt better to their negative mood following depressive episodes. These epidemiological findings parallel the existence of gender differences in hypothalamo–pituitary–adrenal (HPA) axis function in response to stress, where females show a more robust response than males. The HPA axis is a well-described neuroendocrine circuit that begins when a cluster of cells in the hypothalamus receive signals that convey stress-related information (A). Such cells, located in the paraventricular nucleus of the hypothalamus, produce corticotrophin-releasing hormone (CRH). The secretion of CRH into the hypothalamo–hypophyseal portal vasculature occurs in response to both physical and psychological stressors. In turn, CRH binds to specific receptors on pituitary corticotroph cells, which causes them to release adrenocorticotropic hormone (ACTH). ACTH is then transported through the general circulation to its target organ, the adrenal gland. The adrenal glands, located atop the kidneys, respond to ACTH by increasing the secretion of the steroid hormone cortisol (in rats, the primary adrenal steroid is corticosterone). The release of cortisol initiates a series of metabolic responses aimed at alleviating the harmful effects of stress. Additionally, a negative feedback mechanism is targeted to both the hypothalamus and the anterior pituitary as well as other brain sites such as the hippocampus. This reduces the concentration of ACTH and cortisol in the blood once the state of stress subsides. A gender difference in the HPA axis response to stress has been well characterized in the rat model and is depicted graphically in (B). In general, rats

For neurobiologists, the simplest way to think about sex differences is usually to consider mechanisms in the forebrain or the anterior pituitary gland. However, McCormick also found certain sex differences in a protein in the blood that binds corticosterone: corticosteroid binding globulin (CBG) (Fig. 8.2). While such differences in blood proteins do not always bear directly on hormone/behavior relations, they should always be considered.

The magnitude and direction of sex differences in response to stress may depend on the species examined and the type of stress imposed. Siberian hamsters form monogamous male/female bonds, to the extent that long-term separation from a mate is a form of social stress. During chronic separation, male hamsters have shown behavioral changes, according to the report of the Bosch lab at Arizona State University, reminiscent of depression, including elevated levels of the stress hormone cortisol. Females showed decreased levels. In response to acute restraint stress, male cortisol elevations were robust, but females showed no stress response.

Differences between male and female animals in aggressive responses are widespread across mammalian species. Males are usually more aggressive than females, except for situations in which the mother is defending her nest and her young. Some of the hormonal underpinnings of aggressive responses were mentioned in Chapter 3 and are illustrated there.

C. Sex Differences in Responses to Pain

All of the sex differences in aggression and in responses to stress and fear referred to earlier could be influenced by male/female differences in responses

---

Figure 8.1 (*Continued.*)
(whether male, female, or castrated animals) have similar levels of stress hormones when in a non-stressed state. Following stress, both male and female rats respond with a rapid increase in circulating ACTH and corticosterone. This response is much more pronounced in females than in males. Castration of male or female rats extinguishes this difference, suggesting that the sex difference is at least partly due to circulating steroid hormones. Results from such a study are shown graphically in (C). Castrated male rats, devoid of circulating testicular hormones, are compared with castrated male rats treated with either testosterone or estrogen. Castrated males treated with testosterone and then stressed display a reduced ACTH and corticosterone response, typical of intact males. On the other hand, estrogen-treated males display a greater ACTH and corticosterone response to the stressor, more characteristic of females. It appears therefore, that testosterone can act to inhibit HPA function, whereas estrogen can enhance HPA function. Although further research is required to fully elucidate the impact of gender differences in response to stress, an increased understanding of such hormonal effects allows us to appreciate more fully how estrogen and androgen interact in the development of pathologies such as depression, potentially leading to an improvement in preventive and/or pharmacological approaches to mental health. (See Wilhelm *et al.*, 1998; Handa and McGivern, 2000; Suzuki *et al.*, 2001.)

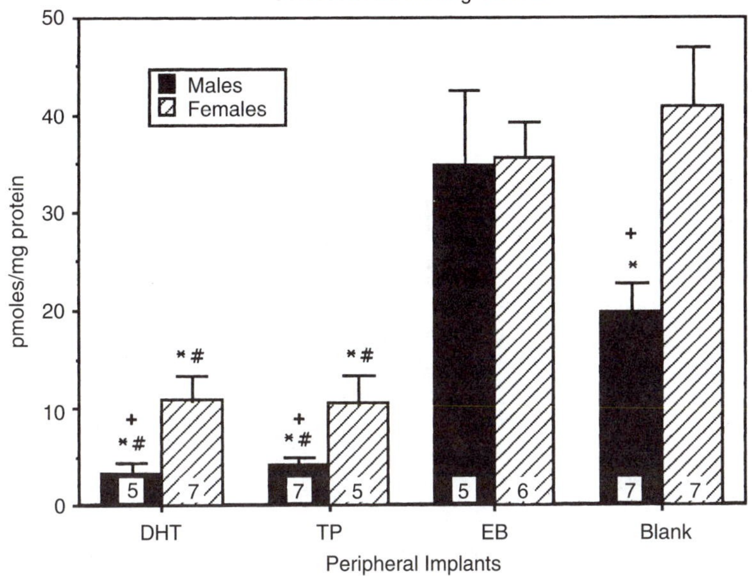

FIGURE 8.2. Some of the sex differences in hormonal responses to stress could be due to circulating hormones. McCormick and her lab at Bates College have gonadectomized male and female rats and studied the effects of specific hormones. As expected, even without hormone treatment there was a sex difference (see Blank). Note also the sex differences following dihydrotesterone (DHT), a reduced testosterone metabolite, and testosterone propionate (TP). (From McCormick, C.M. *et al.*, *Stress*, 5(4): 235–247, 2002. With permission.)

to pain. In animals, as well as in human beings, mechanisms in the brain that suppress responses to pain are called *antinociceptive responses*, or *stress-induced analgesia*. One mechanism underlying these phenomena is a straightforward sex difference in a specific type of opioid analgesia. Males are much more sensitive to pain suppression exerted at the level of the midbrain by opioid peptides that work through μ-opioid receptors. These results in experimental animals have their application in clinical medicine (see below). A completely different set of pain control mechanisms also contributes to male/female differences. Jeff Mogil *et al.*, in Montreal, used a forced, cold-water swim to cause a reduction in the stress-induced analgesia pain response. Females were much less sensitive to this analgesia being blocked by the neurotransmitter *N*-methyl-D-aspartate (NMDA), implying that females had a special pain-control system not possessed by the male experimental animals. The female-specific system is amplified by estrogens. This pain control mechanism in females appears to depend both on sex differences in κ-opioid receptors and on a female-specific action of a different

	Altered behaviors of Estrogen Receptor Knockout (ERKO) mice	
	In the female	In the male
Aggressive behaviors	↑	↓
Sex behaviors	↓	↓

FIGURE 8.3. Two points can be made from the analysis of mice whose classical estrogen receptor gene (ER-α) has been knocked out. (1) The effect of an individual gene on a specific behavior can depend on the gender in which the gene is being expressed. ERKO females are more aggressive than wildtype female littermate controls, whereas ERKO males are less aggressive than male controls. (2) Normal sex behaviors in both genders depend on the integrity of the ER-α gene. (See Ogawa *et al.*, 1996, 1998a,b.)

system, the melanocortin type 1 receptor. Thus, not only do multiple neurochemical systems contribute to controlling the sensation of pain, but the sex differences in such responses also have multiple origins.

D. Genes and Sex Differences in Behavior

The contributions of specific genes to sex differences in behavior have also begun to be explored. Perhaps the most dramatic experiment so far revealed the phenotypes of female and male mice in which the gene for the classical estrogen receptor-α had been knocked out (Fig. 8.3). With respect to certain social behaviors, the sex roles of these animals were reversed. Females were very aggressive; they behaved like males in agonistic encounters showing, for example, vicious offensive attacks with biting; and, most importantly, were treated like males by other males. In contrast, the males were not aggressive. These phenotypes resulted from changes in the brain, not just in abnormal levels of hormones circulating during behavioral tests. This was shown by surgically removing the gonads, replacing equal levels of sex steroids appropriately in the experimental animals, and observing that one could still observe the reversals in sex roles.

E. Mechanisms

Some of the differences in hormone/behavior relations between males and females may be due to differences in sex hormone effects on gene expression

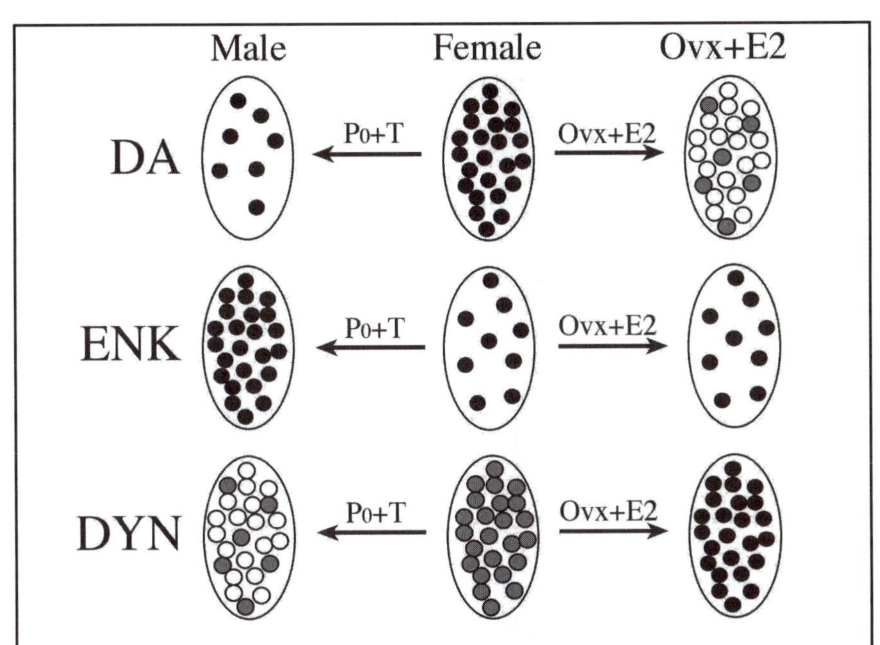

FIGURE 8.4. Sex steroid hormones direct differentiation of hypothalamic neurons. Schematic diagram to illustrate the influence of sex steroid hormones on dopaminergic (DA) or dynorphin (DYN) and enkephalin (ENK) peptide-containing neurons in the anteroventral periventricular nucleus of the hypothalamus (AVPV). The AVPV is a sexually dimorphic nucleus that plays a critical role in controlling gonadotropin secretion and ovulation. Treatment of newborn (P0) female rats with testosterone results in a male pattern of neurotransmitter expression (left). Treatment of ovariectomized adult females with estrogen increases cellular levels of DYN, decreases DA, and does not affect ENK (DeVries and Simerly, 2002). Sex steroid hormones also direct development of neural inputs to the AVPV from sexually dimorphic parts of the limbic region of the forebrain, as well as regulate gene expression within these same pathways in adults (Simerly, 1990, 2003). Such divergent patterns of neuronal development and gene expression illustrate the cell-type specificity that is characteristic of hormonal regulation of neuroendocrine circuits and provide a glimpse into the complexity of neurobiological events underlying physiological differences between males and females. The precise cellular and molecular mechanisms utilized to specify these hormone-dependent cellular phenotypes remain to be discovered (Simerly, 1990, 2002; DeVries and Simerly, 2002).

in specific brain regions. The molecular neurobiological data demonstrating sex hormone effects on opioid peptide genes only in the female, for example, could obviously contribute to the determination of female sex behavior not shared by male rodents; however, other molecular findings could bear on a wider range of social behaviors and even on sex differences in neuronal controls over hormone secretions from the anterior pituitary gland (Fig. 8.4).

F. Intrauterine Environment

It is also interesting to note that the position of a developing XX female in the uterus (*e.g.*, located between other females or between males) can affect her anatomical development, her neuroendocrine status, and her eventual behavior. Female mice who had been between two males are less attractive to males, are more aggressive, more likely to defend food sites, display less lordosis behavior, and do mount other females. This set of effects is thought (by vom Saal at the University of Missouri and others) to be due to the females' prenatal exposure to androgenic steroid hormones from their neighbors. (See also Chapter 9.)

II. CLINICAL EXAMPLES

Throughout the previous discussion on sex differences in animal behavior and their mechanisms, the term *gender differences* has not been used. This is because, for an appreciation of the full range of biological and behavioral differences between men and women and their social manifestations, different terms must be employed to keep things straight. Consider that the neuroscientist is dealing with (1) chromosomal differences (XY versus XX), (2) epigenetic differences in the chromatin, (3) hormonal differences, (4) consequent differences in organ and tissue structure and function, (5) differences in frank sexual behaviors, (6) differences in mate choice, (7) differences in psychological gender self-identification, and (8) differences in social roles. Of course, the most elaborate expressions of all of these variations between XY and XX individuals can be observed in the behaviors of human beings.

In other chapters we refer to the so-called *organizational* versus *activational* effects of hormones. Organizational effects are those that are exerted early in life, during brain development, and which affect later responses to stimuli and hormones in adulthood. Activational effects are exerted in adulthood and more directly facilitate or repress specific classes of behaviors in animals and humans. As a result, as noted above, genetic sex and social gender roles are usually consonant, but in modern human societies all combinations are possible.

In the simplest mechanisms elucidated by neuroendocrinologists, following early exposure to androgens one could see a "masculinized" brain in phenotypic females; also, if, early in development, biochemical events contravened normal androgenic hormonal actions, one could see a "feminized" brain in a phenotypic male. As one specific example, androgen insensitivity in humans is imposed by an enzymatic alteration, a 5-alpha reductase deficiency in steroid hormone metabolism (studied by Imperato-McGinley *et al.*; see also Chapters 7 and 9). Other biochemical mechanisms for human are recounted in endocrinology texts, such as that by DeGroot and Jameison. A countervailing point of view about human sex differences—which was portrayed as embodying an alternative and opposed set of explanations—was suggested in the work of Money and colleagues at Johns Hopkins. This body of work emphasized the importance of how the child is treated by parents and other adults in determining whether the child eventually feels and acts like a man or a woman. "Sex of rearing" was thought of as being an important factor. In sum, despite the fact that biochemical versus behavioral influences have historically been portrayed as opposing camps in medical science, they are not mutually exclusive.

The incidence of major depression is twice as great in adult women as in adult men. Prior to puberty, the incidence of major depression is about the same in girls and boys, and it is not until puberty and adulthood that women have a twofold greater risk of this illness. This implies a hormonal factor that increases risk among sexually mature women. Understanding sex differences in central nervous system (CNS) neurotransmitter function should provide information relevant to major depression, because antidepressant treatments alter the dynamics of several neurotransmitters, including serotonin and acetylcholine. Serotonergic function is known to be affected by sex hormones, and recent evidence suggests cholinergic function also may show a sex difference. For example, Fig. 8.5 shows plasma adrenocorticotropic hormone (ACTH) responses to cholinergic stimulation by low-dose physostigmine (PHYSO, 8 μg/kg), a cholinesterase inhibitor that blocks the hydrolysis of acetylcholine, in 12 women with normal estrogen levels and 8 men of similar age with major depression, and in 12 female and 8 male control subjects individually matched to the patients on age, race, and body surface area, as well as menstrual status for the women.

Several aspects of the data are noteworthy. First, sufficient baseline samples were taken through an indwelling venous catheter to achieve stable ACTH concentrations prior to saline or PHYSO administration at 6:00 p.m. Second, ACTH, a stress-responsive hypothalamo–pituitary–adrenal cortical (HPA) axis hormone, was not affected by the stress of an additional venipuncture for saline administration, providing a control condition against which the effect of PHYSO could be compared. Third, normal men had a greater ACTH response to PHYSO than did normal women, whereas depressed women had a greater ACTH response than did depressed men—a significant sex by diagnosis interaction. These findings suggest that estrogen may have augmented already hypersensitive CNS cholinergic systems in the depressed women, which may bear some relationship to the increased incidence of this illness

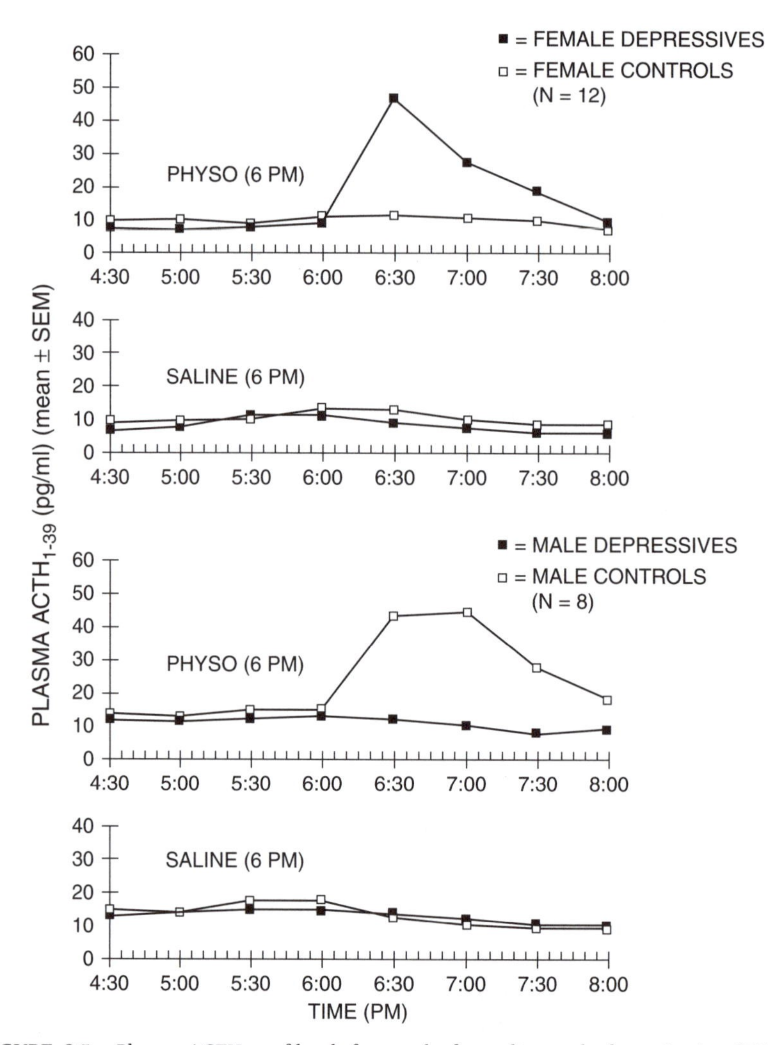

FIGURE 8.5. Plasma ACTH profiles before and after saline and physostigmine (PHYSO) administration at 6 p.m. in 12 premenopausal women and 8 men with major depression and in 12 premenopausal female and 8 male normal control subjects. (Adapted from Rubin, R.T., *et al.*, *Psychiatry Res.*, 89: 1–20, 1999.)

in women. Estrogen enhances CNS cholinergic neurotransmission, and sex differences have been reported for virtually all cholinergic markers. The net effect of estrogen on cholinergic stimulation of the HPA axis is likely a result of a complex set of factors, including estrogen's influence on specific steps in acetylcholine synthesis and metabolism, the role of sexually dimorphic areas of the brain that result from the organizational and activational effects of estrogen and testosterone (as discussed in other chapters), and, in the depressed women, the overlay of a psychiatric illness that itself may involve abnormalities of CNS cholinergic neurotransmission. Gaining an understanding of these factors is an important task for future research.

A few pages ago we briefly summarized differences between males and females in the power of μ-opioid agonists to reduce pain. With modern scanning techniques applied to the human brain, Zubieta and colleagues at the University of Medicine compared men and women with respect to μ-opioid receptor dynamics. Men showed significantly larger magnitudes of μ-opioid receptor activation in response to sustained pain, than women, in the thalamus, in the basal forebrain, and in the amygdala. In contrast, women actually had reductions of μ-opioid activation during pain in the nucleus accumbens, a forebrain cell group closely connected to both negative and positive reward. These differences, both in magnitude and direction of μ-opioid responses in specific regions of the brain, therefore, could contribute to differences in pain experience between men and women.

III. OUTSTANDING CLINICAL AND BASIC SCIENTIFIC QUESTIONS

1. Regarding the so-called premenstrual syndrome, is there any equivalent in men? Is there an andropause? (See Chapters 3 and 11.)
2. Given that some of the mechanisms for sex differences in animals depend on gene expression in the brain altered by hormones in one sex but not the other, what are the causal routes beyond gene expression—beyond the nascent transcript?
3. Even though some behavioral sex differences clearly originate in nerve cells, especially in the hypothalamus and limbic system, what are some roles served by glial cells? Based on the findings of Margaret McCarthy's lab at the University of Maryland, are differences between neonatal male and female animals in glial morphology originating purely in glial cells *per se*, or do they depend on glial/neuronal interactions?
4. What are the most impressive and functionally significant sex differences outside the hypothalamus and limbic system? For example, are

there sexual dimorphisms in the lower brainstem circuitry that mediates arousal?

5. Regarding the most sophisticated and cerebral differences in gender-role behaviors in humans, to what extent do they depend on the individual's appreciation of biological states and changes of state in his or her body? By analogy to the James–Lange theory of emotion (expressed by the great American psychologist William James), a person may reach an emotional state because of the brain's recognition of changes in the viscera. Is there an equivalent situation with sex differences in behavior among humans?

Hormone Actions Early in Development Can Influence Hormone Responsiveness in the CNS During Adulthood

I. BASIC EXPERIMENTAL EXAMPLES

A. Stress Hormones

Stress hormones impacting the fetus either indirectly, from the mother across the placenta, or directly can alter stress responses later in adult life. Nathanielsz (1999) described experiments in which pregnant rats were exposed to restraint stress, which resulted in elevated levels of corticosterone compared to mothers not restrained at all. The babies, allowed to grow to young adulthood, then had a greater hormonal response to restraint stress themselves, compared to the babies of unstressed mothers. Direct elevation of stress hormones soon after birth by removing the baby from the nest will also reprogram the animal's stress responses permanently. Nathanielsz's theory of how this comes about is that set-points in the brain–pituitary–adrenal axis will have been altered by the neonatal stress hormone elevation, thus altering the animal's stress responses even throughout adulthood. Wiegant's experiments with rats tested this suggestion. Indeed, following neonatal treatment with a synthetic glucocorticoid, adult brain, pituitary, and adrenal hormonal responses were blunted. Nathanielsz (1999) speculates that this type of early hormone exposure not only contributes to the formation of temperament but also affects the likelihood of depression and other mood disorders (cf. endocrine data in Fig. 9.1).

The notion that stressful (*i.e.*, adrenal cortex activating) social experiences early in life can lead to a tendency toward depression finds support in the work of Cameron, at the University of Pittsburgh. Among rhesus monkeys, even relatively brief separation from the mother was followed by a marked tendency toward social isolation. The endocrine concomitants of this important behavioral change are under investigation. Moreover, Schneider *et al.*,

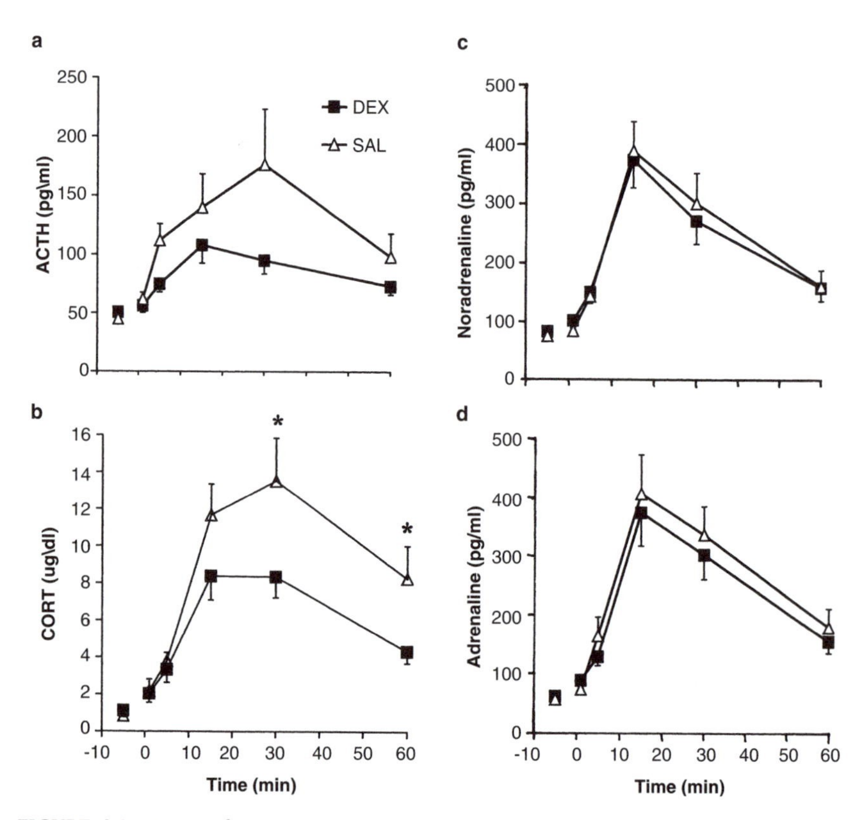

FIGURE 9.1. Neonatal treatment with a synthetic stress hormone, dexamethasone (DEX), compared to neonatal vehicle control treatment (SAL) affected responses to stress as measured by (a) ACTH and (b) corticosterone, but neither (c) noradrenaline nor (d) adrenaline. (From Kamphuis, P.J. *et al.*, *Neuroendocrinology*, 76(3): 158–169, 2002. With permission.)

at the University of Wisconsin, also worked with rhesus monkeys and found that prenatal stress caused reduced attention span and neuromotor ability in the offspring. Prenatally stressed monkeys were less exploratory and showed what the authors interpreted as disturbance behaviors.

B. Mechanisms

One clear possibility for a causal route from earlier stress to later behavioral change is an alteration in the stress-hormone-receiving systems themselves in the brain. Mohammed *et al.* (1993) at the Karolinska Institute, used an environmental enrichment program that might be conceived of as altering stress hormone levels. Compared to controls, rats who earlier had undergone environmental enrichment (Fig. 9.2) had higher glucocorticoid levels in their hippocampal neurons. Both the behavioral and the neurochemical consequences of early glucocorticoid administration have been illustrated widely (see Fig. 9.6a,b). It has not escaped our attention that these phenomena of early stress hormones affecting mechanisms related to later responses could provide mechanisms for the early experience "familial" effects mentioned in Chapter 7.

C. Effects of Early Sex Hormones

In mammalian fetuses, before hormonal differentiation, Wolffian and Müllerian ducts are present in both sexes. The ureters and renal collecting system are derived from the Wolffian ducts. The Y sex chromosome carries a gene or genes coding for the production of testis-determining factor, enabling an XY individual to become phenotypically male. Perhaps the most startling effects of neonatal hormone exposure are in this realm of sex. In rats, the presence of high levels of circulating androgens during the first 2 or 3 days after birth largely determines whether the neuroendocrine and behavioral tendencies of the animal during adulthood will fit those expected of a normal female or those expected of a normal male. If testosterone is injected into the newborn female, she will lose the ability to ovulate during adulthood, as well as the ability to perform courtship and sex behaviors, such as lordosis, in response to adult sex hormone treatment. Conversely, if the male is deprived of circulating testosterone by surgical castration on the day of birth, he will gain the ability to organize an ovulatory luteinizing hormone (LH) surge in response to adult estrogens and progesterone and, correspondingly, will be able to exhibit normal, female-like lordosis behavior. Thus, the most visible features of adult sexual life, endocrinologically and

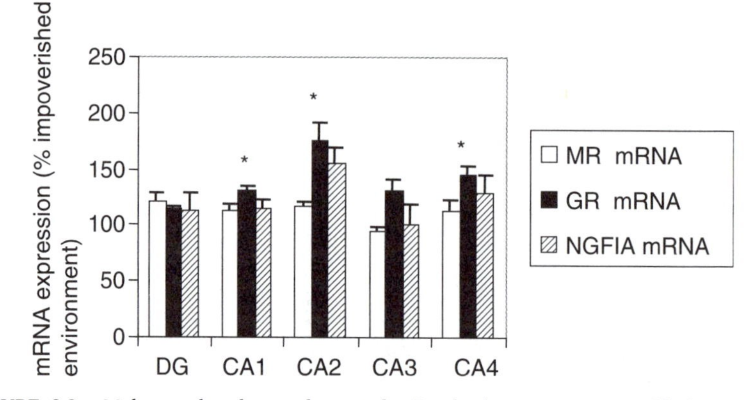

FIGURE 9.2. Mohammed and coworkers at the Karolinska Institute in Stockholm measured transcript levels for glucocorticoid receptors (GRs), mineralocorticoid receptors (MRs), and nerve growth factor receptors (NGF1As) in the hippocampus. One group of rats had been exposed to an enriched environment early in life, while the others had been restricted to an impoverished environment. The enriched group had significantly greater levels of mRNA for GRs in specific parts of the hippocampus (*e.g.*, Ammon's Horn pyramidal cells in area 2; CA2). (From Mohammed, A.H. *et al.*, *Behav. Brain Res.*, 57: 183–192, 1993. With permission.)

	Ability to generate ovulatory surge of LH release?	Sex Behaviors?	
		Male-like?	Female-like?
Normal male	NO	YES	NO
Neonatally castrated male	YES	NO	YES
Normal female	YES	Minimal	YES
Female given testosterone neonatally	NO	Increased	NO

FIGURE 9.3. Removal of testicular androgens from the neonatal male rat permits his neuroendocrine system to generate a pulse of GnRH sufficient for ovulation (if he has been transplanted with ovaries) and to do female sex behaviors such as lordosis. Conversely, injection of testosterone neonatally into the female rat abolishes her ability to ovulate and to respond normally to estrogens with lordosis behavior. Male-like sex behaviors change, some less dramatically, in exactly the opposite direction from female behaviors.

behaviorally, in response to adult sex hormone administration are essentially determined by the neonatal androgen exposure (Fig. 9.3).

In a normal XY animal, the testis secretes testosterone and Müllerian-duct-inhibiting hormone. In the male, in response to testis-determining factor, the Wolffian duct forms the epididymis, seminal vesicle, and vas deferens. In response to Müllerian-duct-inhibiting hormone, these ducts disappear; the testes descend into the scrotum, and the kidneys ascend into the upper abdomen. In a normal XX animal, the Müllerian ducts form the ovaries, fallopian tubes, and uterus. This is the default phenotype, in that the female phenotype will develop in the absence of the specific factors necessary to produce the male phenotype.

Sex-specific behaviors normally are consonant with genotypic and phenotypic sex. Experimentally, these aspects can be dissected, in order to understand the specific components and timing of developmental milestones. For example, in female rats, testosterone treatment on postnatal day (PD) 4, during the critical period of sexual differentiation in females, results in a low number of hypothalamic preoptic area spine synapses, acyclic LH secretion, and low female and high male sexual behaviors. In contrast, testosterone treatment on PD 16, after the critical period for sexual differentiation, has no effect. Correspondingly, in males, castration on PD 1 during the critical period of male sexual differentiation produces a high number of preoptic area spine synapses, cyclic LH secretion, and low male and high female sexual behaviors. Castration on PD 7, after the critical period, has no effect.

The hormonal reduction of same-sex behavior is considered demasculinizing or defeminizing, whereas the hormonal induction of opposite-sex behavior is feminizing or masculinizing. For example, castration of newborn male rats results in a lack of normal male mounting behavior (a demasculinizing effect) and display of lordosis behavior (a feminizing effect) in these animals as adults.

Rodents are altricial (*i.e.*, helpless, naked, and blind at birth), and their central nervous system (CNS)-critical periods are still occurring after birth. In contrast, primates are precocial (*i.e.*, more developed at birth), and their CNS-critical periods already have occurred *in utero*; therefore, in humans, the critical milestones for normal, concordant genotypic and phenotypic sex development, including sexually dimorphic development of CNS regions that influence neuroendocrine function as well as sexually diergic (functionally different) behavior patterns, occur before birth.

D. More Effects of Early Adrenal Hormones

In Chapter 7, we dealt with early deprivation of maternal care and also with neonatal handling of experimental animals. Both are presumed to have long-term effects due to the neonatal alterations in stress hormone release causing later changes in the same neuroendocrine system. Prenatally, also, the hypothalamic–pituitary–adrenal cortical (HPA) axis is very sensitive to programming by high maternal glucocorticoid levels, whether caused naturally or experimentally. Elevated adrenal hormone levels in the prenatal period result not only in increased basal corticosterone levels during adulthood but also in increased stress responses. Perhaps as a consequence of these, there is a greater stress sensitivity in the sympathetic nervous system and altered learning capacities (as reviewed by Jon Seckl at Edinburgh). One possible mechanism connected with a specific example is that prenatal stress hormones may increase neuronal vulnerability to oxidative stress. Ceccatelli and coworkers at the Karolinska Institute exposed rats prenatally to high levels of a synthetic stress hormone, dexamethasone. Their neurophysiological measure in adulthood was the auditory brainstem response (ABR) to sound. Following acoustic trauma in adulthood, the rats treated prenatally with stress hormone showed little recovery of the ABR, whereas normal controls did recover (Fig. 9.4). Additional neurochemical experiments showed that the prenatal stress hormone treatment reduced the ability of the adult CNS to deal with oxidative stress.

II. CLINICAL EXAMPLES

A. Consequences of Abnormal Fetal Hormone Exposure in Humans

Gender identity (*i.e.*, one's sense of self as female or male) and sex-specific behaviors normally develop concordantly with genotypic and phenotypic sex. Several genetic abnormalities can alter this course of events. For example, crossing-over can occur in a homologous region of the X and Y chromosomes, such that testis-determining factor becomes translocated on the X chromosome. If this occurs in a male gamete (sperm), the offspring could be either an XY female (having no Y-chromosome testis-determining factor region) or an XX male (having an abnormal X-chromosome testis-determining factor region).

Genetically determined deficiencies of enzymes involved in sex-steroid production and genetically determined deficiencies in steroid hormone receptor function can produce disconnects between genotypic and phenotypic sex. An example in XX girls is virilizing congenital adrenal hyperplasia (CAH), in which genetically based deficiencies in enzymes involved in glucocorticoid synthesis lead to a shunting of steroid precursors into androgenic pathways. The most common of the congenital adrenal hyperplasias results from a deficiency in 21-steroid hydroxylase, the

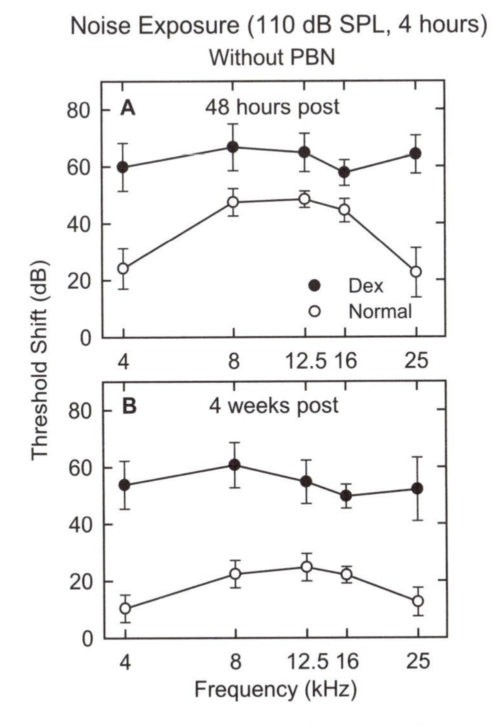

FIGURE 9.4. Fetal exposure to high levels of glucocorticoids, even those coming from the mother, can alter nervous system development. In this experiment by Ceccatelli and her colleagues, rats were treated prenatally with a synthetic glucocorticoid (dexamethasone, DEX) or control (Normal). The recovery of response of brainstem auditory neurons following damage of the auditory system due to high amplitude noise exposure was harmed in DEX-treated rats compared to controls tested as young adults; that is, the thresholds for electrophysiological response to a wide variety of frequencies of auditory stimuli were higher in the DEX rats either 48 hours (A) or 4 weeks (B) after damage. SPL = Sound Pressure Level. (From Canlon, B. *et al.*, *Eur. J. Neurosci.*, in press. With permission.)

enzyme that converts 17-α-hydroxyprogesterone to 11-deoxycortisol. 17-α-hydroxyprogesterone instead is shunted to androstenedione and then to testosterone. Excess testosterone *in utero* results in female pseudohermaphroditism and precocious puberty (see Chapter 7), with an enlarged phallus, a common urogenital sinus at the base of the phallus, and fused and pigmented labioscrotal folds. Fortunately, this condition is almost always recognized at or shortly after birth and can be successfully treated with glucocorticoids, which suppress adrenal androgens, and with mineralocorticoids, which are also deficient and are necessary for salt and water balance, as well as reconstructive surgery to normalize the female genitalia. These

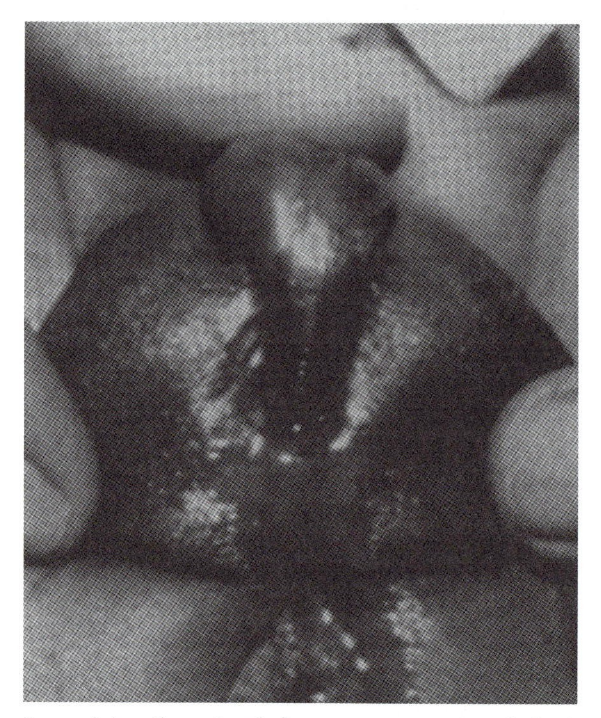

FIGURE 9.5. Some of the effects of early hormone exposures—or lack of same—on adult behavior can be due to peripheral anatomical effects of such exposure. Shown here are the genitalia of an XX female infant with 21-hydroxylase deficiency. The phallus is enlarged, below which is a common urogenital sinus. The labioscrotal folds are pigmented and fused. (From Forest, M.G., in *Endocrinology*, 4th ed., DeGroot, L.J. and Jameson, J.L., Eds., Philadelphia, PA: W.B. Saunders, 2001, p. 1997. With permission.)

children should be raised as girls, because with treatment they will develop normally and have a female gender identity and a normal reproductive life. Figure 9.5 illustrates the external genitalia of an XX female infant with virilizing congenital adrenal hyperplasia. Berenbaum *et al.*, 2000, has reported that girls with CAH play more with boys' toys, and during adolescence their interests and activities are intermediate between the distributions typical of unaffected girls and boys.

An example in XY boys of a disconnect between genotypic and phenotypic sex is complete androgen insensitivity, a genetically determined failure of androgen receptor function (see also Chapter 7). All patients are XY, their testes are capable of normal testosterone secretion, and they undergo normal fetal regression of Müllerian derivatives. However, because of the absence of androgen receptor function, their external genitalia remain female. These genetic males come to medical attention in infancy because of undescended, inguinal testes or at puberty because, as phenotypic girls, they do not develop menstrual cycles. As adults they have a paucity of pubic hair, normal female breast development secondary to unopposed estrogen action, and generally tall stature. They are always raised as girls and can have satisfactory vaginal intercourse without surgery, but, lacking ovaries, they are sterile. Figure 7.3 in Chapter 7 illustrates an XY male with complete androgen insensitivity.

B. Sex, Gender Identity, and Behavior are Usually Consonant But May Not Be

Consistent differences in gender role behaviors across cultures appear to be related to whether or not the brain has been masculinized during development. As indicated earlier, in the absence of male hormones, estrogens promote female CNS development. In addition to *in utero* masculinizing factors and postnatal androgen-organizing influences on the CNS, parents' and others' interactions with the infant based on appearance of the infant's external genitalia, developing self-awareness of testes and male external genitalia, environmental influences such as sex of rearing, and societal norms for sex-appropriate behavior all are important in human gender identity and sexual behavior.

Verified normal sex differences in behavior include the following: Males are more aggressive than females across cultures and from childhood through adulthood, more so in children. Boys are more active and display more playful aggression. For toys, boys prefer vehicles, weapons, and building toys, whereas girls prefer dolls, kitchen accessories, and cosmetics. As adults in most cultures, women are more interested in parenting than men are.

Men are more likely to be left-handed than women, and women are less likely to have exclusive left cerebral language dominance. Whereas there is no sex difference in general intelligence, some differences in specific cognitive measures have been found. Females have slightly better overall verbal ability. Males are slightly better on analogies, whereas females are slightly better in speech production and verbal fluency. From childhood through adulthood, males rotate mental images more accurately and rapidly than do females. Females solve mathematical problems slightly better in childhood, whereas males' problem-solving ability is slightly better in adulthood. There is no sex difference in comprehension of mathematical concepts at any age. Finally, females show better perceptual speed and accuracy than do males, but the magnitude of the difference has been declining since the 1940s, likely related to evolution of the tests used to measure this performance characteristic.

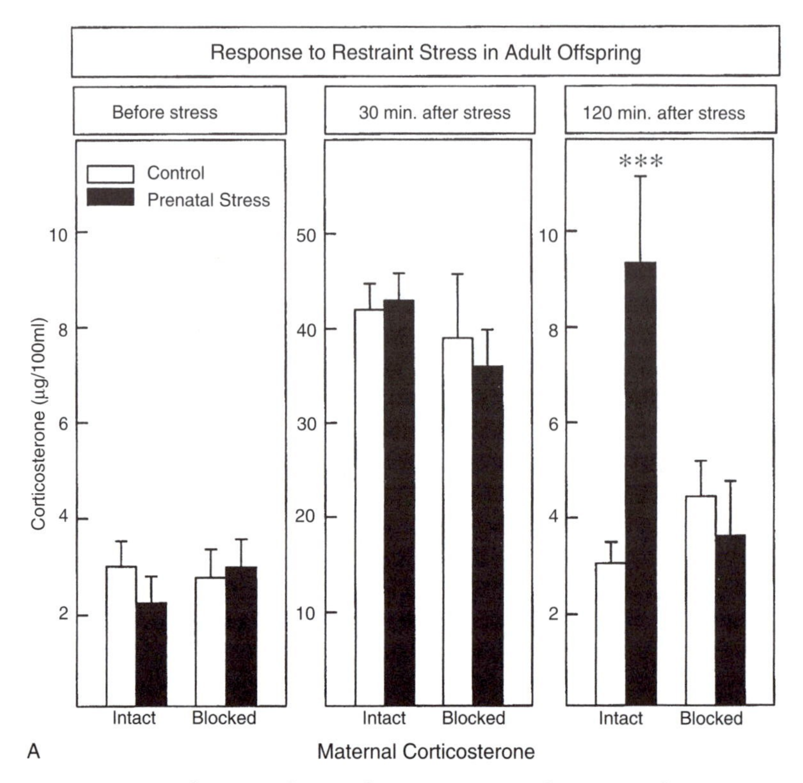

FIGURE 9.6. Prenatal stress and prenatal exposure to stress hormones influence neuroendocrine and behavioral responses in later life. (A) Prolonged responses to restraint stress (right panel, black bar) were evident in the adult offspring of mothers who had been intact and secreted stress hormones during gestation, but not in adult offspring of mothers whose gestational stress secretions had been blocked (comparison on far right). The inability to turn off the stress response has been linked with various behavioral maladies. (From Barbazanges, A. *et al.*, *J. Neurosci.*, 16(12): 3943–3949, 1996. With permission.) (B) Prenatal exposure to the synthetic stress hormone dexamethasone (DEX) throughout gestation (DEX 1–3) or only in the last week of gestation (DEX 3) influenced certain anxiety-related behavioral responses after these baby rats had grown up. Number of rears in a 12-minute open field test (black bars) were very sensitive as behavioral indicators, as were open-arm entries in the elevated plus maze assay (open bars). Less sensitive was an assay that is supposed to model depression: time spent floating in a forced-swim test (striped bars). (From Seckl, *J. Neuroendocrinology*, 2001. With permission.)

As illustrated by the above examples of congenital adrenal hyperplasia in XX females and complete androgen insensitivity in XY males and the example of 5-alpha-reductase deficiency in Chapter 4, there can be hormone-induced discordances among sex, gender identity, and sexually diergic cognitive and emotional behaviors. XY individuals with complete androgen

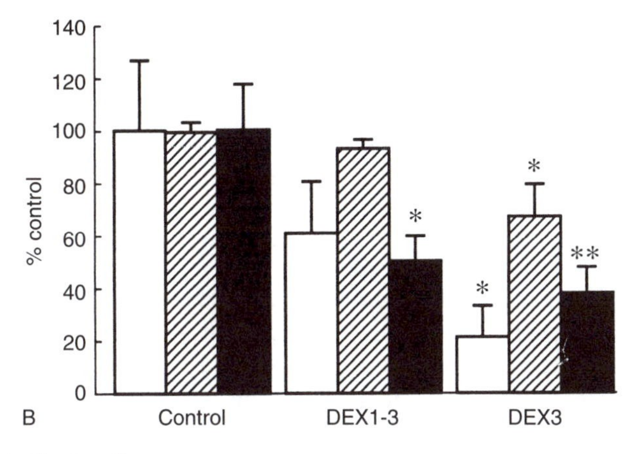

FIGURE 9.6. (*Continued*).

insensitivity have concordant female phenotypic sex, sex of rearing, and gender identity. In contrast, as indicated in Table 4.1, XY individuals with 5-alpha-reductase deficiency have female-appearing or ambiguous genitalia at birth, such that some have been raised as girls. At puberty, however, due to an overriding increase in testosterone production, many have developed male secondary sex characteristics and have experienced a male-gender-identity awakening. More difficult to explain on an endocrine basis is male transsexualism, where there is a pervasive female gender identity in spite of unambiguous male genital sex and sex of rearing. No abnormalities of CNS morphology or function and no endocrine abnormalities have been verified in either transsexual males or females.

XX individuals with congenital adrenal hyperplasia have masculinized external genitalia at birth, but, as indicated previously, are now usually quickly recognized and treated. Nevertheless, they may have higher than usual energy levels and some tomboyish behaviors. As well, they may have higher than usual homosexual fantasies and arousability, but not necessarily more homosexual activity. Some older individuals who were raised as boys because of the extent of masculinization of their external genitalia have successfully maintained a male gender identity.

C. Mechanisms

Some of the mechanisms by which exposure to high levels of glucocorticoid hormones, or stress hormones, during fetal life or soon after birth could influence later hormone/behavior relationships are just beginning to be explored. One set of possible alterations reside in the lower brainstem.

Lucion's laboratory at the University of Sao Paulo reported altered numbers of neurons in the locus ceruleus. This is potentially important because neurons there respond strongly to salient stimuli (including, of course, stressful stimuli) and then tell the rest of the brain about them through long-ranging adrenergic projections. Lucion imposed the mild stress of neonatal handling of baby rats. As a result, both in males and females, there were fewer neurons in the locus ceruleus at later developmental stages than in non-handled control animals.

Neuronal mechanisms in the forebrain are also candidates for remembering the effects of early stress. Maccari and colleagues in Lille, France, injected the synthetic glucocorticoid dexamethasone into pregnant female rats for each of the last 5 days of gestation, then they measured the consequences for hypothalamic control of the pituitary (adrenocorticotropic hormone, or ACTH) adrenal axis. Compared to controls, prenatal stress hormone injection not only reduced ACTH levels in the neonates but also, in the paraventricular nucleus of the hypothalamus, reduced expression of CRF, the neuropeptide that causes ACTH release as well as glucocorticoid receptors and vasopressin. Decreased biosynthetic activity of hypothalamic neurons controlling pituitary and adrenal stress-related hormone release could help to explain the behavioral effects of early stress, mentioned above and in Chapter 7. Even farther forward in the brain, in the prefrontal cortex, lie other molecular mechanisms that could participate in altered hormone/behavior relations. Riva, in Milano, found that chronic maternal stress during pregnancy significantly reduced gene expression for brain-derived neurotrophic factor (BDNF) specifically, without affecting other telencephalic regions. In contrast, postnatal stress imposed in the form of maternal separation caused a decrease in BDNF expression in the adult hippocampus. These changes in BDNF expression are potentially important because its role as a neurotrophic molecule could underlie permanent behavioral adaptations.

D. Thyroid Hormones

Clearly, if a child develops without circulating thyroid hormones for prolonged periods, his adult performance will be marked by disastrous cognitive failures (see material in Chapter 5). Abnormalities in morphological brain development accompany and, in fact, can explain the cognitive problems. These comprise the very definition of cretinism. What about brief or subtle decreases in thyroxine? Might one of the implications of early thyroxine loss be an abnormality in later responses to this and other hormones?

Thyroid function in the fetal rat, a useful experimental animal, begins at about embryonic day 17.5; therefore, losses of thyroxine after that point are of special practical significance. As reviewed by Zoeller at the University of Massachusetts, even relatively subtle changes in thyroxine in pregnant women can have consequences for their children's neurological development. Zoeller and Rovet (2003) have reviewed evidence that later psychological performance can be affected, including the well-documented significantly lower IQ. The precise type of behavioral deficit depends on the time of thyroid hormone loss: Early in pregnancy, problems with visual attention and processing have been reported. Later in pregnancy, decreased visuospatial skills and slower motor responses may result. These latter can include deficits in eye–hand coordination, motor imitation, and object manipulation. With thyroid insufficiency extending until after birth, problems in linguistic and memory skills are evident. Zoeller and Rovet (2003) reflect that these clinical observations far exceed our neurological understanding. By exactly what mechanisms do these losses come about? Do some of the causal routes involve later responses to thyroxine (T4)?

Dowling and Zoeller (2000) used the differential display technique to examine the genomic consequences of acute T4 administration to the mother just before the normal onset of fetal thyroid function. Several of the thyroxine-sensitive genes discovered were expressed selectively in brain areas that contain thyroid hormone receptors (TRs). One of them, transcription factor Oct-1, shows how interesting these genes of potential importance for cognitive development can be. In turn, how are the effects of early hormonal changes stored in such a way that the individual's later responses, genomically and behaviorally, to both hormones and environment are changed permanently? In this era of discerning causal relations between gene expression and behavior, a new set of possible mechanisms has emerged. Epigenetic mechanisms are defined as those closely related to gene expression with changes in the primary nucleotide sequences themselves. We know now that DNA can be methylated, and that such a chemical alteration changes the sensitivity of a gene to transcriptional enhancers throughout the life of the animal. Further, changes in the chromatin can be long lasting. Histone acetylation or methylation provide additional possible mechanisms by which prenatal or neonatal hormone exposures could alter the animal's hormone responsivity for the rest of its life.

III. SOME OUTSTANDING QUESTIONS

1. With respect to early effects of sex hormones, why is core gender identity more consonant with genetic sex than is sexual partner preference in humans?

2. The incidence of transsexualism, in which gender identity is discordant with phenotypic sex and sex of rearing, is about 1:25,000 in males and 1:75,000 in females. These ratios are reasonably consistent across cultures. In contrast, the incidence of bisexuality/homosexuality, in which there is concordance among gender identity, phenotypic sex, and sex of rearing but same-sex partner preference, is about 1:25 in males and 1:20 in females. Are there identifiable CNS determinants or correlates of male or female transsexualism and male or female homosexuality?
3. To what extent do the effects of early exposure to high and prolonged levels of stress hormones such as cortisol relate to the psychological effects of low socioeconomic status early in life? Do they represent some of the mechanisms by which early feelings of deprivation and loss have their eventual behavioral effects?
4. Considering the effects of environmental toxins on the mother, what are some circumstances where these act through hormonal mechanisms, be they thyroidal, sexual, or adrenal?

CHAPTER 10

Puberty Alters Hormone Secretion and Hormone Responsivity and Heralds Sex Differences

Instead of dividing this chapter into discussions of basic and clinical examples, we have treated the material in a way more convenient to its nature, first illustrating some phenomena and then exploring the mechanisms.

I. PHENOMENA OF PUBERTY

In the lifetime of changes in the endocrine environment of the brain and the pituitary gland, puberty represents one of the more obvious transitions. Over a period of several years in humans, changes in the gonads and other organs associated with reproduction and secondary sexual characteristics offer the clinician clear developmental stages to track (Table 10.1). In regard to behavior, hormonal changes during puberty impose some of the most dramatic changes imaginable. A critical period for the organization of social behaviors and the neural circuits that underlie them, puberty is being increasingly recognized in Western societies as a set of years in which boys and girls are vulnerable to some of the most delicate and sometimes unfortunate developmental changes.

TABLE 10.1 Human Development Stages

Female development			
Breast stages	Description	Pubic hair stage	Description
Bl	Prepubertal: elevation of the papilla only	PHI	Prepubertal; no pubic hair
B2	Breast buds are noted or palpable with enlargement of the areola; this is quite subtle and often missed on examination	PH2	Sparse growth of long, straight, or slightly curly minimally pigmented hair, mainly on the labia; this stage is very subtle and sometimes missed on cursory examination
B3	Further enlargement of the breast and areola with no separation of their contours	PH3	Considerably darker and coarser hair spreading over the mons pubis
B4	Projection of areola and papilla to form a secondary mound over the rest of the breast	PH4	Thick adult-type hair that does not yet spread to the medial surface of the thighs
B5	Mature breast with projection of papilla only	PH5	Hair is adult in type and is distributed in the classic inverse triangle

Male development			
Genital stages	Description	Pubic hair stages	Description
Gl	Preadolescent	PHI	Preadolescent; no pubic hair
G2	The testes are more dian 2.5 cm in the longest diameter excluding the epidydimus; and the scrotum is thinning and reddening	PH2	Sparse growth of slightly pigmented, slightly curved pubic hair mainly at the base of the penis; this stage is very subtle and may be missed on cursory examination
G3	Growth of the penis occurs in width and length and further growth of the testes is noted	PH3	Thicker, curlier hair spread laterally
G4	Penis is further enlarged and testes are larger with a darker scrotal skin color	PH4	Adult-type hair that does not yet spread to the medial thighs
G5	Genitalia are adult in size and shape	PH5	Adult-type hair spread to the medial thighs

Modified from Marshall and Tanner (1969, 1970).

Clearest are the sexual changes. During puberty, the hypothalamus and pituitary in females are first able to manage the positive feedback effects of estrogens in order to generate an ovulatory surge of luteinizing hormone (LH). Boys undergo a rise in testosterone levels. Libidinous interests in both sexes undergo a marked rise over a period of several years during which hormonal changes are instituted.

At least for girls, this period of marked hormonal change can occur during a time that can already be considered stressful. Jacqueline Eccles at the University of Michigan has documented the harmful effects of stressful transitions for girls when they coincide exactly with the occurrence of pubertal alterations. For boys, the years of puberty are just as delicate, but with different social consequences. Marked elevations in aggressive behavior are seen under a wide variety of circumstances. The easiest form of aggression to quantify is the rate of homicides. Wilson and Daly, at McMaster University in Canada, have charted murders of unrelated persons of the same sex as a function of the age of the killer (Fig. 10.1). Two facts are immediately clear. First, males kill much more frequently than females. Second, the homicide

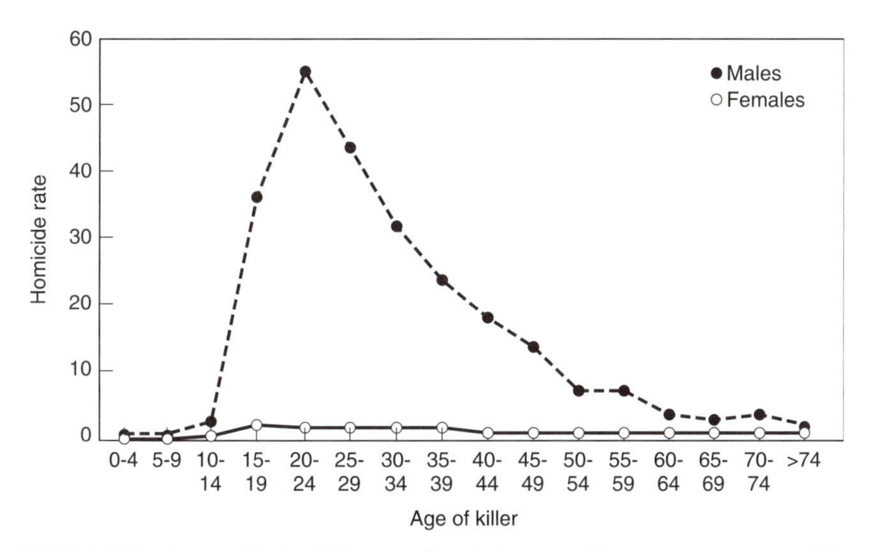

FIGURE 10.1. Age-specific homicide rates (homicides per million persons per annum) for men and women who killed an unrelated person of the same sex in Canada, 1974–1992 (upper panel), and in Chicago, 1965–1989 (lower panel). Data include all homicides known ot police in which a killer was identified. (From Wilson, M., Daly, M., and Pound, N. (2002). An evolutionary psychological perspective on the modulation of competitive confrontation and resk-taking. *Hormones, Brain, and Behavior*, Vol. V, Pfaff, D. W., ed., San Diego: Elsevier, p. 389. With permission.)

rate rises tremendously at puberty and falls gradually after the early 20s. Wilson and Daly's conclusions transcend national boundaries and are not limited to Western societies. Clearly, from the correlations Wilson and Daly have made with circulating testosterone levels and from decades of mechanistic work in animals, one can conclude that androgenic hormones themselves are part of the problem. As a result of the actions of testosterone in the brain—as well as androgen metabolism to estrogens—changes in serotonergic systems in the forebrain facilitate aggressive behaviors, and other neurotransmitters have not been ruled out. Genetic influences are clear. Surprisingly, with respect to the special vulnerability of the pubertal animal, male mice with the gene for estrogen receptor-β knocked out show high levels of vicious biting attacks when tested at puberty but not later in life.

We hasten to point out that even though hormonal, genetic, and neurochemical influences on aggressive behavior obviously can play important roles in the violent behaviors exhibited by pubertal boys, they are not the entire story. James Gilligan, of Harvard Medical School, has drawn together evidence for the important roles played by socioeconomic status, size of schools (smaller being preferred), initiation rites, and gun control. Just as crucial is the treatment of the child during an earlier critical period, the neonatal period (see Chapter 9). Deprivation at that time, sometimes associated with low birth weight, predicts later trouble. All of these social and psychological factors (acting in a manner consistent with a focus of Chapter 19) must be considered together with androgenic hormone action for a total understanding of violence exhibited by teenaged boys.

Pubertal years also are the developmental period during which gender differences in psychiatric states emerge. In several nations, the predominance of diagnoses of depression in women, compared to men, has been tabulated. As mentioned in Chapter 8, in the United States the ratio of depression in women to men is about 2:1; in Denmark, about 6:1. Mechanisms for the peripubertal appearance of this set of mood states are unknown.

II. MECHANISMS OF PUBERTY

Puberty is occasioned by huge changes in the synthesis and secretion of hormones associated with growth and with reproduction (Fig. 10.2). In both cases, the pituitary hormones growth hormone (GH) and (for reproduction) luteinizing hormone (LH) and follicle-stimulating hormone (FSH) must be secreted in pulsatile form in order to be effective (see related material below). Their increased biological actions during puberty depend upon greater pulse amplitudes, not pulse frequency.

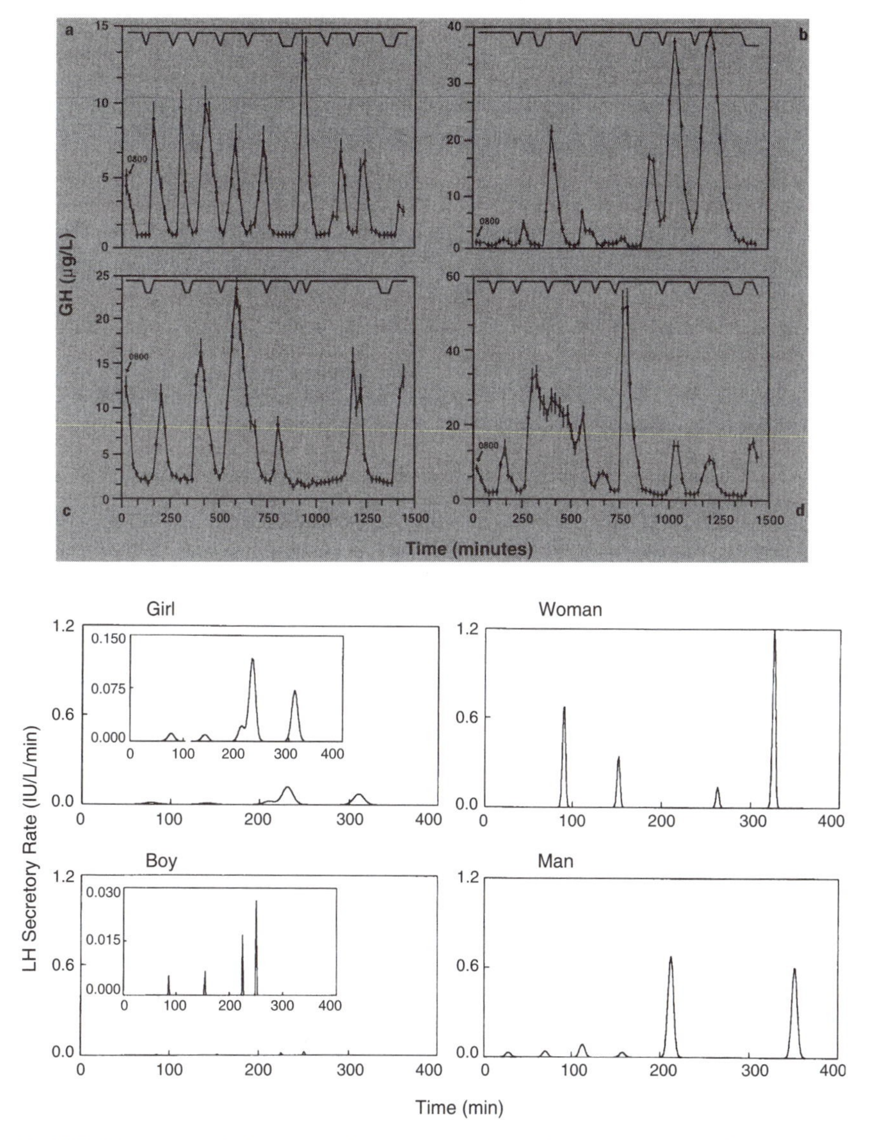

FIGURE 10.2. Pulsatile hormone release during puberty. (Top panel). Growth hormone (GH) secretory profiles of 3 boys. (a) A 14 year old boy, prepubertal. Note the pulses. (b) Another 14 year old boy going through puberty. Note the larger numbers on the Y-axis. (c) A hypogonadal 14 year old boy before testosterone therapy. If a boy is allowed to grow up hypogonadal, there are serious implications for his level of libido. (d) The same boy as in c, after testosterone administration. Note the larger numbers on the Y-axis. (From Mauras *et al.*, 1987.) (Bottom panel). Luteinizing hormone (LH) secretory profiles. Note the larger Y-axis numbers for the postpubertal woman and man, compared to the girl and boy, respectively. (From Clark *et al.*, 1997.)

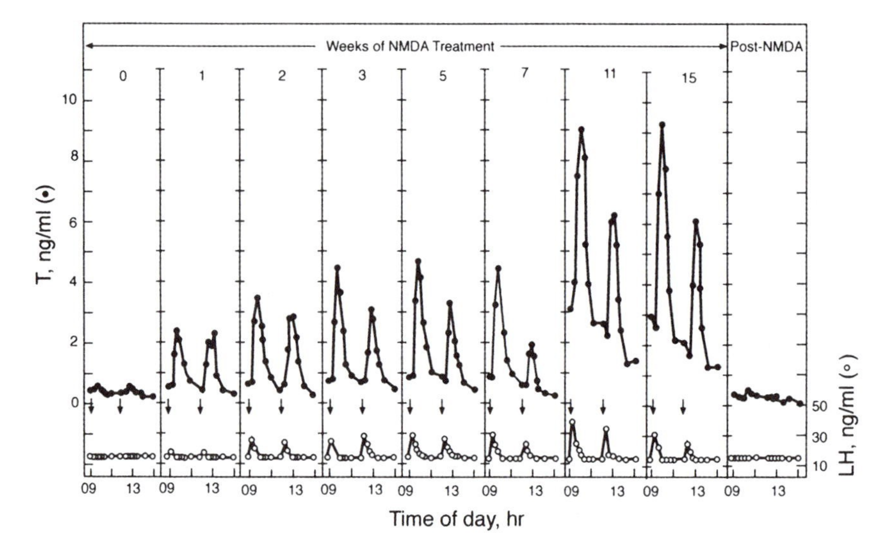

FIGURE 10.3. Premature activation of the hypothalamic–pituitary–testicular axis in immature male rhesus monkeys induced by repetitive stimulation of the hypothalamus with NMDA, an excitatory amino acid receptor agonist, administered once every 3 hours for 15 weeks. The treatment was initiated 1.5–2 years before the normal age of puberty in this species. The arrows indicate the time of injection. T, testosterone; LH, luteinizing hormone. (From Plant *et al.*, *Proc. Natl. Acad. Sci. USA* **86**, 2506–2510 (1989).With permission.)

What are the neuroendocrine mechanisms that allow these dramatic transitions at the level of the pituitary gland? The initiation of puberty requires increased secretion of the neuropeptide gonadotropin-releasing hormone (GnRH) (also known as LHRH) from the hypothalamus to the pituitary. The deposition of GnRH needed for puberty is associated with more excitatory transmitters impacting GnRH neurons and less inhibitory transmitters. Plant and colleagues worked with rhesus monkeys and showed that they could advance puberty by treatment with *N*-methyl-D-aspartate (NMDA), which stimulates a subset of receptors for the excitatory transmitter, glutamate (Fig. 10.3). Likewise, there is decreased release of the inhibitory transmitter γ-aminobutyric acid (GABA) (Fig. 10.4). Further, blockade of synthesis for a synthetic enzyme for GABA induces a pubertal-like GnRH release in monkeys. In turn, blocking GABA receptors causes precocious puberty (see Chapter 7). Most interesting is that Ojeda and colleagues in Oregon have determined that these pubertal mechanisms are not limited to neurons. Glial cells (astrocytes) control GnRH secretion through growth factor pathways, including transforming growth factor-alpha (TGF-α) and the neuregulins, both of which act through tyrosine kinase receptors such as erbB-1 and erB-4. The full complexity of the mechanistic steps involving astrocytes can be appreciated

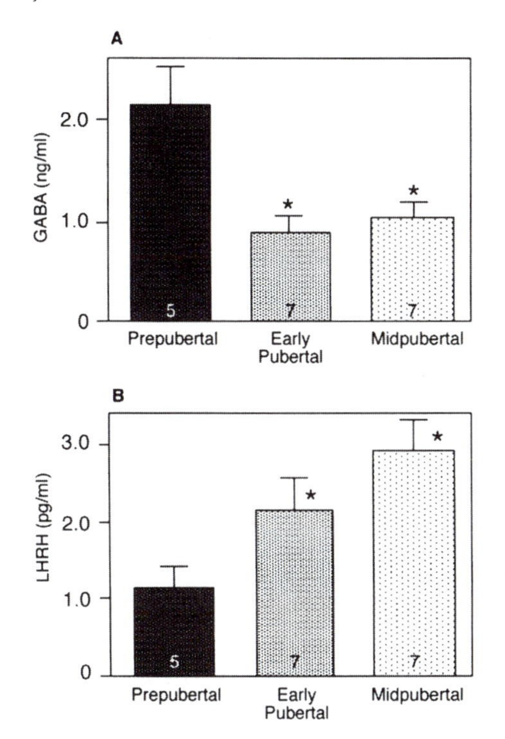

FIGURE 10.4. Decrease in GABA release (A) from the rhesus monkey hypothalamus at the time of puberty. The decrease coincides with the pubertal increase in LHRH release (B). Bot GABA and LHRH were measured in conscious animals, via push-pull perfusion of the stalk-median eminence (S-ME) of the hypothalamus. Numbers inside each bar represent number of animals per group. *Significantly different from prepubertal value. (From Mitsushima *et al.*, *Proc. Natl. Acad. Sci. USA* **91**, 395–399 (1994). With permission.)

by looking at Fig. 10.5. A simplified view of the several hierarchical levels involving the various mechanisms important for pubertal changes in behavior is shown in Fig. 10.6.

A. Growth and Feeding in Puberty

Puberty is a period of dynamic changes in body shape, size, and composition. The changes are sexually dymorphic, leading to body composition changes in the distribution of fat and growth. The adolescent growth spurt and alterations in body composition depend on the release of gonadotropins, sex steroids, growth hormone, and leptin. Bone mass is maintained by

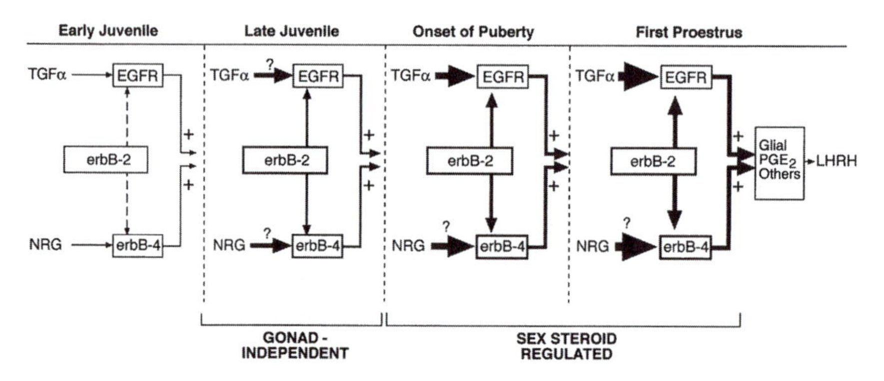

FIGURE 10.5. Changes in hypothalamic erbB receptor expression during juvenile and peripubertal development of the female rat. The first increase in synthesis occurs at the end of the juvenile period in the absence of changes in gonadal steroid secretion. This first increase is, therefore, gonad-independent and is postulated to be a component of the centrally activated, gonad-independent process that sets in motion the initiation of puberty. Once puberty is initiated, erbB receptor synthesis is further increased first by the rising estrogen levels, and then by the concerted action of estrogen plus progesterone on the day of the first preovulatory surge of gonadotropins. Whether there is a sex-steroid-independent increase in TGFα and/or NRG gene expression during juvenile development is not yet known. (From Ojeda *et al.*, *Recent Prog. Horm. Res.* 55, 197–224 (2000). With permission.)

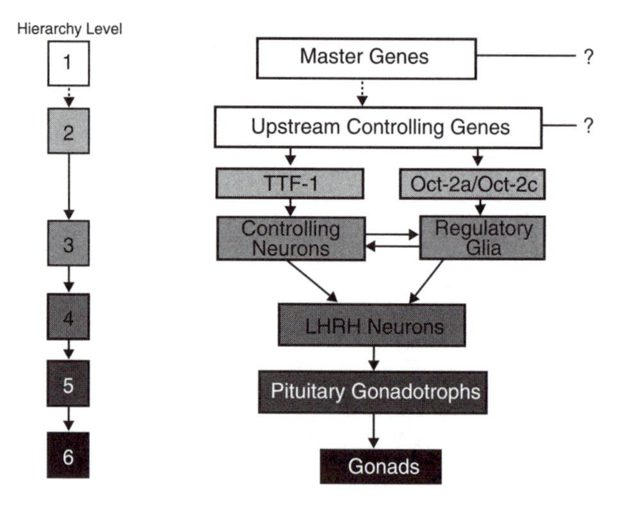

FIGURE 10.6. Postulated hierarchical levels of control in the hypothalamic-pituitary-ovarian axis at the time of puberty. 1 represents the highest level of control, 5 the most subordinated. TTF-1 and Oct-2 are two homeodomain genes postulated to act as upstream regulators of the process. (From Ojeda *et al.*, *Recent Prog. Horm. Res.* 55, 197–224 (2000). With permission.)

a constant yin/yang relationship between osteoclasts, which reabsorb bone, and osteoblasts, which form bone. In addition, there are cartilaginous plates at the growth ends of bone; closure of these plates as they become ossified by the actions of osteoblasts defines the length of the bone. Testosterone plays a role in the growth spurt by increasing bone size and closing the growth plates. Nutrition, which provides both energy and specific nutrient intake, is a major determinant of pubertal growth, as poor nutrition inhibits linear growth.

An important factor in nutrition is leptin, which is probably one of the hormones centrally involved in determining the onset of puberty. Leptin-deficient mice (*Ob/Ob*) or leptin-receptor-deficient mice develop high bone mass. Infusing leptin into the third ventricle of the brain of the *Ob/Ob* mice decreases bone mass and bone formation. This finding places bone formation under hormonal control from the brain, probably the hypothalamus. Bone formation responds to testosterone levels that are under the control of LH from the hypothalamus and are influenced by the level of leptin. Leptin, as we saw in Chapter 1, is the hormone of fat cells. As fat cells increase, leptin is released into the circulation and exerts its anorexigenic effects on neurons in the brain. Leptin acts on at least two nuclei in the hypothalamus: the ventromedial hypothalamic nucleus (VMH) and the arcuate nucleus (ARC), as revealed by the fact that both have a high density of neurons with the leptin receptor. Neurons in the ARC can be specifically destroyed by treating young animals with monosodium glutamate (MSG), a toxin that overexcites cells by an action on the glutamate receptor (Olney, 1969). Mice treated with MSG had normal bone volume, but leptin was unable to decrease body weight. When the VMH neurons were specifically destroyed by gold thioglucose, leptin infusion decreased the body weight but did not affect bone mass. This experiment by Takeda *et al.* (2002) showed that leptin has an anti-osteogenic function, mediated through the VMH, that is differentiated from its anorexigenic function mediated by the ARC. To achieve high bone mass, the mice needed not only low leptin but also activation of β-adrenergic receptors. The implication of this study is that, during puberty, good nutrition is important and high metabolism is necessary to burn off fat. Reduced fat reduces leptin so that appetite is increased and high bone mass is formed.

III. SOME OUTSTANDING QUESTIONS

1. What are the interactions among social and hormonal influences on libidinal changes during puberty?

2. How do we relieve anorexia nervosa, a problem more common in teenaged girls than boys?
3. What mixture of social and biological tactics will work to reduce violence by teenaged boys?
4. What accounts for the rise in depression diagnosed in girls during and after puberty, but not in boys?
5. What factors lead to stunted growth resulting from the lack of a pubertal growth spurt?

CHAPTER 11

Changes in Hormone Levels and Responsiveness During Aging Affect Behavior

Questions of changed hormone/behavior relations during aging and behaviorally relevant effects of hormones on diseases of the aged are taking on new dimensions because of our steadily increasing life span (Fig. 11.1). Under circumstances where animals would die at a young age due to predation, inanition, accident, or disease, it would not be important to answer either of the classes of questions above. This chapter documents that there are, indeed, changes during aging and reviews some of the initial investigations. Because of the plethora of clinical examples, we have concentrated, in this chapter, on material of importance for aging humans.

I. CLINICAL EXAMPLES

Aging in humans has been considered by some to begin at birth. While this might be a somewhat overstated perspective, it is true that some tissues can begin the process toward senescence early in the life cycle. For example, exposure of skin to excess ultraviolet radiation, (*i.e.*, sunburn) in childhood can induce changes that result in prematurely aging skin and skin cancer in adulthood. With reference to endocrine influences on behavior, it is

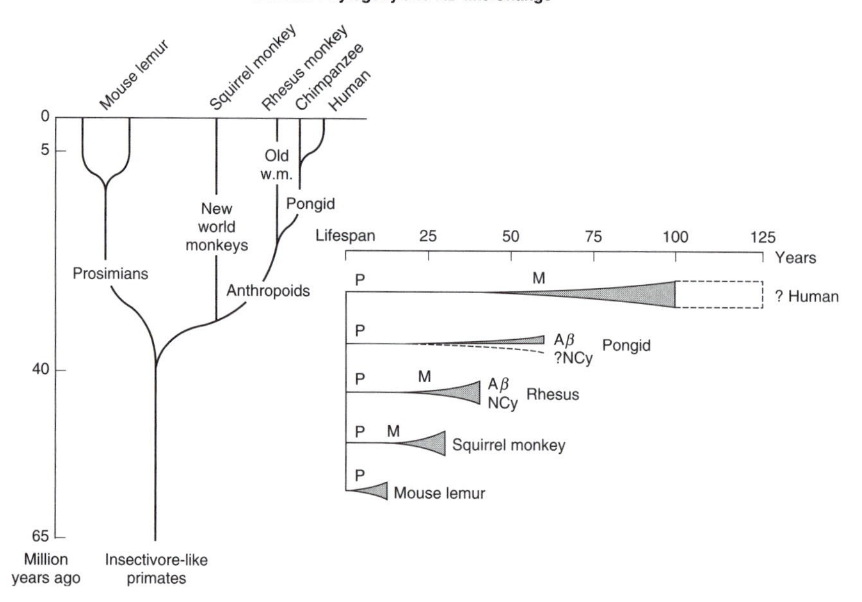

FIGURE 11.1. Behaviorally relevant effects of hormones on the diseases of the aged. In this case, effects of hormones include those on menopause (M) and the production of the protein fragment Aβ, crucial in Alzheimer's Disease (AD).

important to recognize that both neurotransmitter and neuroendocrine systems continually change during the life cycle, at different rates and in very different ways.

The human life span is increasing, such that we now need to consider elderly people as the "old" and the "oldest old," the latter including those in their 80s and 90s and even over 100 years of age. These individuals comprise an ever-increasing percentage of the population; it has been estimated that half of the female children born in industrialized countries in the 21st century will live to be 100 years of age. Thus, studying the physiology of the oldest old, in particular their endocrinology, becomes increasingly important as this sector of the population grows.

Several neurotransmitter systems in the central nervous system (CNS) become less active with advancing age, including the actions of norepinephrine, serotonin, dopamine, and acetylcholine. Decreases in neurotransmitter synthesis, uptake, receptor binding, and responsiveness to stimulation have been noted. Neuroendocrine activity in general also declines with age, and an underlying factor may be a decline in CNS hypothalamic catecholamine neurotransmitter activity. Loss of hormone receptor sensitivity and a reduction in the output of target endocrine glands also may be contributing factors. Following are examples of endocrine changes in later life that can have major influences on behavior.

A. Hypothalamic–Pituitary–Gonadal Function: Menopause

Perhaps the most profound endocrine change to occur with aging is the female menopause. The pool of ovarian follicles becomes exhausted, estrogen and progesterone secretions by the ovary decline, and gonadotropin (luteinizing hormone [LH] and follicle-stimulating hormone [FSH]) secretion from the pituitary increases, due to the lack of gonadal steroid feedback to the hypothalamus and pituitary. The events occur at several levels of the hypothalamo–pituitary–gonadal (HPG) axis, including early changes in the secretion of LH and FSH coincident with the onset of hot flashes.

Behavioral disturbances can include impaired sleep, reduced energy, fatigability, difficulty concentrating, increased sensitivity to pain, and mood changes, most often depressive in nature. The overall incidence of major depression, however, does not seem to be increased after menopause. Measurable cognitive changes are often minor, although estrogen replacement can result in improved memory function. Estrogen plus progestin replacement in normal postmenopausal women has salutary effects on hot flashes, night sweats, disturbed sleep, bone mineral density, and a general sense of well-being, but at the possibly increased risk of hormone-sensitive malignancies (*e.g.*, breast), blood clots, and stroke. Estrogen replacement by itself, without a progestin, may have a different risk/benefit ratio and is still being studied. As well, selective estrogen receptor modulators (SERMs) are being designed to maximize beneficial effects, such as increasing bone density, while not activating receptors associated with malignancies. For example, both tamoxifen and raloxifene prevent bone loss, but tamoxifen increases the risk of uterine cancer, and raloxifene is an estrogen antagonist in the uterine lining; however, raloxifene does not alleviate hot flashes. SERMs with a broad spectrum of efficacy and with few or no unwanted side effects are an area of active drug development.

Although estrogen is not an effective antidepressant by itself, it may be a useful adjunctive therapy in postmenopausal women who are only partially responsive to standard antidepressant drugs. Dementias in the elderly are becoming an increasing public-health problem as the old and oldest old become an ever-increasing percentage of the population, and estrogen therapy also may have a role in addressing these illnesses. Alzheimer's disease accounts for about 50% of dementias in the elderly, and its incidence rises dramatically after age 70. Its clinical course is unremittingly downhill. A prominent neurochemical change in Alzheimer's disease is the loss of cholinergic neurons in the brain. Currently available treatments are mainly directed toward enhancing the neurotransmitter effect of whatever acetylcholine is still being produced by these neurons. Estrogen has significant neuroprotective effects, so that an important question is whether estrogen replacement in postmenopausal women can prevent the onset and slow the progression of Alzheimer's disease. Although research data are still being gathered, evidence to date suggests that estrogen replacement may be protective against disease onset as well as being effective in slowing progress of the disease when used as an adjunct to treatment with drugs that enhance cholinergic neurotransmission.

B. Androgen Decline in Aging Men

The relatively abrupt and profound neuroendocrine changes associated with menopause do not occur in men; rather, throughout their adult life men experience a gradual decline in circulating testosterone concentrations and blunting of its circadian rhythm with advancing age. Because protein binding of testosterone increases somewhat with age, the net result is a faster decline in free (unbound) testosterone, the bioactive fraction, in blood than in total testosterone. Properly termed *androgen decline in aging men* (ADAM), this slow decrease in testosterone with age also has been referred to as "andropause," implying a male counterpart to menopause (*i.e.*, occurring universally in men and requiring some degree of androgen replacement). In fact, the decrease in androgen with age varies from person to person, and only men with documented subnormal circulating testosterone and who have symptoms attributable to low testosterone should be considered for replacement therapy.

Symptoms associated with low testosterone include reduced sex drive, reduced ability to maintain erections, low energy level, fatigability, sleep disturbance, difficulty concentrating, and depressive mood. These symptoms can respond dramatically to hormone replacement. Many of these nonspecific symptoms, however, also occur in both young and elderly patients with major depression who have normal androgen levels. Therefore, these symptoms alone cannot be used as indications for hormone replacement therapy; rather, abnormally low testosterone must be documented in order to justify androgen replacement. As with estrogen replacement in postmenopausal women, there are risks associated with testosterone replacement in elderly men, including accelerating the development of prostatic cancer and increasing circulating red blood cells, blood viscosity, and the risk of stroke.

C. Aging Women

It is clear that at least in a subset of older women, estrogen administration can protect against some cognitive or emotional decline evident during aging. Some of these health-preserving actions may be on mechanisms of associative memory themselves. Others may be on more primitive cognitive functions such as arousal, alertness, and attention. Still others may be emotional effects, which in turn could affect cognitive performance. Importantly, Henderson and Reynolds (2002) have reviewed how, under some circumstances, estrogen treatment can delay some of the symptoms of Alzheimer's disease. From a clinician's point of view, the number of possibilities is large (none of them mutually exclusive and some of them synergistic) as to how estrogens could protect mental functions in the aging brain (Table 11.1). Detailed mechanistic investigations have been carried out in experimental animals and are best illustrated by Foster's work at the University of Kentucky (Fig. 11.2). In this series of experiments, virtually every estrogenic action is in a direction opposite to that evident during aging. Further, a reasonable possibility for one component of the anti-aging effects of estrogen is that estrogen reduces the consequences of vascular accidents in the brain, large or small, whose cumulative effects over time would reduce cognitive performance. An impressive body of data demonstrates that chronic estrogen administration can reduce cerebral damage following strokes (see below). Thus, through genomic and metabolic actions on positive anabolic functions (*e.g.*, neuronal growth; see review by Woolley and Cohen, *Hormones, Brain and Behavior*, 2002) and through the reduction of negative effects, estrogens can, in easily understandable ways, delay degeneration in the aging brain.

D. Estrogens and Neuroprotection

In general terms, we now appreciate that estrogens are pleiotropic gonadal steroids that exert profound effects on plasticity and cell survival of the adult brain. Over the past century, the lifespan of women has increased dramatically from 50 to over 80 years, but the age of onset of menopause has remained constant at approximately 50 years. This means that women may now live over one-third of their lives in a hypo-estrogenic, postmenopausal state. The impact of prolonged hypoestrogenicity has many repercussions, as these women may suffer an increased risk of cognitive dysfunction and neurodegeneration diseases, including Alzheimer's disease and stroke. Accumulating evidence from both clinical and basic science

TABLE 11.1 Protective Actions of Estrogen

Short-term protective effects	Preventive effects	Miscellaneous effects
Oxidative stress	β-Amyloid precursor	Neurotrophism
Excitatory neurotoxicity	Protein metabolism	Synaptic plasticity
β-Amyloid toxicity	Apolipoprotein E	Neurotransmitter systems:
Hypoglycemia	Vascular system	acetylcholine, noradrenaline, serotonin, dopamine, etc.
Ischemia		
Apoptosis		
Hypothalamic–pituitary–adrenal axis reactivity		
Inflammation		
Cerebral blood flow		
Glucose transport		

From Henderson, V.W., and Reynolds, D.W. (2002). Protective effects of estrogen on aging and damaged neural systems, *Hormones, Brain, and Behavior*, Pfaff, D.W., ed., Vol. 4, ch. 80. San Diego: Elsevier, pp. 823. With permission.

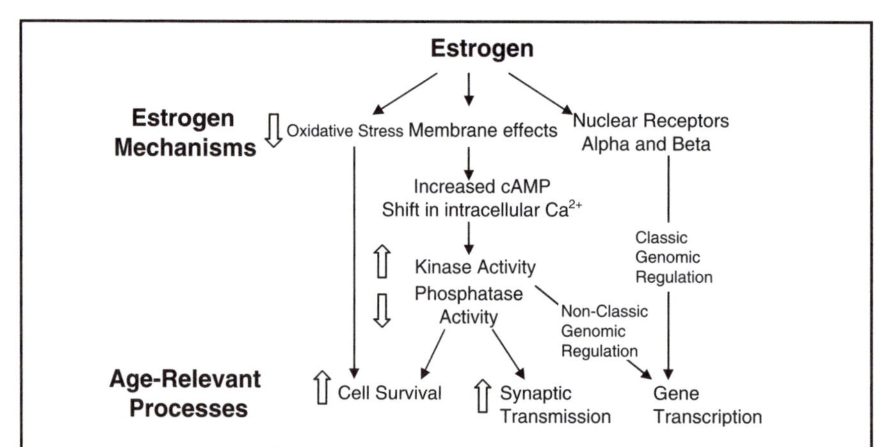

FIGURE 11.2. A growing body of research indicates that estrogen can influence hippocampal anatomy and physiology, lending support to the idea that estrogen may influence memory function. Interest in estrogen effects on memory and the hippocampus has increased due to evidence indicating that hormone replacement can delay the progression of Alzheimer's disease and reduce memory decline observed during normal aging. The beneficial effects of estrogen may involve protection from oxidative stress, trophic support, and from the of maintaining Ca^{2+} homeostasis. It is clear that estrogen can act on neural tissue through several distinct mechanisms, including activation of estrogen nuclear receptors and membrane interactions that influence second-messenger signaling cascades. Classical genomic effects depend on estrogen binding

studies indicates that estrogens exert critical protective actions against these conditions.

Studies using animal models of cerebral ischemia provide strong evidence that estradiol is a neuroprotective factor that profoundly attenuates the degree of ischemic brain injury. Experiments conducted by Phyllis Wise's group (Fig. 11.3) have found that low physiological concentrations of estradiol replacement exert dramatic protective actions in the brains of young and middle-aged female rats as well as in cultured explants of the cerebral cortex. The results emphasize that estradiol attenuates delayed, apoptotic cell death by acting at multiple levels to simultaneously decrease cell death and promote cell survival. It enhances the expression of genes that attenuate apoptosis, diminishes the expression of genes that promote apoptosis, decreases activation of caspases, increases activation of signaling pathways that promote cell survival, promotes the proliferation of neurons in the region of the brain that may migrate to the cortex, and attenuates the immune response. Together, these actions can dramatically decrease the extent of brain damage

Figure 11.2. (*Continued.*)
to either the alpha or beta estrogen receptor; however, the hippocampus exhibits relatively low expression of estrogen receptors, which is limited to a small number of cells. Nevertheless, most hippocampal cells exhibit rapid changes in several physiological processes upon estrogen exposure, indicating powerful estrogenic influences through non-genomic mechanisms. Memory-enhancing effects of hormone replacement can be observed under specific experimental conditions, particularly in animals whose learning ability has been pharmacologically impaired.

The non-genomic estrogen effects are rapid, independent of RNA or protein synthesis, and usually require the continued presence of estrogen. Interestingly, many of the rapid effects are diametrically opposite to changes observed in aged, memory-impaired animals. For example, aging is associated with modifications that reduce transmission through the hippocampus (Foster, 1999). In contrast, estrogen rapidly increases cell excitability and the strength of synaptic transmission (Kumar and Foster, 2002; Sharrow *et al.*, 2002). In addition, estrogen adjusts synaptic plasticity processes, facilitating the induction of long-term potentiation and impairing the induction of long-term synaptic depression, forms of Ca^{2+}-dependent plasticity that are modified by aging and thought to play a role in the establishment and maintenance of memory (Sharrow *et al.*, 2002).

The exact mechanism for non-genomic estrogen effects is not established; however, research indicates that estrogen can influence processes that are regulated by G-protein activity and intracellular Ca^{2+}. A shift in these signaling cascades changes the phosphorylation state of proteins by increasing and decreasing protein kinase and phosphatase activity, respectively. An estrogen-mediated shift in the balance of kinase/phosphatase activity is involved in regulating synaptic plasticity and contributes to cell survival (Sharrow *et al.*, 2002). Moreover, estrogen-induced changes in the phosphorylation state of transcription factors, such as CREB, regulate gene transcription and provide a non-classical mechanism for genomic regulation. Importantly, the efficacy of hormone replacement therapy may be reduced if therapy is not initiated until several years after the onset on menopause or after the symptoms of Alzheimer's disease are manifest. Thus, it will be important for future research to understand age-related changes in estrogen signaling pathways

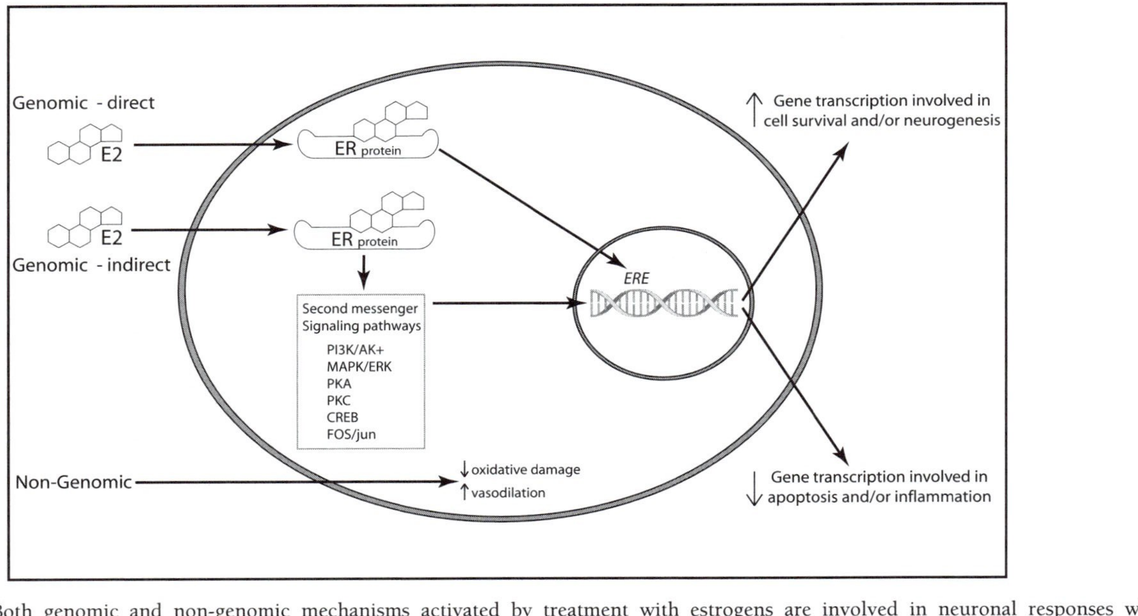

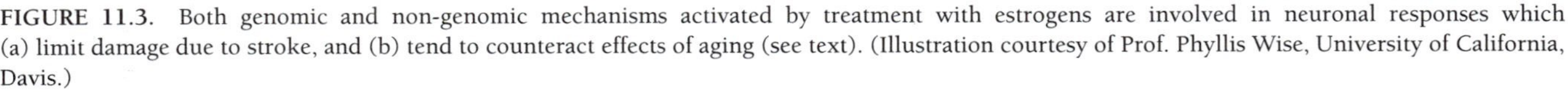

FIGURE 11.3. Both genomic and non-genomic mechanisms activated by treatment with estrogens are involved in neuronal responses which (a) limit damage due to stroke, and (b) tend to counteract effects of aging (see text). (Illustration courtesy of Prof. Phyllis Wise, University of California, Davis.)

after stroke injury. These protective actions of physiological levels of estradiol replacement absolutely require the presence of estrogen receptor-α, as estradiol was unable to protect in estrogen receptor-α knockout mice. Intriguingly, this receptor is not normally expressed in the adult cortex, but only in the developing cortex. Indeed, after injury in the adult, it reappears and plays a pivotal functional role in neuroprotection. Pharmacological concentrations of estradiol, on the other hand, appear to protect against neurodegenerative agents by very different mechanisms that do not require estrogen receptors. At these concentrations, estrogens act on the vasculature to influence blood flow and endothelial cell function, are effective antioxidants and free radical scavengers, and can alter ion channel biophysics (Golden *et al.*, 1999).

It is important to emphasize that estrogens do not always protect. Recent clinical studies suggest that under some circumstances, hormone replacement therapy does not afford protection and may increase risk. Together, these clinical studies would suggest that estrogen replacement therapy affords effective primary prevention of many neurodegenerative disease processes, but it does not effectively protect against or reverse a disease process that has already been initiated. Furthermore, these studies reveal that the effects of hormone replacement depend greatly on the formulation of the therapy, whether estrogen is combined with progestin replacement, how the hormone is administered, and the doses that are used. Clearly, much more work is necessary before we gain a more complete understanding of the spectrum of estrogen actions and understand the multiple interacting and diverse mechanisms that underlie its actions.

While the implications of estrogenic treatments for the reduction of stroke damage and the delay of Alzheimer's symptoms would seem to apply to men as well as women, a more controversial subject deals with whether or not there is a clearly defined, functional andropause. As noted above, testosterone levels, on the average, decline steadily, by about 1% per year, in aging men. The autonomic nervous system equivalent of the hot flash is not observed. Nevertheless, the decreasing physical and mental energy and reduced libido call attention to the possibility of hormonal therapy to deal with behavioral changes as well as other physiological declines. Transdermal treatments with androgens in gels, testosterone, or its reduced metabolite dihydrotestosterone certainly can improve the ability to initiate and maintain penile erections in aging men; however, reports of improved feelings of general well-being have been clouded by subsequent discoveries of significant placebo effects. While positive effects of systemic androgen treatment on sexual function and physical energy may occur, the chance of exacerbating prostate cancer growth in aging men requires that these behavioral benefits be pronounced, quite certain, and worth the risk.

II. MECHANISMS

The field of neurochemical and cell biological examination of how hormones could be used to combat aging of the brain has been growing rapidly. One extensive set of cellular and molecular investigations was referred to in Fig. 11.2; in addition, non-reproductive hormones must be considered.

A. Hypothalamic–Pituitary–Adrenal Function

In contrast to the marked changes in the pituitary–gonadal axis, plasma glucocorticoids change little with age. Some studies indicate no change, whereas others suggest a slight increase. As well, there may be a blunting of the glucocorticoid circadian rhythm and a shift in its timing. The increase in circulating glucocorticoids with age appears to be based on the reduced sensitivity of hippocampal negative feedback on corticotropin-releasing hormone (CRH) and adrenocorticotropic hormone (ACTH). Some investigators have proposed a glucocorticoid-induced loss of hippocampal glucocorticoid receptors, such as from early stress-induced increases in hypothalamic–pituitary–adrenal cortical (HPA) axis activity and continuing stress activation of this axis throughout adult life. In older individuals, stress and other provocations of the HPA axis result in a prolonged cortisol response, in contrast to younger individuals. This, in turn, could contribute to the major problem of depression among the elderly.

From a human behavior standpoint, one current mechanistic hypothesis of major depression is that increased CRH, through its anxiogenic properties, produces some of the symptoms, and increased glucocorticoids, which impair hippocampal functioning, contribute to other symptoms. Drugs that interrupt these neuroendocrine processes, such as CRH receptor blockers, glucocorticoid receptor blockers, and glucocorticoid synthesis inhibitors, have been proposed as treatments for major depression, but none has yet shown sufficient therapeutic activity to be clinically useful. As well, the hypothesis upon which these treatments is being developed (*i.e.*, increased HPA axis activity in major depression) is weak, in that the majority of such patients, including elderly depressives, have normal HPA axis activity.

B. Hypothalamic–Pituitary–Thyroid Function

In the elderly, the half-life of thyroxine (T4) is decreased, but circulating T4 usually is within the normal range, because its production is correspondingly reduced. The production of triiodothyronine (T3) may be reduced as well, but its half-life is not increased, leading to somewhat reduced circulating T3. Basal thyroid-stimulating hormone (TSH) concentrations are more variable, some being below and some above the normal range for younger individuals, perhaps representing the early stages of hypo- and hyperthyroidism, respectively. What is important is that the symptoms of both hypo- and hyperthyroidism are frequently more muted in the elderly, so that clinicians need to carefully screen their geriatric patients for thyroid abnormalities.

C. Water Balance in the Elderly

In elderly patients, disorders of water balance are common (Phillips *et al.*, 1991). The capacity of aging kidneys to concentrate urine is reduced. Thirst mechanisms are also impaired in healthy elderly people. Elderly subjects show reduced thirst and water intake following dehydration, whether or not water or some other beverage (non-alcoholic) is available. There appears to be a resistance to arginine vasopressin (AVP) in the kidney, because the kidneys do not produce as concentrated urine as those of younger people. Although thirst is impaired, AVP responses are appropriate for the level of osmotic increase in the plasma; therefore, this reduction in appropriate thirst is not due to an impaired osmoreceptor neuron in the hypothalamus leading to AVP secretion.

Plasma atrial natriuretic peptide (ANP) levels may also change with drinking. Ten healthy old men (64–76 years of age) and ten young men (20–32 years of age) were deprived of water for 24 hours and the effects were investigated. After 24 hours of water deprivation, plasma, sodium, osmolality, and AVP were increased similarly in both groups. Plasma ANP levels were consistently higher in the older group. Although thirst increased in both groups, it was significantly less in the elderly. Unfortunately, additional data are lacking. The increase of ANP in the elderly might have an inhibitory effect on angiotensin II (see Chapter 2). Therefore, we can only speculate that the reduced thirst in the elderly might reflect an inhibition of angiotensin or a reduction of angiotensin receptors in the critical sites around the third ventricle.

D. Steroid Receptors and Memory Loss

Aging is commonly accompanied by mild hypercorticism, with implications for both stress responses and fluid balance. Due to the catabolic effects of elevated glucocorticoids, glucocorticoid receptors (GRs) may be lost in hippocampal neurons. This parallels the concept of Sapolsky (2000) that stress, which causes increased exposure to glucorticoids, produces irreversible loss of glucocorticoid receptors in the hippocampus. In addition to glucocorticoid receptors, there are also mineralocorticoid receptors (MRs), and a strong structural homology exists between GRs and MRs (see Chapter 4). This allows for MRs and GRs to recognize identical hormone responsive elements within the promoter region of genes activated by corticoid steroid receptor complexes. Thus, there is a binary hormone response system for one single hormone, namely corticosterone. The loss of excitability and even the loss of neurons in the hippocampus could, if sufficiently critical in number, lead to losses in memory because clearly hippocampal neurons are involved in memory functioning. The hippocampus contains the highest concentration of MRs. In aging, the number of hippocampal MR binding sites is reduced, and the resulting imbalance between GR and MR impairs their capacity to control neuroendocrine feedback so that corticosterone levels are lower with aging.

III. SOME OUTSTANDING QUESTIONS

1. What is the optimal clinical paradigm for postmenopausal female hormone replacement?
2. The usual combination therapy of estrogen plus a progestin has shown increased risk of breast cancer and stroke. (a) Will estrogen alone be sufficient to reduce hot flashes and night sweats, to increase bone density, and to have other salutary effects without these risks? Studies are underway to address this question. (b) Will SERMs prove to have the desired spectrum of beneficial effects without detrimental side effects? Drug development in this area is ongoing. (c) What is the usefulness of dietary phytoestrogens in this regard (*e.g.*, from soybean products)? They have weak estrogen-like activity, and some may antagonize the actions of estrogen itself.
3. What is the optimal clinical paradigm for androgen replacement in aging men?
4. As indicated above, circulating testosterone declines with age, and normal levels of free and total testosterone exist for groups of men at different ages, including the elderly. These are averages, however, and cutoff points for low testosterone that may require androgen replacement therapy are based on these averages. The problem remains of the elderly man who has symptoms suggestive of low testosterone, who is not suffering from major depression, but who has circulating

testosterone above the established cutoff point for hypogonadism in men his age. What cannot be established is what that individual's starting testosterone concentration was as a young man and how much it has fallen in the intervening decades. The percentage decline may have been severe, even though the current concentration is above the average cutoff point. Strategies for addressing this dilemma need to be developed.

5. How can we best address the issue of hormone therapy for the prevention and treatment of dementias such as Alzheimer's disease?
6. Does a possible hormonal preventative therapy for Alzheimer's disease increase the risk of other complications, such as cancer, heart disease, or stroke, similar to hormone replacement therapy in menopausal women?

SECTION III

Time: Hormonal Effects on Behavior Depend on Temporal Parameters

CHAPTER 12

Duration of Hormone Exposure Can Make a Big Difference: In Some Cases Longer is Better; In Other Cases Brief Pulses are Optimal for Behavioral Effects

I. BASIC EXPERIMENTAL EXAMPLES

A. Sex Steroids

For large numbers of steroid sex hormone effects on a wide variety of behaviors, longer duration hormone administrations make for greater behavioral increases. A case in point is the effect of testosterone on male-typical sex behaviors in male rodents. Typically, for increased mounting behavior and especially for pelvic thrusting, penile insertions, and ejaculation, many days or several weeks of continuous testosterone treatment are required for high levels of response. Female sex behaviors are more complicated. Consider first the role of estrogenic hormones. A long "priming" period is useful to produce high levels of lordosis behavior. If the female is cycling normally or has been exposed recently to substantial levels of estrogens, then 48 hours is enough. The longer the female is without ovarian estrogens, the longer the priming period required for female sex behavior. A molecular interpretation of the mechanism operating in this case is the following: Prolonged absence of estrogens allows the decline of nuclear co-activator protein levels (see Chapter 18) which are needed to transduce the nuclear binding of estrogen receptors into behaviorally relevant transcriptional facilitations. A surprising development with respect to long priming actions came from biochemical work with the uterus by Jack Gorski, at the University of Wisconsin, who found that two brief exposures to estrogen, suitably timed, could be substituted for a priming period of 24 hours. Parsons, at Rockefeller, followed up this finding for the central nervous system (CNS) and behavior. Again, an estrogen priming of 24 hours could be replaced by two 1-hour exposures, the first from hour 0 to 1 and the second beginning between 4 and 13 hours after the first. This pulse schedule was effective for inducing both PR and lordosis behavior (Fig. 12.1).

In females, estrogen priming must be followed by progesterone for optimal behavioral facilitation (refer back to Fig. 3.4). The progesterone, if timed correctly, amplifies the estrogenic effect. The required temporal parameters for progesterone administration are much different from those for estradiol, as there is a biphasic action of progesterone both on pituitary release of luteinizing hormone (LH) and on behavior. In female rats, for example, 2 to

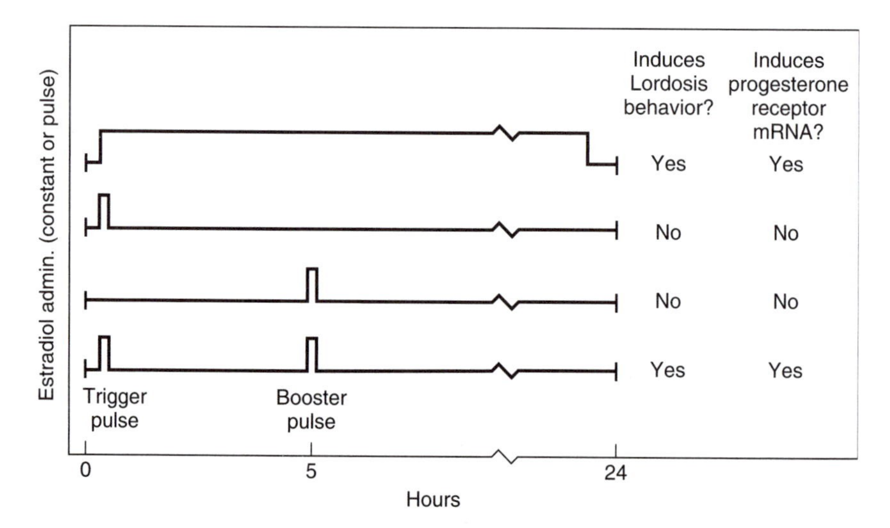

FIGURE 12.1. (Top line) Prolonged high levels of estrogens are sufficient for both the female reproductive behavior, lordosis, and for the induction of progesterone receptor mRNA (whose protein product, in turn, permits progesterone to amplify estrogen action). (Middle lines) A single brief of estradiol is sufficient for neither. (Bottom line) A trigger pulse followed by a booster pulse is sufficient for both. For the lordosis behavior test, the estrogen action has been amplified by progesterone treatment on the second day of the experiment. (Adapted from Parsons *et al.*, 1981, 1982).

5 hours after progesterone injection, LH release and lordosis behavior are facilitated mightily. But, the continued presence of progesterone, several hours later, actually inhibits both the endocrine output (LH) and the behavioral output (lordosis).

B. Protein Hormones from Pituitary

In dramatic contrast to the situation with testosterone and with estrogenic priming, protein hormones secreted by the pituitary gland not only (1) do not require long priming periods, but also (2) actually require rapid pulsatile exposure for maximum effectiveness. The classic example is the sudden pulse of gonadotropin-releasing hormone (GnRH, also known as LHRH), which is necessary to trigger the LH surge, itself an obligatory pulse for ovulation (Fig. 12.2). In addition to the LH surge, a surge of brain angiotensin II lasting 30 minutes occurs 1 hour before the LH surge, and there is also a surge of prolactin. Also, a steady buildup of neuropeptide Y (NPY) begins some hours before the LH surge. Thus, several events, possibly arranged in an ordered sequence, are necessary to prime the sudden release of large amounts of LH in the ovulatory surge.

C. Angiotensin Exposure at a Young Age

In contrast to the angiotensin surge that we previously discussed, and in agreement with the points in Chapter 9 (regarding early hormone effects), prolonged exposure to a hormone during development can lead to a lifelong change in physiology. To study the differences in the population of people who are likely to have hypertension in adult life, researchers have resorted to animal models, particularly the spontaneously hypertensive rat (SHR). If blood pressure is measured in these rats from the first week after birth up to 20 weeks, a distinct curve of the development of the hypertension is seen. Up to 5 weeks, there is little change in blood pressure. Between 5 and 8 weeks, a dramatic increase in blood pressure is observed that reaches a level that continues to increase more slowly over time. The question then is what happens during this developmental period between weeks 5 and 8? In hypertension treatment, angiotensin-converting enzyme (ACE) inhibitors are frequently given. When the mother SHRs were given the ACE inhibitor captopril, the baby rats were weaned on the inhibitor. When they stopped weaning, their water was doctored with ACE inhibitor. In this way, these animals were exposed to inhibition of angiotensin formation from birth.

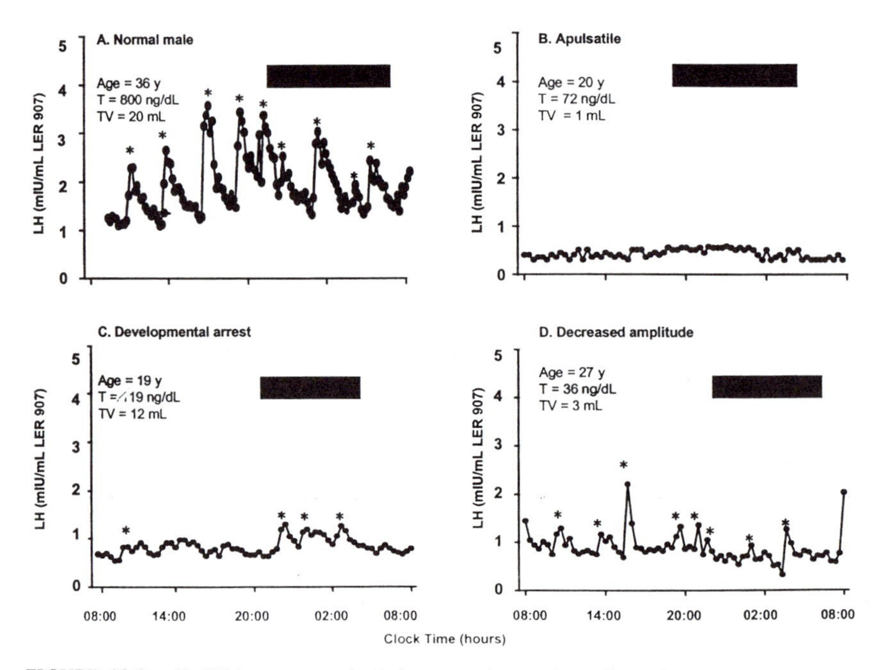

FIGURE 12.2. GnRH is necessary both for normal LH release from the pituitary gland and for normal sexual behaviors. The effective mode of GnRH application is not a constant administration but is in pulsatile form. In fact, the authors of this figure (led by Dr. William Crowley at Harvard Medical School) can bring infertile men into a state of fertility by pulsatile GnRH administration. In turn, the normal mode of LH release is pulsatile. (A) Pulsatile LH release in a normal male; (B) a patient who lacks normal GnRH and LH secretion; (C) a patient with a developmentally arrested, low amplitude pattern; and (D) a patient who has the same type of pulses as the normal man in (A) but whose amplitudes are significantly decreased. The asterisk (*) signifies a statistically significant pulse. (From Seminara, S.B. *et al.*, *Endocr. Rev.*, 1998. With permission.)

The result was that these SHR pups never developed hypertension (Wilson *et al.*, 1988). The study has been repeated using a more sophisticated approach utilizing a retrovirus containing an antisense to angiotensin type 1 receptor (AT_1R) (Reaves *et al.*, 1999), but the result was the same. If the animals are not exposed to angiotensin during that developmental period, they do not develop hypertension. When we analyze the angiotensinogen levels in these SHRs at different time periods, we see a distinct significant peak of angiotensinogen in the SHR at 3 weeks of age. Thus, the peak of angiotensin occurs before the rise in blood pressure seen at 5 weeks and later. Besides showing the importance of a peak in hormone concentration, these data illustrate how exposure to a hormone at a very young age can permanently affect the physiology in adulthood.

II. CLINICAL EXAMPLES

Gonadotropin-releasing hormone (GnRH) is famous for its pulsatile release. On the one hand, it is fascinating that populations of identical GnRH neurons can manage pulsatile output, an outcome that, as it turns out, is made comprehensible through the use of computer simulation. On the other hand, pulsatile outputs of GnRH are absolutely necessary for the pituitary to

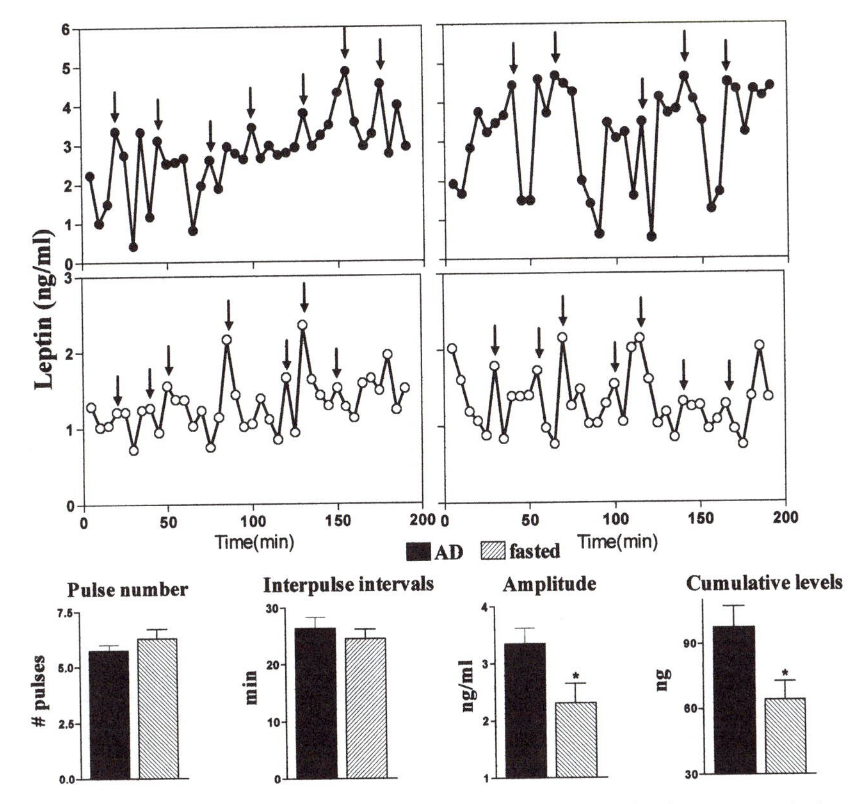

FIGURE 12.3. Pulses of leptin release had higher average amplitudes, causing higher cumulative levels in fed rats than in fasted rats. Top of the figure shows raw data from individual animals. Filled circles, fed; open circles, fasted. Note differences in numbers on the *y*-axis. Bottom of the figure shows average values. (From Kalra, S.P. *et al.*, *Regul. Pept.*, 111(1–3): 1–11, 2003. With permission.)

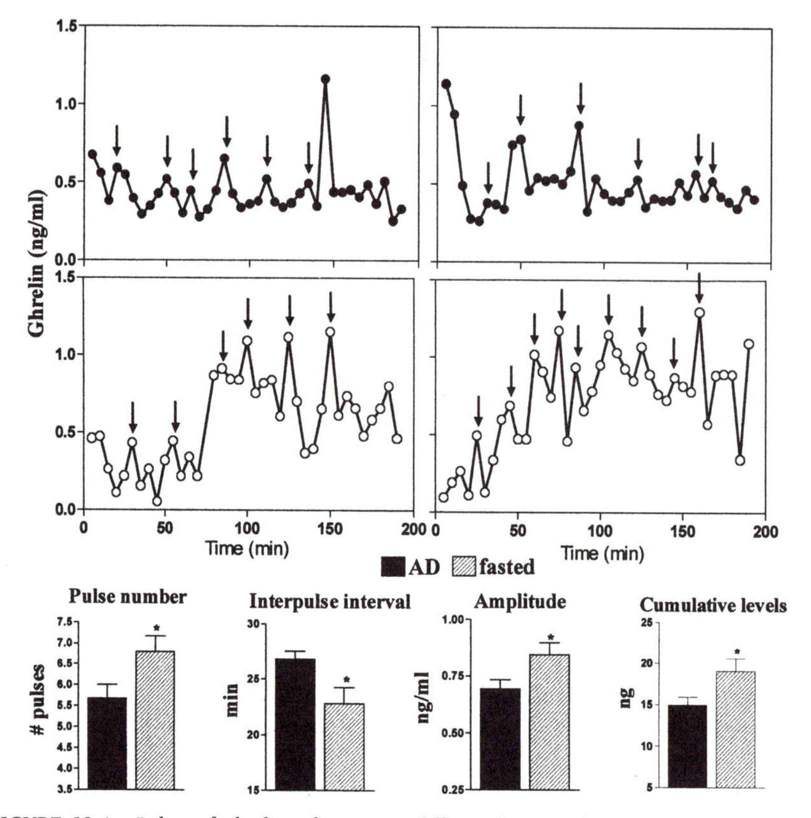

FIGURE 12.4. Pulses of ghrelin release were different between fed and fasted rats. Top of the figure shows raw data from individual animals. Filled circles, fed; open circles, fasted. Bottom of the figure shows average values. Ghrelin pulse numbers were higher in the fasted animals, and their interpulse intervals were reduced. Amplitudes were, on the average, higher in the fasted animals, leading to higher cumulative levels. See Chapter 13 for additional material. (From Kalra, S.P. *et al.*, *Regul. Pept.*, 111(1–3): 1–11, 2003. With permission.)

respond with substantial gonadotropin release into the blood. In fact, a steady high level of administration of GnRH actually turns off the gonadotropin system and thus can be used either (1) as a birth control device or (2) to ramp down the system in case of gonadal cancers. Some of these phenomena are beginning to be understood at the molecular level. The GnRH promoter is activated in an episodic fashion in GnRH neuronal cultures. A specific 410-base region of the GnRH promoter is required for pulsatile GnRH promoter activity. Within that region a small 5-base site that represents a binding site for the transcription factor Oct-1 is likewise required. In sum, order to achieve fertility, LH has to be given in a pulsatile fashion.

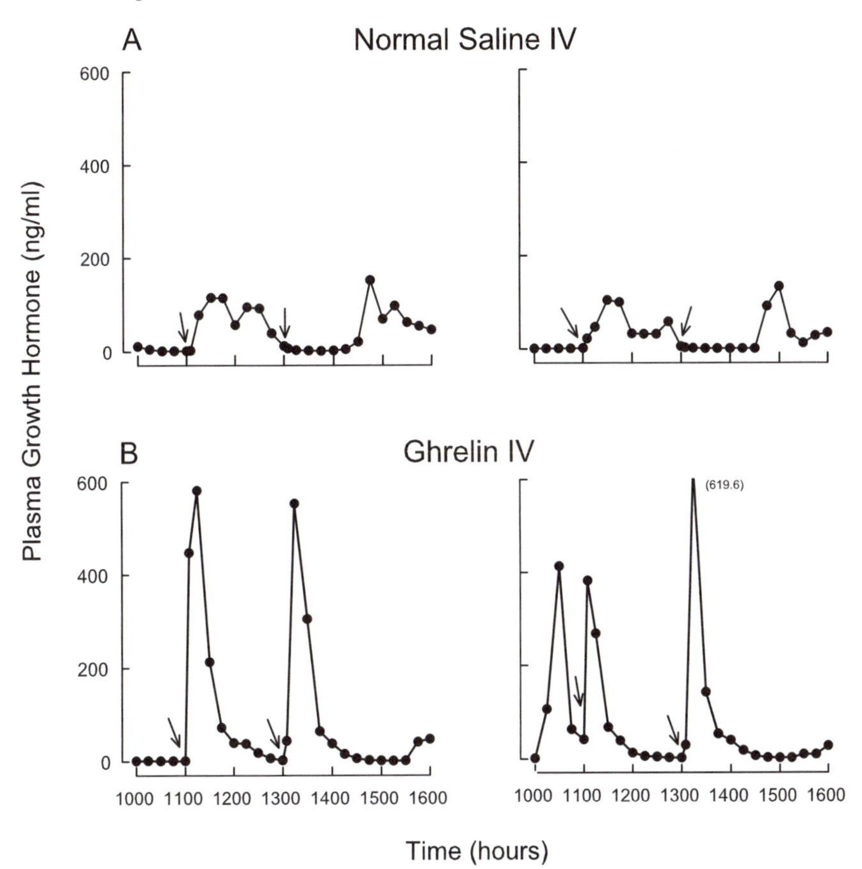

FIGURE 12.5. Pulsatile release of growth hormone (GH) from the pituitary gland. (A) Normal saline given intravenously cannot stimulate large pulses of growth hormone from somatotroph cells of the anterior pituitary. (B) The stomach hormone ghrelin given intravenously is able to stimulate significant GH pulses. (From Tannenbaum, G.S. *et al.*, in *Brain Somatic Cross-Talk and the Central Control of Metabolism*, Kordon *et al.*, Eds., Springer-Verlag, Heidelberg, 2002. With permission.)

Referring to the ability of leptin to inhibit energy intake, what is the optimal temporal schedule for its application? It turns out that plasma leptin levels in adult rats are themselves pulsatile (Fig. 12.3), as are the levels of another hormone related to food intake—ghrelin (Fig. 12.4 and Chapter 13). In turn, ghrelin effects could help to account for the pulsatility of growth hormone (GH) release (Fig. 12.5), an effect proven by the use of a GHRH antibody that operates in the CNS through GHRH (Fig. 12.6). In males, leptin pulses are low amplitude and high frequency. In females,

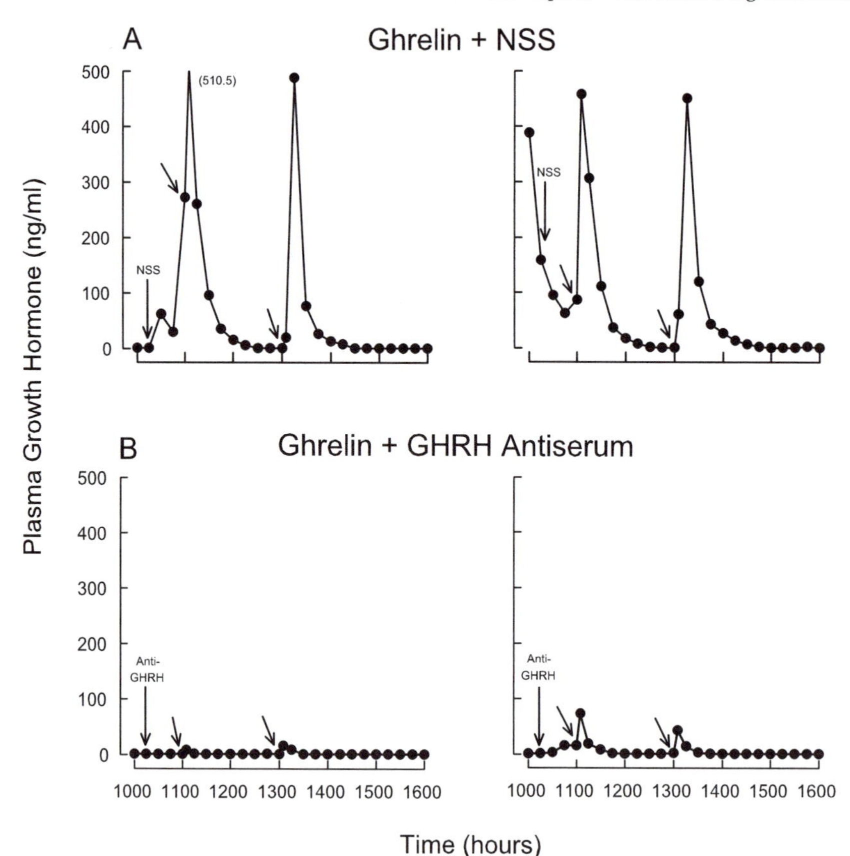

FIGURE 12.6. One mechanism for pulsatile release of growth hormone (GH). (A) The stomach hormone ghrelin stimulates pulsatile release of GH from the pituitary, and the effect is not blocked by the normal saline control (NSS). (B) Ghrelin apparently works through GH-releasing hormone (GHRH) because a GHRH antiserum blocks its effect. (From Tannenbaum, G.S. *et al.*, in *Brain Somatic Cross-Talk and the Central Control of Metabolism*, Kordon *et al.*, Eds., Springer-Verlag, Heidelberg, 2002. With permission.)

high-amplitude pulses dominate, leading to a higher average concentration of leptin in the plasma in females. Moreover, removal of the ovaries reduces leptin pulse frequency and amplitude, perhaps explaining the higher body weights of ovariectomized female rats (also see Chapter 13).

An important feature of medical practice in the endocrine clinic is the possibility of undoing therapeutic effects by overly long exposure to the hormone in question. One mechanism for this complication is the downregulation of hormone receptors by constant exposure to ligand.

This phenomenon finds its medical use in the application of long-acting LHRH agonists to suppress LH and testosterone secretion in the treatment of prostate cancer in men and in the treatment of endometriosis in women. A second type of mechanism is the actual damage of constant tissue exposure to glucocorticoids, in Cushing's syndrome (see Chapter 5).

III. OUTSTANDING NEW BASIC OR CLINICAL QUESTIONS

1. Are there circumstances in the application of hormone replacement therapy (for older women) or androgen supplementation (for older men) where fluctuating hormone levels would be more effective than constant exposure?
2. When two pulses of hormone treatment can substitute for a longer constant exposure, can the time period be broken up even more finely into three or four well-timed pulses as a route to finer dissection of the underlying mechanisms? For example, it is already known that a first pulse of estrogen, acting at the membrane, can augment the action of a later pulse, acting in the nucleus.
3. When long-term hormone priming is required, what molecular mechanisms are changing during that long time period? Conversely, when, as in older adults, there is a prolonged absence of a certain hormone, what cellular components are declining, to account for reduced medical effectiveness of hormone treatment?
4. What are the detailed molecular and biophysical mechanisms that allow a neuroendocrine population of cells to generate a pulse of secretion?

CHAPTER 13

Hormonal Secretions and Responses are Affected by Biological Clocks

I. BASIC SCIENTIFIC EXAMPLES

In a wide variety of experimental animals and in humans, very regular 24-hour fluctuations in circulating hormone levels can be observed (Fig. 13.1). In addition, given certain hormone levels, some behavioral responses to hormone administration vary in amplitude according to time of day. As summarized by Silver and colleagues (*Hormones, Brain and Behavior*, 2002), such 24-hour rhythms are compound functions of endogenous circadian clocks and entrainment by daily patterns of light intensity. In mammals, the brain's circadian clock is manifest in the activity of neurons in a small portion of the hypothalamus, the suprachiasmatic nucleus (SCN). The "master clock" of the body, SCN governs slave clocks in most peripheral cells that are synchronized to the SCN. This system accounts for the daily oscillations and circadian rhythms of hormone levels and behavioral habits. To keep the clock on time, there must be external cues (*zeitgebers*, from the German word for "time givers"). The major zeitgeber is light. The location of the SCN above the optic chiasm permits a retinohypothalamic tract to keep the SCN in tune with

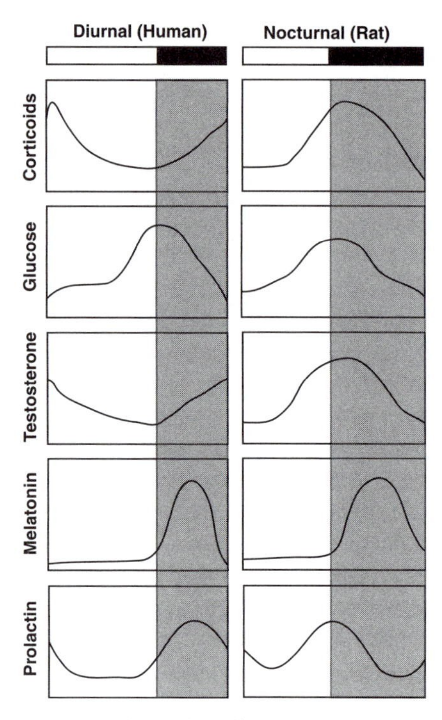

FIGURE 13.1. There are very regular 24-hour fluctuations in circulating hormone levels.

light and dark cycles outside the body. The major neurotransmitter in this process is glutamate. If animals or humans are kept in constant darkness or constant light, the accuracy of the master clock is impaired, and there is a shift in the circadian gene expression of the clock and the clock-controlled genes in other tissues.

Glucocorticoids can trigger circadian gene expression in cells, and dexamethasone, a glucocorticoid receptor agonist, can cause phase shifting for peripheral oscillators. Recently, the role of glucocorticoid hormones in controlling feeding time has been elucidated. Feeding time is a zeitgeber for gene expression in liver, pancreas, liver, and heart. Food metabolites, or hormones eliciting feeding, are the normal cues for peripheral (slave) oscillators. Controlling the feeding time depends on glucocorticoid hormones. Glucocorticoid hormones are secreted by the adrenal gland in a circadian and episodic manner (see Chapter 5). The SCN controls circadian corticoid steroid levels via the release of adrenocorticotropic hormone (ACTH). Nocturnal animals feeding in the day or day-feeding animals eating at night causes increased amounts of glucocorticoid to be released. Probably the same thing happens when we fly from the United States to Europe. It takes several days to adapt to the change in phase. It is about a day for every hour of flight. Such a change affects our sleeping time and appetite. Experiments by Schibler (2001) in Geneva, Switzerland, indicate that glucocorticoids are the reason why the adaptation takes so long. Secretion of glucocorticoids uncouples the SCN influence over the individual cells so that they become desynchronized. Although light is a major zeitgeber, feeding time also plays a critical role. Thus, we have a system where the SCN is situated to control hypothalamic–pituitary–adrenal cortex secretion, and the glucocorticoids are released episodically to control the rhythm of peripheral cells. The glucocorticoid receptor is expressed in all cell types. Thus, glucocorticoid is a zeitgeber for peripheral clocks, although it is probably not the only one.

II. MECHANISMS

This is an exciting and dynamic time with respect to discovering mechanisms involved in circadian biology and in the impact of sleep mechanisms on hormone/behavior relationships. For example, Yau at Johns Hopkins Medical School reports that a novel type of neuron, melanopsin-containing cells in retina, controls pacemaking activities in the SCN of the hypothalamus. In turn, daily regulation of calcium currents in the SCN neurons, which comprise the most important CNS biological clock, were found by workers at the Netherlands Institute of Brain Research to transduce molecular rhythms into electrophysiological firing rate rhythms.

Transcriptional feedback loops have been unveiled as essential components of circadian clocks. These have been discovered in a series of brilliant experiments using fruit flies (Fig. 13.2) and mice (Fig. 13.3). The amazing similarity of these two sets of molecular mechanisms demonstrates the conservation of CNS clock mechanisms, a feature typical of self-conserved neuroendocrine systems. Further, the importance of these transcriptional feedback loops for clinical medicine and human hormone/behavior relations can be illustrated by a *per* gene mutation in a patient who, as a result, gets up very, very early each morning and goes to bed in the afternoon (Fig. 13.4). Moreover, sleep physiology itself lays bare molecular mechanisms related to hormone effects on behavior. At Rockefeller University, a microarray study (using DNA chips) showed that gene expression for the synthetic enzyme that produces a prominent sleep-inducing substance, prostaglandin D, is significantly reduced by estrogen administration, in a basal forebrain region in the preoptic area that governs sleep. It was inferred that estrogens would, by this route, reduce the drive for sleep, thus increasing opportunities for courtship behaviors and sexual arousal. All of these data are related to changes in behavior throughout the 24-hour day.

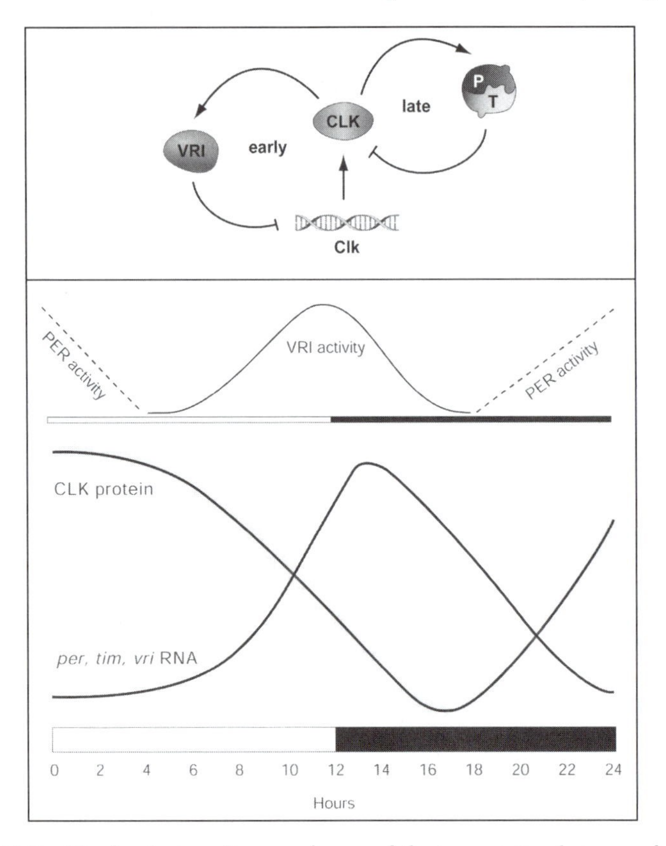

FIGURE 13.2. The fly clock; a diagram of some of the transcriptional steps underlying the circadian clock in *Drosophila*. Notice the conceptual similarity and conservation to a mammalian clock (see Fig. 13.3). (Top) Negative feedback cycles with a time delay make a rhythm. P, *period* gene; T, *timeless* gene. (Middle) High levels of VRI activity block the expression of the genes for *period* (*per*) and *timeless* (*tim*). Entrainment by the light cycle is observed. (Bottom) During the day, clock (CLK) mRNA and protein levels decline, whereas VRI mRNA and protein accumulate. Because T is quickly degraded by light, the absence of T after the lights go on allows CLK to switch on its repressor VRI. This simplified summary does not include every autoregulatory link conceivable from current data. In particular, it puts emphasis on transcriptional feedbacks but does not fully represent posttranscriptional steps. (Adapted from Young and Kay, 2001.)

A. Seasonal Rhythms

A different kind of rhythm is the seasonal rhythm, classified into three distinct types by Irving Zucker at the University of California at Berkeley (Fig. 13.5). As reviewed by Zucker and Nelson (*Hormones, Brain and Behavior*, 2002), certain seasonal rhythms are dependent on melatonin, the secretion of which

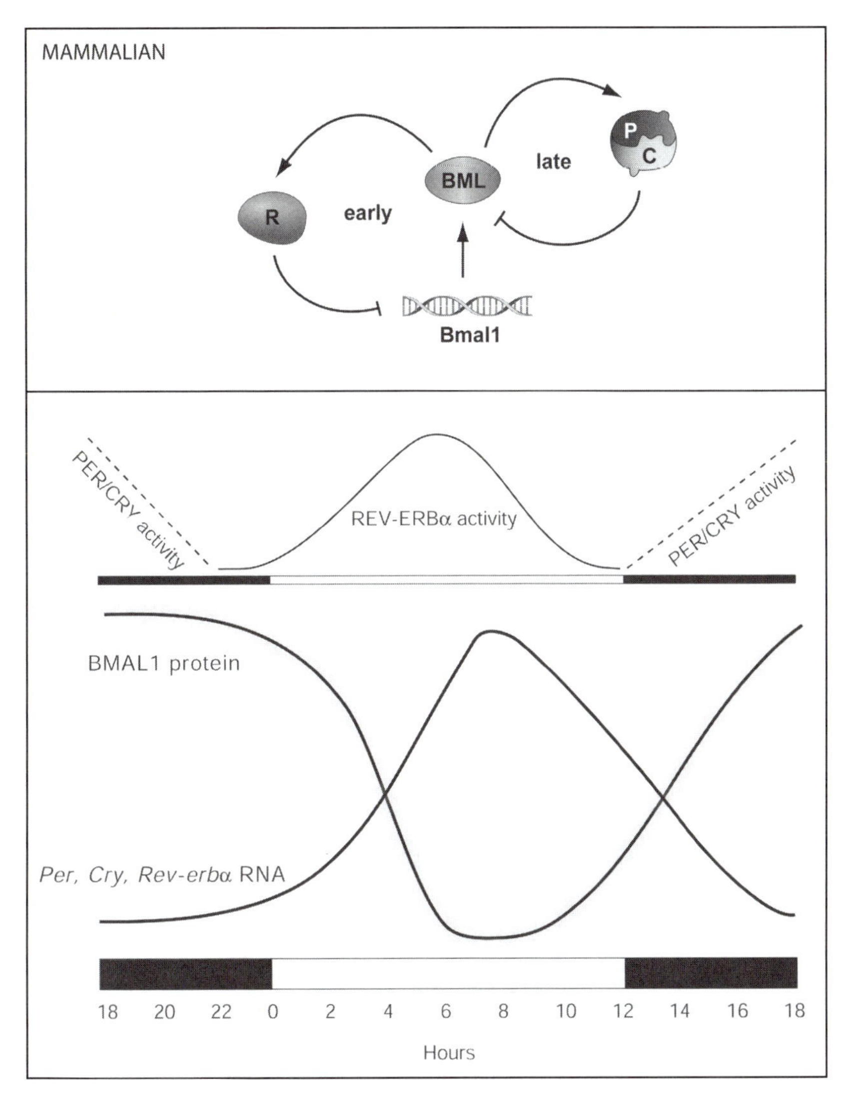

FIGURE 13.3. The mouse clock. Note similarity to sketches of the fly clock (see Fig. 13.2); some of the abbreviations are repeated. (Top) Negative feedbacks at the transcriptional level, coupled with time delays, make a rhythm. R, orphan nuclear receptor REV-ERB-α. (Middle) During the day, REV-ERB-α protein activity restricts the BML-protein product function, thus keeping PER/CRY activity low. (Bottom) In the morning, high BMAL protein levels induce a wave of Per (P), Cry (C), and REV-ERB-α transcription. Accumulation of Per and Cry proteins will be delayed until after dark. Note that other autoregulatory steps are already suggested by data extant, and that this set of sketches emphasizes the transcriptional mechanisms in the mammalian circadian clock. (Adapted from Young, Neuron, 2002.)

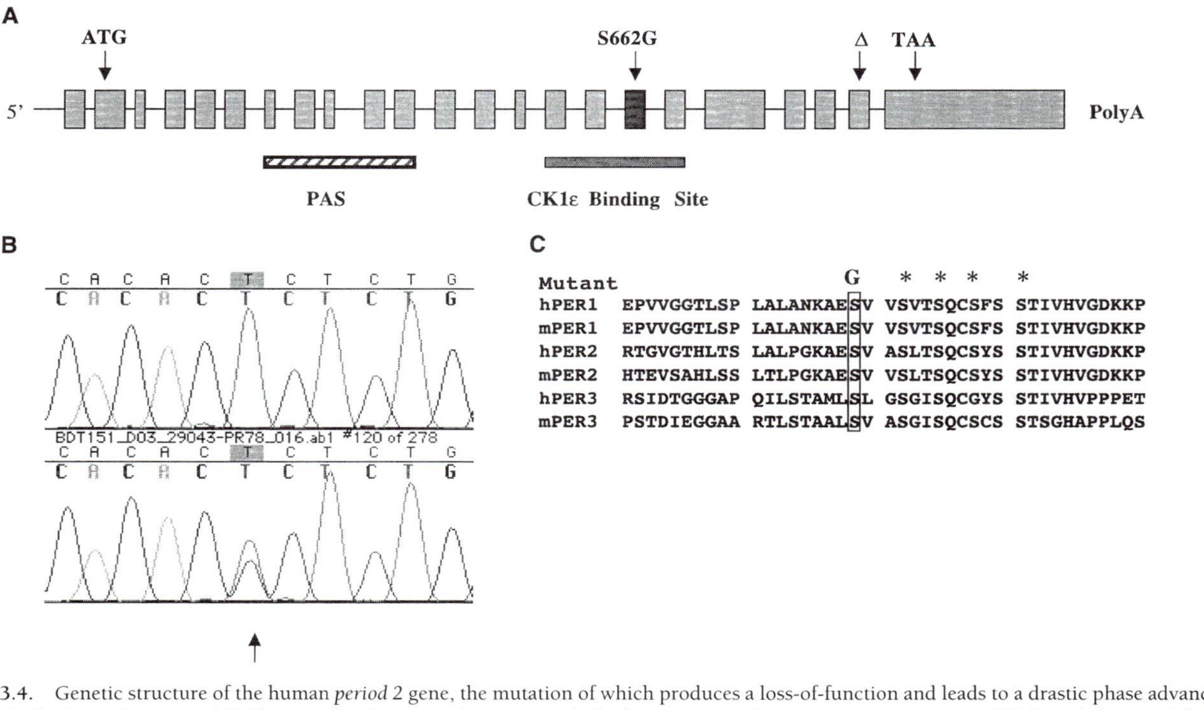

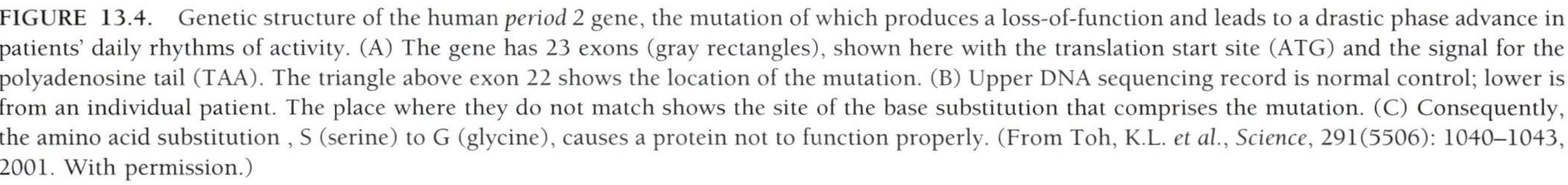

FIGURE 13.4. Genetic structure of the human *period* 2 gene, the mutation of which produces a loss-of-function and leads to a drastic phase advance in patients' daily rhythms of activity. (A) The gene has 23 exons (gray rectangles), shown here with the translation start site (ATG) and the signal for the polyadenosine tail (TAA). The triangle above exon 22 shows the location of the mutation. (B) Upper DNA sequencing record is normal control; lower is from an individual patient. The place where they do not match shows the site of the base substitution that comprises the mutation. (C) Consequently, the amino acid substitution , S (serine) to G (glycine), causes a protein not to function properly. (From Toh, K.L. *et al.*, *Science*, 291(5506): 1040–1043, 2001. With permission.)

Type I Rhythm
Mixed: Driven by endogenous and exogenous components
(e.g., Syrian hamster reproduction)

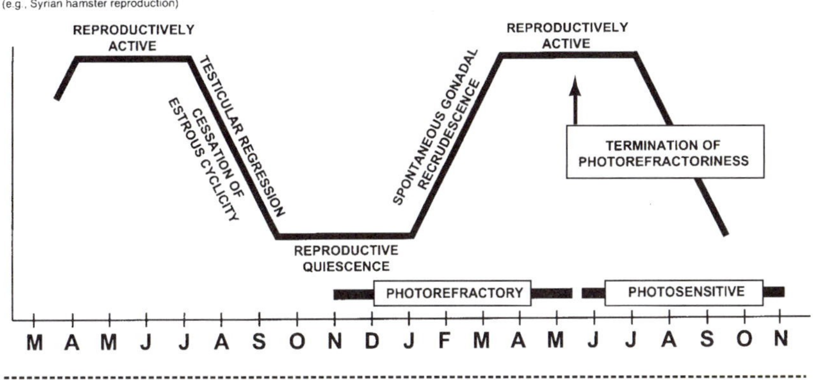

Type II Rhythm
Endogenous: Product of circannual clock(s)
(e.g., ground squirrel body weight)

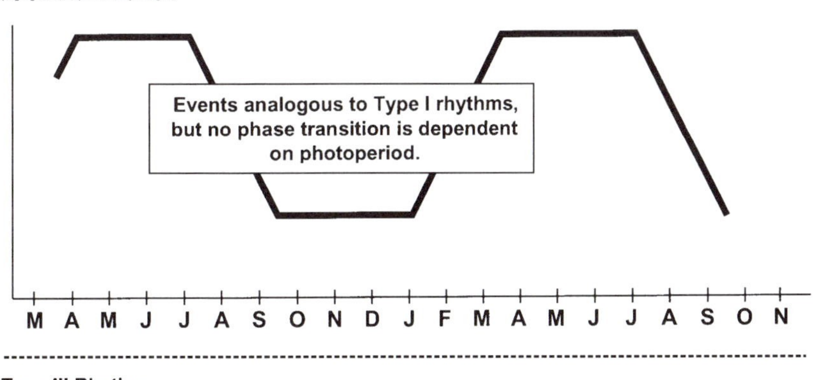

Type III Rhythm
Exogenous: Environmentally triggered
(e.g., hay fever)

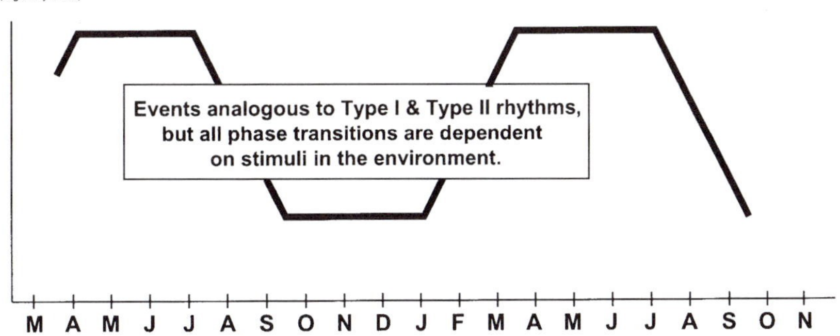

FIGURE 13.5. Three distinct types of seasonal rhythm. (Zucker and Nelson, *Hormones, Brain and Behavior*, 2002.)

is controlled by the light/dark cycle. For example, the duration of the melatonin signal is important in regulating circannual reproductive cycles in many mammals (*e.g.*, deer, hamsters, and sheep). In the winter, longer dark periods make for longer periods of melatonin secretion from the pineal gland, which entrain 12-month rhythms of pituitary secretion of hormones controlling reproduction such as luteinizing hormone (LH). In turn, gonadal steroid hormones dependent on LH are more or less available to drive sex behaviors, which therefore fluctuate seasonally. Experimental infusions of melatonin have tested this idea successfully; high melatonin levels achieved during the correct phase of the 12-month year can reduce LH. Thus, both the duration and the annual phase of the melatonin signal are important for transducing season ⇒ melatonin ⇒ sex hormone ⇒ behavior.

While tremendous efforts have recently been made to elucidate the mechanisms of daily rhythms, the melatonin fluctuations mentioned above are, so far, carrying the causal burden for seasonal rhythms. With respect to what Irving Zucker has identified as being the three types of seasonal rhythms (endogenous, environmentally set, and stimulus-triggered, depending on specific signals present at a particular time of year), physical and environmental variables driving seasonality would include availability of food and water, day length, and environmental temperature. In small animals, whose bodies have high surface-to-volume ratios, the cold of winter is no time to reproduce. In some such animals, the solution to this problem is simple: Simply shut down hormone secretion from the pituitary, which stimulates the gonads to release behaviorally relevant steroid hormones. The hormonal mediator of the "winter effect" includes long-duration releases of melatonin from the pineal gland. Thus, a winter decline can be observed in testis volume and hormonal secretion, for example. On top of that, short day lengths can decrease responsiveness to sex hormones (Fig. 13.6). In other cases, the solution is more complex. Randy Nelson, at Ohio State University, found that hamsters inhibit their reproductive systems in a manner dependent on thyroid hormones. This seasonal regulation seems to be carried out by fluctuations in thyroxine-binding proteins.

B. Feeding Rhythms

Throughout the animal kingdom, feeding occurs in a rhythmic fashion. The seasons, the day/night cycle, and the availability of food all affect feeding, but there is an underlying rhythm controlled by hormones and neural mechanisms. The fundamental unit of feeding is meal size. Some animals, such as rats, feed by small meals but with high frequency. Humans, by contrast, tend to consume three meals a day during the daytime and early hours

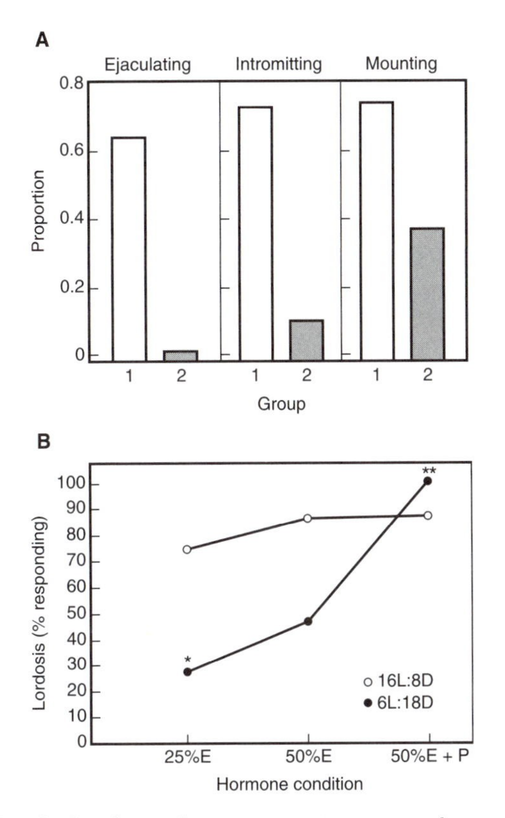

FIGURE 13.6. Short day lengths can decrease responsiveness to sex hormones. (I. Zucker *et al.*, *Hormones, Brain and Behavior*, 2002.)

of the night. These rhythms can be upset by time changes, such as when one travels from one time zone to the other, or by tumors or lesions in the hypothalamus, but the basic regulation of feeding is rhythmic. Measurements of the hormones involved in feeding also show a rhythmic increase and decrease over time. Lesions to the hypothalamus result in the loss of these rhythms. For example, lesioning the ventromedial hypothalamus (VMH) results in animals feeding continuously and becoming extremely obese because of it. They completely lose their normal patterns of feeding episodes. In mice in which the leptin gene has been knocked out, continuous feeding and overt obesity are also observed. This abnormality can be reversed by giving leptin replacement (Friedman, 2002). Also, mutating the leptin receptor has the same effect of producing constant feeding and obesity (Cohen *et al.*, 2001). There is clearly a circadian rhythm for circulating leptin levels (Bagnasco and Kalra, 2002a,b; Xu *et al.*, 1999; Kalra *et al.*, 1997). In rodents, blood levels of leptin rise in the dark phase (when the animals primarily feed). While this pattern of leptin secretion appears to be linked to the light/dark cycle, it can be altered by feeding patterns. When rats are allowed to eat only during the light hours, the timing of leptin hypersecretion changes to coincide with the initiation of feeding (Xu and Kalra, 1999). Thus, the plasma leptin rhythm is actually a feedback from feeding. However, in adipocytes, leptin synthesis increases before the onset of nocturnal feeding (Fig. 13.7). Leptin receptor mRNA in the hypothalamus increases at the beginning of the dark phase also. Thus, there appears to be an independent controller of the natural rhythm of leptin that is turned on before exposure to light that is not correlated with feedback from feeding. In humans, circadian fluctuations of leptin in the plasma peak during sleep. So, if the leptin levels are highest when there is no feeding, as occurs during the sleep/wake cycle, is the onset of feeding triggered by low levels of leptin? There are, of course, other orexigenic hormones, such as ghrelin, which can trigger feeding when injected either centrally or peripherally. Within the day/night cycle are ultradian rhythms of circulating leptin levels. These are low-amplitude, high-frequency pulses and are related to gender. In females, the amplitudes of the pulses are of slightly lower frequencies than for males, but overall the baselines are significantly higher in females than in males. Removal of sex steroids (testosterone or estrogen) decreases the pulse amplitude and frequency in both sexes. Kalra *et al.* (2003) have proposed that there are circadian and ultradian cycles in the principal hormones of feeding and energy homeostasis (*i.e.*, leptin and ghrelin) that provide feedback to neuropeptide Y (NPY) in the hypothalamus. The ultradian (less than a day) leptin secretion may be innate to adipocytes and subject to increase by steroid-mediated mechanisms (Bagnasco *et al.*, 2002). A connection exists between leptin and insulin, and Song *et al.* (2000) have reported a pulsatile insulin secretion in human subjects; therefore, the source of the leptin rhythm could be its connection with insulin (as the major "adiposity signal") or it could be due to hypothalamic pattern generators dictating pulsatile leptin secretion from adipocytes via the autonomic nervous system. When leptin levels are at their lowest, upregulation of peripheral ghrelin occurs.

Ghrelin is primarily produced in the oxintic glands of the stomach. It is released into the blood and activates the NPY cells of the arcuate nucleus (ARC) in the hypothalamus to initiate feeding. The cells that contain NPY, and also agouti-related peptide (AGRP) and γ-aminobutyric acid (GABA), have receptors for ghrelin. If ghrelin is given continuously or repeatedly, there is an increase in weight gain and fat mass. NPY is released in the hypothalamus and acts on Y1 receptors (Y1Rs) in the paraventricular nucleus (PVN), and AGRP acts by tonically inhibiting melanocortin-4 receptors

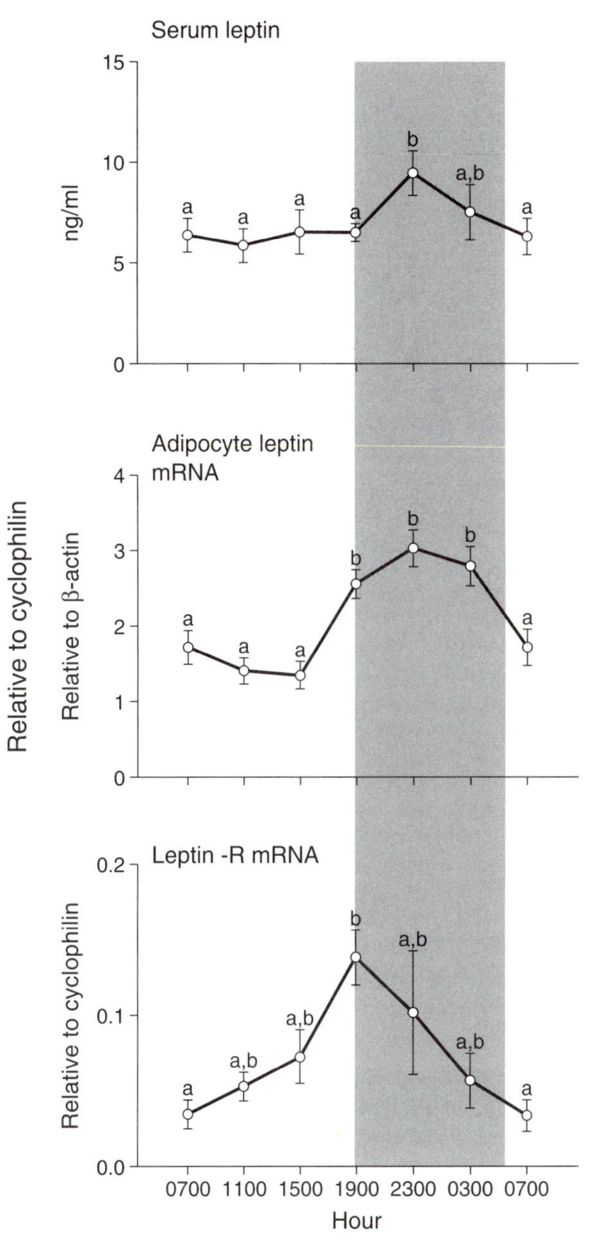

FIGURE 13.7. Day/night fluctuations of leptin; production of higher levels of leptin transcripts (middle panel) and serum leptin levels (top panel) during the dark part of the daily light cycle in rats. These are preceded by a daily peak in leptin receptor mRNA levels (bottom panel). (From Kalra, S.P. *et al.*, *Regul. Pept.*, 111(1–3): 1–11, 2003. With permission.)

(Mc4Rs) in the PVN. Thus, feeding is the result of Y1R stimulation and blockade of Mc4R.

Ghrelin is secreted in anticipation of a meal. There is little evidence to suggest that this is due to low glucose levels or some other indication of energy levels. It is more a conditioned event or a rhythm imposed physiologically. When rats are maintained on an *ad libitum* diet, ghrelin is secreted in a pulsatile fashion (see Fig. 12.4). Fasting increases ghrelin secretion with high-amplitude pulses. This occurs at the same time as a decrease in leptin pulses and coincides with an increase in NPY rhythmic discharge in the PVN. In fasting, leptin levels are low, so it appears that leptin normally restrains ghrelin secretion. This decrease of leptin allows ghrelin to be secreted in high-amplitude, high-frequency pulses to stimulate feeding, which may account for hunger pangs.

Importantly, GABA acts synergistically with NPY, which acts on GABA-A receptors to inhibit the proopiomelanocortin (POMC) neurons release alpha-melanocyte-stimulating hormone (α-MSH) in the PVN. Thus, NPY both activates the neural circuits for feeding, and inhibits the tonic inhibition of feeding. NPY is secreted in a rhythmic pattern (Kalra *et al.*, 1991). Before a meal the rhythm and amplitude of NPY increase, and during the meal the rhythm and amplitude of NPY gradually decrease. If food is not available at mealtime, NPY is hypersecreted. Thus, the picture seems clear. NPY is secreted in high-amplitude, high-frequency episodes when energy levels are low, while secretion decreases when energy levels are normal. The situation for leptin is the reverse, as it is secreted in high amounts when energy levels are high and in low amounts when energy levels are normal. Thus, there is a leptin–ghrelin–NPY feedback loop. High leptin restrains the release of ghrelin from the stomach and blocks its central action by inhibiting NPY release. When this leptin restraint is reduced, feeding can occur by the release of ghrelin.

In the hypothalamus, the SCN is the master clock controlling circadian rhythms in the ARC, VMH, and other hypothalamic areas. Lesions of the SCN abolish the circadian patterns of feeding behavior; therefore, the overall rhythm of ingested behavior is regulated by SCN. However, it can be fooled. By limiting food availability to a few hours, the patterns of hormone secretion adapt to the time that the food is available.

Hormones that regulate food intake involve those that act rapidly (such as ghrelin), those that act slowly (such as leptin and insulin), and those that act in between slowly and rapidly. For short-term meals, ghrelin increases appetite, and the meal is terminated by cholecystokinin (CCK), which decreases appetite. The frequency of meals is related to the levels of leptin and insulin, which affect meal size and frequency depending on the energy needs. $PYY_{3\text{-}36}$ is a member of the neuropeptide Y protein family that is secreted from endocrine cells in the small bowel and colon. It is secreted in response to food. In humans and rodents, doses of PYY_{3-36} can inhibit eating for up to 12 hours. PYY_{3-36} binds to the Y2 receptor, which results in inhibition of the NPY/AGRP neurons that accelerate feeding and holds it in abeyance until the level of PYY_{3-36} secretion is reduced.

As noted earlier with regard to how the CNS manages these various rhythms, the master clock in SCN controls slave clocks in most peripheral cells. As mentioned, glucocorticoids can trigger circadian gene expression in cells and dexamethasone, a glucocorticoid receptor agonist, can cause phase shifting for peripheral oscillators. Recently, the role of glucocorticoid hormones in controlling feeding time has been elucidated. Feeding time itself is a zeitgeber for gene expression in liver, pancreas, liver, and heart. Food metabolites, or hormones eliciting feeding, are the normal cues for peripheral (slave) oscillators. Controlling the feeding time depends on glucocorticoid hormones. Glucocorticoid hormones are secreted by the adrenal gland in a circadian and episodic manner. The SCN controls circadian corticoid steroid levels via the release of ACTH. Nocturnal animals feeding in the day or day-feeding animals eating at night causes increased amounts of glucocorticoid to be released. The experiments by Schibler, mentioned above, indicate that glucocorticoids control the time course of adaptation to imposed phase changes as we travel from time zone to time zone.

III. IN HUMAN BEINGS

In humans, as in experimental animals, daily changes in hormone secretions and behavior are governed by at least two types of forces: (1) light-entrained, 24-hour rhythms in which the underlying circadian mechanisms, referred to earlier, are synchronized with the environment by visual stimuli; and (2) the drive for sleep, which can be gated by light-entrained rhythms but which works by different mechanisms. With respect to daily changes in hormone secretions by human volunteers, for example, Czeisler and colleagues at Harvard Medical School have documented very clear rhythms in cortisol, growth hormone, and thyroid-stimulating hormone (Fig. 13.8).

Also, in humans, especially those living far from the equator, seasonal changes in moods can be quite striking. Depression, alcohol use, violence against women, breast feeding, food consumption, and other behaviors associated with hormone-influenced neurochemical systems in the midbrain and forebrain during winter months all tell us that day-length-dependent and temperature-dependent mechanisms in the human brain are invoked as factors in hormone-controlled behaviors.

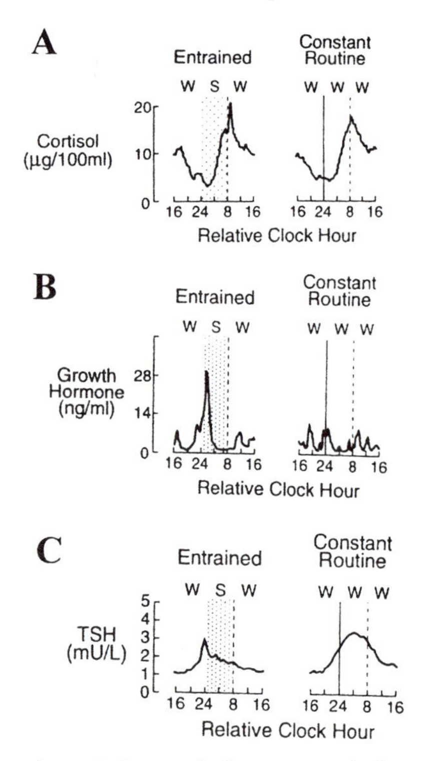

FIGURE 13.8. Rhythms in corisol, growth hormone, and thyroid-stimulating hormone (Czeisler *et al.*).

IV. SOME OUTSTANDING QUESTIONS

1. What is the relationship of entrainment of hormonal biological clocks by light/dark cycles to the development of behavioral pathology in humans (*e.g.*, seasonal winter depression)?
2. Can seasonal or daily periods underlying behavioral rhythms be demultiplied to account for shorter rhythms of behavioral change? Molecular experiments by Patrick Chappell and Pamela Mellon at the University of California, San Diego, suggest this is possible.

Space: Spatial Aspects of Hormone Administration and Impact are Important

CHAPTER 14

Effects of a Given Hormone Can be Widespread Across the Body; Central Effects Consonant with Peripheral Effects Form Coordinated, Unified Mechanisms

Many hormones have effects both in the central nervous system (CNS) and in the periphery. The two routes of hormonal actions may correspond precisely to the two ways an animal or human has to cope with challenges to homeostasis—namely, (1) changes in internal management of the physiological system in question, the better to meet the challenge; and (2) the initiation of behaviors that make use of external resources to help with the support of those same internal systems. A clear example in laboratory science and in medicine is offered by the hormone angiotensin in fluid homeostasis. Angiotensin, released from the kidneys, acts both (1) to conserve water the animal or patient already has, and (2) to stimulate drinking behavior.

I. BASIC EXPERIMENTAL EXAMPLES

A. Control of Food Intake

Body weight is tightly regulated over time. Over a year, we burn about 1 million calories and yet, even for those putting on weight, increases in this intake are rarely more than a few thousand calories per year. This balance is achieved by the coordination of many inputs into the brain, both hormonal and neural, that are integrated to produce a unified mechanism for dealing with hunger, satiety, and metabolic rate, as well as thermogenesis and locomotor activity.

The principal sensory neural component arrives in the brainstem through the vagus nerve (Fig. 14.1). The input to the vagus comes from the stomach, the gut, and the liver. As introduced previously, cholecystokinin (CCK) is released from the small intestine into the blood but stimulates contractions in the stomach. These contractions are detected by stretch receptors of afferent vagal nerves that carry their signal to the nucleus tractus solitarius (NTS). Adiposity signals (leptin and insulin) are bloodborne and transported across the blood–brain barrier into the arcuate nucleus (ARC). There they stimulate arculate neurons containing neuropeptide Y (NPY) and agouti-related peptide (AGRP), which activate cells containing melanocyte-stimulating hormone (MSH) and project onto cells in the paraventricular nucleus (PVN) where they act on melanocortin-4 receptors (Mc4Rs; see Chapter 1). In this way, leptin activates a pathway that inhibits food intake. There is then a feedback because eating less reduces the amount of fat; therefore, less amount of leptin is released, which switches off the system. Similarly, insulin decreases food intake. It is also delivered in the blood to the ARC and stimulates the same catabolic pathways through the PVN. The PVN releases corticotropin-releasing hormone (CRH) and oxytocin, and the catabolic effects reduce food intake and increase energy expenditure. Experiments have clearly shown that insulin infused in the brain decreases feeding, while a decrease of insulin in the brain leads to obesity. Neither leptin nor insulin is made in the brain, yet numerous receptors for leptin and insulin are located inside the blood–brain barrier. Therefore, in order for leptin and insulin to reach those receptors from the blood, they have to be taken up by a special transport mechanism in the brain blood vessels or at the circumventricular organs where the blood–brain barrier is lacking. In sum, both peripheral and central hormonal actions limit food intake.

One of the characteristic ways in which the brain coordinates feeding is by the limitation of meal size. Meal size is controlled by ghrelin and CCK, which can act both peripherally and by stimulation of vagal afferents. Meal satiation is limited by satiety hormones, including CCK, peptides of the bombesin family such as GRP and glucagon, enterostatin, an apolipoprotein, somatostatin, and neuromedins. As proof of this, when CCK is given before a meal to rats, it reduces the amount of the meal eaten. Oppositely, inhibitors of CCK increase meal size. It might be thought that one could simply swallow a pill of CCK before a meal to reduce the amount eaten. While this might work, the problem is that hunger does not cease and more frequent meals

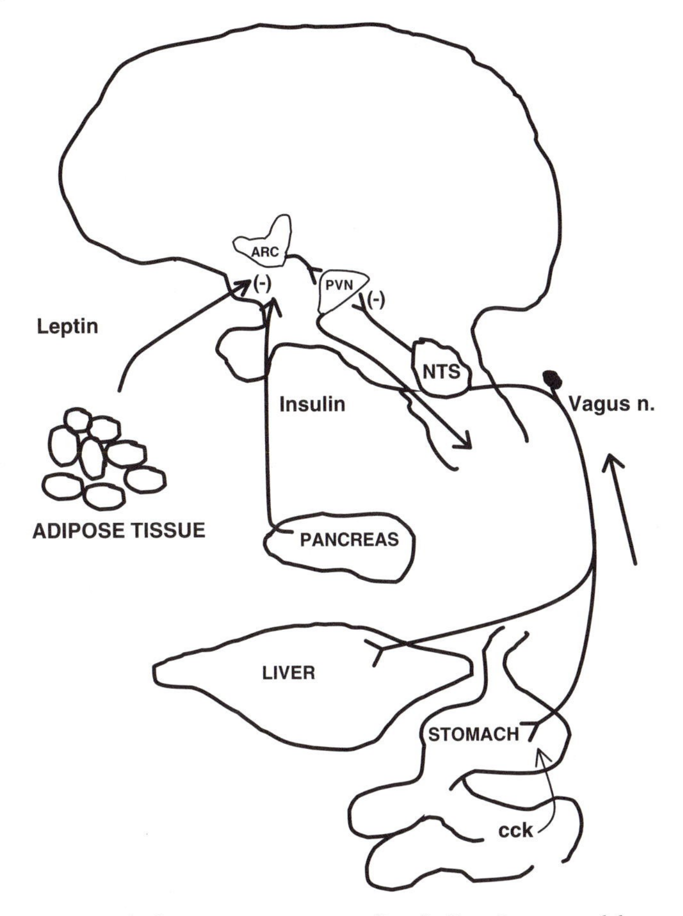

FIGURE 14.1. Brain/body communication regarding feeding. Sensory and hormonal inputs arrive from peripheral organs via the blood and the vagus nerve. The vagus nerve inputs from the liver and the gut and stomach relays stretch, induced by CCK to the nucleus of the tractus solitarius (NTS) in the brainstem. This leads to decreased feeding. Other hormones are released into the blood and must cross a blood–brain timer to enter the brain. ARC, arcuate nucleus; PVN, paraventricular nucleus of the hypothalamus.

would be eaten, leading to no net decrease in calories. Thus, the brain, due to signals being sent directly to the NTS in the lower brainstem and indirectly to the hypothalamus by blood, coordinates the forebrain signals with the hindbrain signals to produce food intake and energy expenditure control.

In addition to this hormonal–neuronal reflex, there are also influences of social factors and memory on food intake. It is not surprising that insulin and CCK have receptors in the hippocampus and elsewhere in the forebrain, presumably to serve these functions. It may be important, in this regard, that insulin applied directly to hippocampus neurons can inhibit electrical activity.

Another hormone with the capacity to integrate behavioral responses with the state of the body is leptin, produced by the *ob* gene, the mutation of which is responsible for an extremely obese mouse. Cloned by the Friedman laboratory at Rockefeller University, it is produced by fat cells, adipocytes. The more fat, the higher the leptin level in the blood. Can leptin affect the brain, thus altering behavioral responses relevant to the acquisition and ingestion of food? Apparently, yes. This fact is slightly surprising because leptin is a 16-kDa protein and by the usual thinking would not be able to cross the blood–brain barrier. Nevertheless, leptin receptors are found in the CNS. Their presence not only in the hypothalamus but also in the brainstem has been demonstrated convincingly. Moreover, neurochemical consequences of leptin administration have been reported. It is known that the Ob-Rb receptor activates a signal transduction pathway featuring the Janus kinase (Jak) protein and, in turn, the transcriptional activator STAT3 (see Chapter 1). Peripheral administration of leptin increases STAT3 phosphorylation not only in the hypothalamus but also in the brainstem; therefore, the neurobiological basis for excellent coordination of food intake and energy expenditure with fat supply, orchestrated by leptin working through the Jak/STAT pathway, is in place. The relevant behavioral experiments are now of great interest.

In summary, food intake controls require inputs to the brain of sight and scent (which may affect appetite, taste, and texture), signals from the gut, and fat signals. When an individual is underweight, the level of adiposity signals is decreased which permits larger meals to be consumed; when individuals are overweight, the level of adiposity signals is increased. Maintaining a balance between food intake and energy expenditure (homeostasis) is accomplished with the help of both peripheral and central actions of hormones.

B. Alpha-Melanocyte-Stimulating Hormone

α-Melanocyte-stimulating hormone (α-MSH) illustrates the principle of this chapter in an interesting way in that it has an unusual route of action. Secreted from the neurointermediate lobe of the pituitary gland, it stimulates glands in the skin that produce pheromones. They, in turn, influence the behavior of the test animal's partner, which then influences the behavior of the test animal. Thus, if the α-MSH of an aggressive hamster is making the animal's flank glands grow and secrete, then the resulting sebaceous gland products will stimulate aggressive responses in the partner, causing even more aggression in the first animal. Peripheral actions of the hormone in the test animal, causing CNS responses in the partner, subsequently alter the behavior of the test animal.

C. Oxytocin

Taking off from the introduction of oxytocin's behavioral effects in Chapter 1, oxytocin (OT) gene expression also illustrates the "unity of the body" concept. Physiological conditions controlling OT gene expression include the organism's stage of development, puberty, the estrus cycle, thyroid hormone levels, plasmid osmolality, and circadian rhythms. Both peripheral and central mechanisms alter this important gene. Regarding development, for example, mRNA for OT is strongly increased in magnocellular anterior hypothalamic neurons around the time of birth. OT mRNA expression continues to increase during postnatal development, and the increase can be eliminated by removal of the gonads. This indicates that it is under gonadal steroid control. During the female estrus cycle, OT mRNA expression is highest at estrus. In pregnant females, OT mRNA is strongly expressed in the magnocellular and parvocellular neurons of the PVN and supraoptic nucleus (SON) toward the end of gestation. The powerful contractions of the myometrium of the uterus during parturition are triggered, in part, by OT.

After birth of the young, a crucial role of OT is the "milk letdown reflex" during lactation. The very first reports of OT gene knockout mice by Young and by Nishimori highlighted the absence of this important function; however, OT also has many other effects, and some of its striking CNS actions, including the facilitation of affiliative behaviors, were covered in Chapter 1.

Returning to the molecular level, we have just begun to sort out the physiological stimuli that can increase either OT or OT-receptor transcript levels. For example, we know that estrogens act on regulatory elements in a complex 5′-flanking region that mediates the genomic effects of several nuclear hormone receptors, including those for estrogens, thyroid hormones, and retinoids (Dellovade *et al.*, 2000). During pregnancy, OT mRNA expression increases in both the hypothalamus and uterine epithelium. Thus, in preparation for birth, the hypothalamus, uterus, and mammary glands are all associated with increased oxytocin and oxytocin receptors (Burbach *et al.*, 1992).

A reflexive release of OT into the blood occurs during coitus and parturition. This reflex is neural and depends on mechanical stimulation activating stretch receptors in the uterus, which are the end organs of afferent nerves leading into the spinal cord. A relatively complex and circuitous pathway then signals to the PVN and SON of the anterior hypothalamus.

OT in the PVN and SON then will be released into the blood via projections to the neurohypophysis (*i.e.*, the posterior pituitary gland).

Not all ovarian hormones pull in the same direction. In both coitus and parturition, estrogens stimulate both OT synthesis and secretion; however, progesterone has the opposite effect and, consequently, as levels of both hormones rise during pregnancy, this release of OT remains fairly constant. Approaching birth, however, progesterone levels fall rapidly, releasing OT from that inhibition so that OT levels dramatically rise. At this same time, OT receptors are greatly increased in the uterus, and subsequent uterine contractions release OT in a feed-forward reflex to help with delivery.

In regard to the mechanisms of lactation, OT causes contraction of the myoepithelial cells of the mammary gland, resulting in the ejection of milk. Prolactin increases OT mRNA in the hypothalamus and is critically important to the synthesis of milk. Thus, lactation is both a hormonal and neurophysiological event, involving electrical activity of the hypothalamus that results in neurohypophysial release of OT.

Finally, with respect to behavior, OT has a large number of effects, in addition to simple reproductive physiology. Associated with the reproductive actions is maternal behavior, which provides a positive feedback on central OT release. OT also increases pain thresholds, probably through direct release of oxytocin onto the β-endorphin cells in the arcuate nucleus. In various species, oxytocin is associated with facilitating sexual behavior in both males and females. As mentioned earlier, OT facilitates friendly social behaviors in some species. For example, in prairie voles, OT is necessary for pair bonding. Infusions of OT increase the tendencies for partnering and parental behaviors.

Overall, from the various fields of work on OT covered here, it is easy to see that the multiple peripheral and central controls over OT or OT-receptor gene expression are matched by the breadth of the physiological actions of OT, both peripheral and central.

D. Relaxin

Relaxin is a hormone secreted during pregnancy from the uterus corpus luteum. It was named for its effects of relaxing connective tissue and thus pubic symphysis during parturition; however, as it circulates in the blood, it also reaches the circumventricular organs of the brain. In one of these, the subfornical organ (SFO), relaxin activates neurons in the same way that angiotensin II does and thus induces the same behavioral response of fluid intake by drinking (see Chapters 2 and 3). It is therefore proposed that relaxin has multiple roles during pregnancy, including an influence on fluid balance controlled in the brain, and multiple effects in the pregnant uterus.

II. CLINICAL EXAMPLES

All of the material presented here on food intake and relaxin pertains to humans as well. As part of the energy balance equations in humans, the widespread actions of leptin (see Chapters 1, 10, 12, and 13) serve to illustrate the main idea of this chapter. A parallel set of examples regarding the role of angiotensin in controlling drinking can also be appreciated.

A. Thyroid

Thyroid hormone systems, frequently altered following thyroid surgery, provide another excellent example. Thyrotropin-releasing hormone (TRH) is a three-amino-acid peptide that has widespread synaptic excitatory actions, short-lived behavioral effects (including reduction of food intake), and a central role in releasing thyroid-stimulating hormone (TSH) from the pituitary. Additionally, it is a prohormone, in that its metabolite cyclo-histidine–proline (CHP) also can reduce food intake. In turn, thyroxine released following TSH action in the thyroid gland affects overall body metabolism and nervousness/irritability of behavior.

B. Corticotropin-Releasing Hormone

As introduced in previous chapters, corticotropin-releasing hormone (CRH) is a classically integrative hormone (Figs. 14.2 and 14.3). In terms of peripheral effects, its neuroendocrine role of releasing adrenocorticotropic hormone (ACTH), which acts in the adrenal to release glucocorticoid hormones, causes a chain of events that radiate throughout the body. Centrally, its overproduction in the amygdala has been linked by Gold's laboratory at NIMH to depression and anxiogenesis, and its receptor system (CRH receptor type 1) plays an important part in controlling anxiety, as shown by Franz Holsboer and his team in Munich. Thus, CRH is an excellent example of the main principle of this chapter.

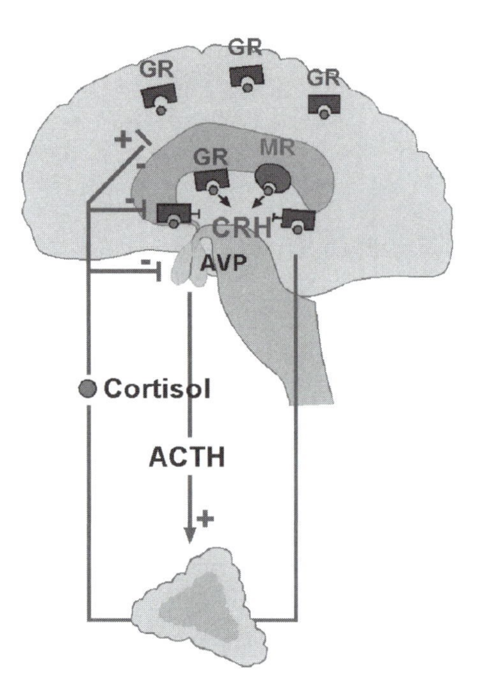

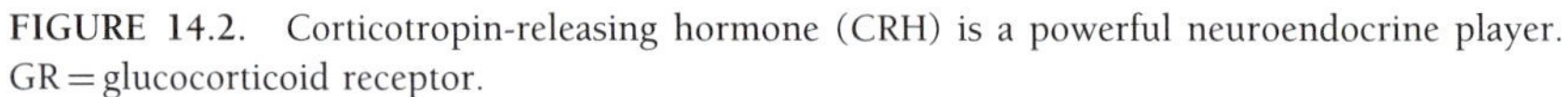
FIGURE 14.2. Corticotropin-releasing hormone (CRH) is a powerful neuroendocrine player. GR = glucocorticoid receptor.

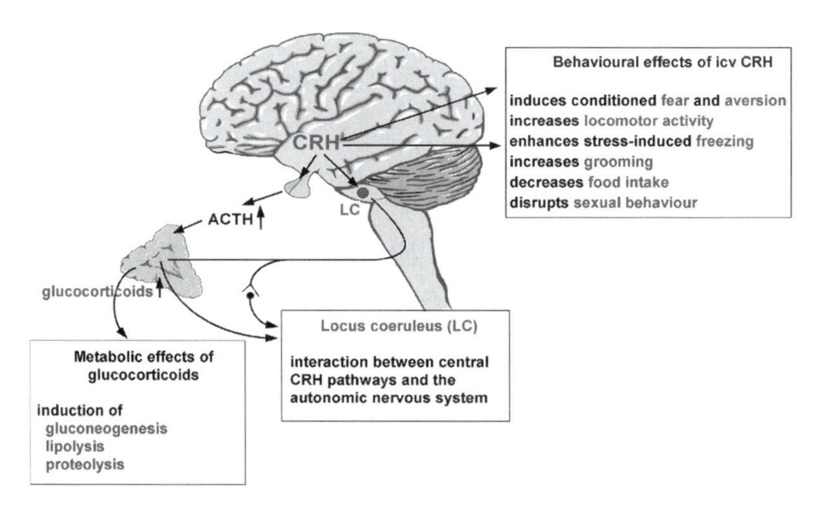

FIGURE 14.3. Corticotropin-releasing hormone (CRH) is a classical integrative hormone.

C. Gonadotropin-Releasing Hormone

The roles for the 10-amino-acid peptide gonadotropin-releasing hormone (GnRH, also known as LHRH) offer examples of bodywide coordination both in experimental and in clinical settings. Schwanzel-Fukuda at Rockefeller discovered in 1989 that these neurons, amazingly, are born in the olfactory pit and undergo a surprising migration from the nasal apparatus into the preoptic area and hypothalamus. This fact probably unites certain aspects of pheromone biology with reproductive biology. In the preoptic area and hypothalamus, GnRH will, after puberty, control reproduction by regulating release of luteinizing hormone (LH) and follicle-stimulating hormone (FSH) from the anterior pituitary gland. Importantly, GnRH also potentiates mating behavior. In many animals, this would be important for the females of the species because it would serve to coordinate sexual responses with ovulation. Such a synchronization of endocrine and behavioral systems is biologically adaptive, because the female need not expend energy or expose herself to predation during courtship and mating unless fertilization would indeed be attempted at the time of ovulation. The discovery of the nose-to-brain GnRH neuronal migration has been replicated and extended to all vertebrate species studied. In humans, Kallmann's syndrome, hypogonadotropic hypogonadism coupled with anosmia, can be linked either to the mutations on the X chromosome or to the autosomes (see also Chapter 17). The X-linked Kallmann's syndrome is due to a failure of migration of GnRH neurons during early development. Elucidation of the syndrome has linked nasal biology with basal forebrain, with pituitary, with blood hormones, with testes, and, hence, with testosterone (Fig. 14.4).

Indeed, testosterone itself makes an interesting case for bodywide alterations during aging. As men grow older, testosterone levels in the blood decline steadily. As a result, not only does their sex drive dwindle, but they also lose muscle mass and bone density. In addition, testosterone can fuel mental energy, so a drop in testosterone levels can result in a decline of mental abilities in later years, thus leading to a greater chance of cognitive performance loss and/or depressive moods.

III. OUTSTANDING NEW BASIC OR CLINICAL QUESTIONS

1. In view of these widespread hormonal effects throughout the body—adaptive physiologically, but often troublesome in terms of therapy—what are the opportunities for development of hormone analogs that

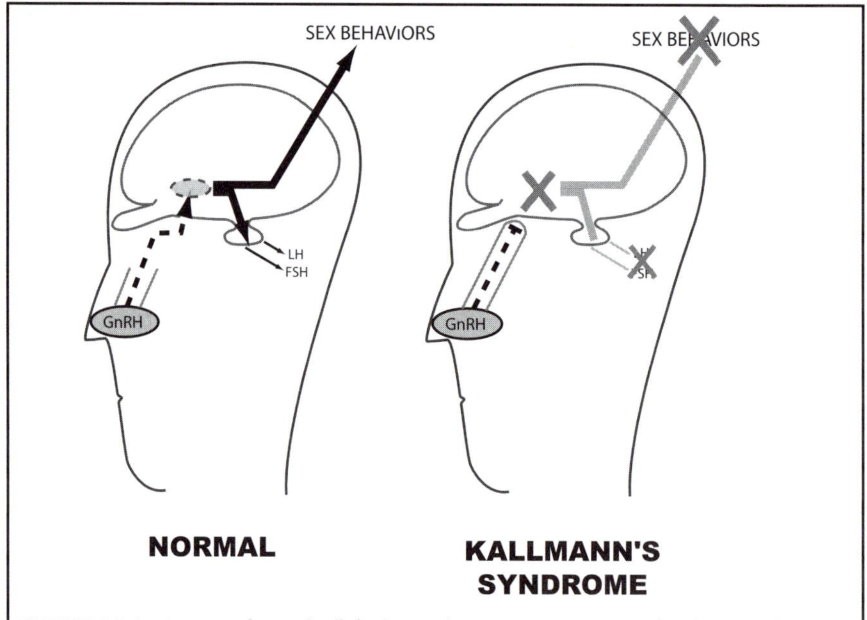

FIGURE 14.4. In normal people (left drawing), neurons expressing the decapeptide GnRH (gonadotropic-releasing hormone) are born on the medial side of the olfactory placode during fetal life; they migrate up the nose, enter the brain, and arrive at their final functional positions in the preoptic area of the basal forebrain and the hypothalamus. This occurs during development in all vertebrate species studied and is illustrated here for humans. In the basal forebrain and hypothalamus, GnRH neurons control release of reproductive hormones LH (luteinizing hormone) and FSNH (follicle-stimulating hormone) from the pituitary gland. These hormones in turn control ovulation and steroidogenesis (*e.g.*, estrogens) in the ovary and spermatogenesis and steroidogenesis (*e.g.*, testosterone) in the testes. GnRH neurons also facilitate sexual behaviors. In Kallmann's syndrome, disruption of a specific gene on chromosome 22 causes the loss of an extracellular matrix protein which in turn prevents GnRH from entering the developing forebrain. Consequently, drastically low levels of steroidogenesis result because of the low release of LH and FSH, and a notable absence of libido is observed. (From Schwanzel-Fukuda, M. and Pfaff, D.W., *Nature*, 338: 161–164, 1989; Schwanzel-Fukuda, M. *et al.*, *Mol. Brain Res.*, 6: 311–326, 1989; Schwanzel-Fukuda, M. *et al.*, *J. Comp. Neurol.*, 366: 547–557, 1996. With permission.)

narrow the effects of a hormone to a single system? Current efforts highlight *selective estrogen receptor modulators*, but discoveries of androgens and stress hormone analogs that have the desired effects in only one organ would also be great steps forward, especially for the aging population.

2. How, exactly, do the effects of leptin in the CNS register as behavioral effects?

3. In thyroid-related systems, how do we conceive of the separate contributions to behavioral controls of TRH versus thyroid hormones themselves?
4. Do receptors for food-intake-related hormones in the telencephalon (for example, insulin receptors in the hippocampus) mediate food-related learning and memory by the hungry subject?

CHAPTER 15

Hormones Can Act at All Levels of the Neuraxis to Exert Behavioral Effects; The Nature of the Behavioral Effect Depends on the Site of Action

Think back to Chapters 1, 3, and 14. Controls over feeding and hunger, important both for animals and humans, are orchestrated by actions of neuropeptide Y (NPY), leptin, alpha-melanocyte-stimulating hormone (α-MSH), cocaine- and amphetamine-regulated transcript (CART), agouti-related peptide (AGRP), melanin-concentrating hormone (MCH), and other hormones. They work (1) through different receptor distributions, and (2) at different CNS sites, sometimes even (3) outside the CNS. These systems illustrate the main principle of this chapter; however, several other examples will also make the case.

I. BASIC EXPERIMENTAL EXAMPLES

A. Angiotensin

Angiotensin has a number of actions (Table 15.1). Some of these are achieved by the peripheral renin–angiotensin system, some by the brain renin–angiotensin system, and some possibly by the interaction of both.

TABLE 15.1 Summary of Effects of Angiotensin by Different Route of Administration

Actions	Central	Peripheral
Drinking	Low dose	High dose required
Blood pressure increase	Slow rise/long duration	Fast rise/short duration
	Complex mechanisms	Local vasoconstriction
Heart rate	Up or no change	Down
AVP release	Prompt	Questionable
Sympathetic nerve activation	Increase	Local changes
Catecholamines	Levels increased in plasma/brain	Prolongs action in periphery
Na^+ appetite	Slow onset/long lasting	Questionable
Na^+ excretion	Natriuresis	Antinatriuresis
Prolactin release	Positive	Questionable
LH release	Positive	Questionable
ACTH release	Positive	High dose required
CRF interaction	Interacts	High dose required
Prostaglandins	Questionable	Increases synthesis
Aldosterone	Indirect release	Direct release

The main role of angiotensin is to retain water in the body. Drinking is clearly a motor effect that involves the brain. It involves coordinating input of sight; memory; recognition; motor effector systems for body positioning, licking, and swallowing; and circuits for thirst. A very-low-dose injection of angiotensin II into the brain produces drinking, whereas a relatively high dose of angiotensin II is required to be infused peripherally to initiate drinking. Van Eekelen and Phillips measured the angiotensin levels in plasma at the moment of drinking in rats given intravenous infusions of angiotensin II. There was a threshold below 50 ng angiotensin II per kg per minute to be reached before drinking occurred. After that, as the levels increased, the drinking increased. Thus, it appears that angiotensin has to reach a threshold level in the plasma before drinking occurs. When that level is compared to levels induced by dehydration, 50 ng angiotensin II per kg equals the level reached after 24 hours of dehydration. This means that in the normal day-to-day control of thirst, peripheral angiotensin probably plays no role; therefore, it is the brain angiotensin that is active normally and probably integrates with meal size as prandial drinking (*i.e.*, drinking with eating).

Angiotensin acts on receptors in the forebrain, and signals from these receptors are transmitted to the motor system, the hippocampus, and the brainstem. In the brainstem, these connections stimulate the sympathetic nervous system. The sympathetic nervous system increases vasoconstriction, which in turn elevates blood pressure. Indeed, when humans or animals drink, their blood pressure goes up (Hoffman and Phillips, 1977). This finding from rats is now used clinically to help bring up the blood pressure of patients with autonomic dysfunction. A glass of water is the treatment (Schroeder *et al.*, 2002).

Based on studies with knife cuts in the brain, three pathways appear to be involved in angiotensin-induced drinking (Fig. 15.1). One is a pathway between the subfornical organ (SFO) and the median preoptic (MNPO) area. Either brain-angiotensin-stimulating receptors in the SFO or high blood levels of peripheral angiotensin stimulate this circuit. Two pathways extend

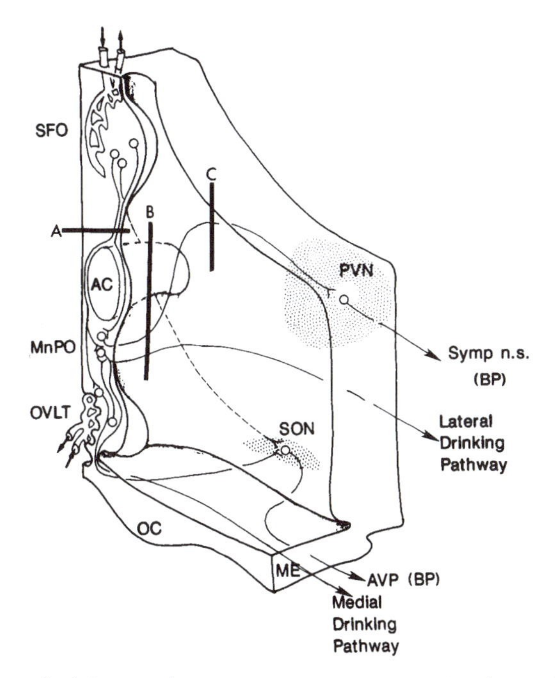

FIGURE 15.1. Multiple brain pathways participating in vasopressin release; schematic drawing of efferents and afferents involved in vasopressin release and pressor and thirst responses to angiotensin II. The subfornical organ (SFO) has efferents to the median preoptic area (MPO) and organum vasculosum lamina terminalis (OVLT). The OVLT projects to the MPO and SFO. Cuts at A, B, and C decrease the pressor response but not the drinking or arginine vasopressin (AVP) release. One explanation is that the OVLT ventral pathways are left intact, but the cuts interrupt a dorsal, caudally curving tract from the MPO–OVLT to the PVN and the locus ceruleus, thereby disconnecting angiotensin II receptors from the sympathetic nervous system. (From Phillips, M.I., in *Circumventricular Organs and Body Fluids*, Vol. III, Gross, P.M., Ed., CRC Press, Boca Raton, FL, 1987, pp. 163–182. With permission.)

from the MNPO: one, a lateral pathway that inhibits drinking when cut, and, the second, a medial pathway that emanates from the lower third ventricular region. In addition, a direct pathway from this median preoptic area to the paraventricular nucleus has afferent connections to the nuclei in the brainstem that activate the sympathetic system. A second connection is to the supraoptic nucleus (SON) from which arginine vasopressin (AVP) is released through the median eminence and the neurohypophysis. Angiotensin receptors are distributed along the forebrain that includes the SFO median preoptic area and organum vasculosum of the laminae terminalis (OVLT). Because these receptors are also part of the circumventricular organs that have no blood–brain barrier, it has been difficult to dissect the relevant roles of central versus peripheral angiotensin. However, the experiment described above, where plasma levels of angiotensin II have to reach the levels found after 48 hours of dehydration, clearly indicate that the frequent bouts of drinking experienced in a normal day are due to central angiotensin. Angiotensin in the brain also releases adrenocorticotropic hormone (ACTH) and luteinizing hormone (LH). It takes extremely high levels of circulating angiotensin II, working by itself, to release ACTH and LH. Vasopressin release is very prompt with central injections of angiotensin II and much more difficult to achieve with infusions of angiotensin II, so there are differences between the peripheral route and the central route (see Table 15.1). Neural inputs can affect central angiotensin. The vagal inputs from the baroreceptors in the heart and blood vessels to the nucleus tractus solitarius (NTS) detect decreases in pressure. The baroreflex increases heart rate when pressure falls. Angiotensin II plays a role in this reflex by directly acting on NTS neurons, but the input from the periphery also stimulates angiotensin II in the hypothalamus. This is vitally important during hemorrhage when there is a loss of pressure and a need to conserve fluid by any means. The pathways running through the anterior ventral third ventricle (AV3V) have given rise to a number of experiments in which that area has been lesioned. This results in a complete loss of drinking behavior. It also prevents hypertension from developing in different models of hypertension, but it does not affect feeding. Thus, although eating and drinking are obviously associated, the circuits that control them in the brain are separated in the hypothalamus.

In addition to angiotensin as a stimulant for thirst, an increase in plasma osmolality will also elicit drinking. Thus, the brain appears to have two separate circuits for thirst induced by angiotensin and for thirst induced by an osmotic change. The latter can be initiated by an injection of a long-acting acetylcholine agonist (carbachol) into the hypothalamus; therefore, the circuits are angiotensinogenergic and cholinergic. The separation is logical because angiotensin responds to hypovolemia by increasing sodium appetite and by releasing vasopressin to increase sodium concentration. Thirst has two opposite roles. In one case (via the osmotic circuit), the role is to reduce the osmotic pressure by diluting the urine concentration. In the other (the angiotensinogenic pathway), drinking is initiated to increase volume. The brain detects small changes in osmotic level (as small as 3 mOSm by osmoreceptors). Studies in different species indicate that the osmoreceptors are localized in special cells in the AV3V region, including the OVLT and the SON and paraventricular nucleus (PVN). Direct application of hyperosmotic solutions in these areas trigger neurons to fire and presumably activate their circuits.

Within the brain, angiotensin interacts with neurotransmitters as part of the signaling pathway. Angiotensin receptors overlap with those of neurons and fibers containing transmitters for serotonin, substance P, glutamate, norepinephrine, and dopamine. Substance P attenuates the drinking response elicited by angiotensin II in the PVN. Evidence also supports the activation of angiotensin II in MDA receptors in the thirst circuit. Serotonin pathways are involved in this circuit and are affected by angiotensin II. One of the sites of interaction between serotonin and angiotensin II is the AV3V region referred to earlier.

B. Sex Hormones

Sex hormones, as well, illustrate how hormones can act at all levels of the neuraxis, with the exact behavioral effect depending on site of action. Working through preoptic neurons, estrogens facilitate locomotor and courtship behaviors and also maternal behaviors. Through medial hypothalamic neurons, the same hormones turn on lordosis behavior. Yet, in the midbrain, with estrogen receptor-β expressed in neurons of the dorsal raphe nucleus, estrogenic effects on serotonergic neurons are of likely importance for mood. Much further posterior in the neuraxis, estrogens act on medullary reticular neurons, which give rise to ascending arousal pathways (Fig. 15.2). In the spinal cord, estrogen binding at specific sites in the dorsal horn (in Rexed layer II) must be of importance for pain. Finally, Raymond Papka and colleagues have shown estrogen receptors in peripheral autonomic neurons: same hormone—different CNS sites—different functional effects.

II. CLINICAL EFFECTS

A. Arginine Vasopressin

Polydipsia is the condition where patients continuously drink copious amounts of liquid and seek fluid to drink even though their volume status is

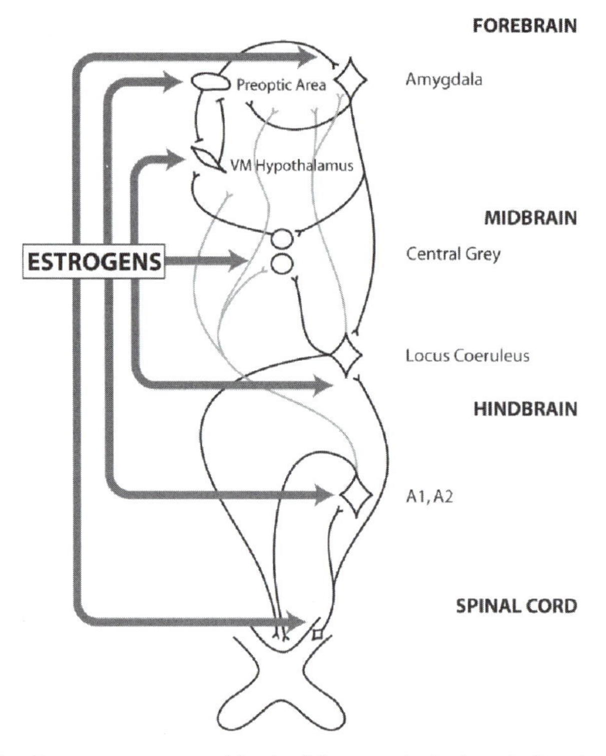

FIGURE 15.2. Estrogens act at several levels of the neuraxis. In the spinal cord and hindbrain, fundamental arousal mechanisms and pain are affected. In the hypothalamus, estrogen-dependent lordosis behavior circuitry is completed in the ventromedial nucleus. In the forebrain, both neuroendocrine cells and emotional states are hormone-dependent. Many other hormones—testosterone, progesterone, stress hormones, and thyroxine—also have widespread targets of actions on the CNS. Estrogens are used here as the example because cellular and molecular mechanisms are understood in greater detail.

normal. Polydipsia, in some cases, is due to diabetes insipidus (DI). This form of diabetes results when there is an absence of AVP release from the brain (central DI) because of a mutation in the AVP gene, or when AVP is released but is not effective in the kidney (nephrogenic DI). We elaborate on this distinction below. Most important, AVP provides the perfect example of actions at multiple levels of the neuraxis. Not only do projections from SON and PVN and other magnocellular anterior hypothalamic cell groups reach the posterior pituitary for release of AVP into the bloodstream, but also AVP-bearing axons project to forebrain cell groups, the lower brainstem, and the spinal cord. In the last case, the implications for spinal cord control over the autonomic nervous system are obvious.

Psychogenic polydipsia in schizophrenics is of unknown cause. These patients are treated with antidopaminergic drugs that correct the underlying schizophrenia. Inhibiting dopamine would interfere with the angiotensin–dopamine circuit.

1. Diabetes Insipidus

Diabetes insipidus is a clinical syndrome involving the lack of vasopressin effects either centrally or in the kidney. As mentioned, the syndrome is characterized by polyuria and polydipsia. It is due to the fact that vasopressin is either not being produced in the brain (central DI) or vasopressin is being released from the brain but the kidney is not responding to it (nephrogenic DI). The most frequent type is central DI, due to an inadequate secretion of AVP. There can be various causes for this, such as tumors in the pituitary stalk, genetic loss of expression of the vasopressin gene, and low amount of vasopressin synthesis in the SON and PVN cells of the hypothalamus. This is the case in an animal model, the Brattleboro rat. The Brattleboro rat was discovered by an observant technician who noted the rats had constant polydipsia and polyuria and reported his observation to a professor, Heinz Valtin, who understood what it meant. When he treated the rats with vasopressin, they stopped overdrinking and urinating and diluted urine became normalized. In the Brattleboro rat, the cause of the DI is a lack of expression of vasopressin in the magnocellular cells of the PVN and SON of the hypothalamus. The gene mutation that causes it is a frameshift mutation that prevents synthesis of the complete prohormone.

2. Central Diabetes Insipidus

In humans, central DI has to be differentiated from psychogenic polydipsia, a condition where patients constantly drink fluid and seize every opportunity to drink, sometimes to the point of fatal water intoxication. In psychogenic polydipsia, vasopressin levels are normal. Restricting water in these patients eliminates the diluted urine, showing that AVP is working. Such patients do not respond to treatment with a vasopressin-like hormone. Polydipsia in dialysis patients, however, is due to increased angiotensin and can be treated with angiotensin-converting enzyme inhibitors (Kuriyama, 1996).

In diagnosing central DI, one must also differentiate it from nephrogenic DI. To determine if the DI is due to a lack of vasopressin, the patient is tested with desmopressin, a synthetic form of AVP. Replacing the vasopressin by treatment with desmopressin restores the antidiuretic effects in the case of central DI. Among the causes of brain lesions that lead to central DI are germinoma, cranial pharyngioma, and sarcoidosis of the central nervous system. In addition, autoimmune and vascular diseases, trauma following surgery, and genetic autosomal-dominated or X-linked recessive traits may lead to a defect in vasopressin biosynthesis.

3. Nephrogenic Diabetes Insipidus

If, however, desmopressin does not work, then attention turns to the kidney. The kidney may not respond because it lacks either vasopressin V_2 receptors or V_2 effects on cells. In nephrogenic DI, mutations occur in the arginine vasopressin receptor type 2 gene. Genetic defects may also occur in the signal transduction pathway after V_2 receptors have been stimulated. Normally, V_2 receptors are linked to cyclic adenosine monophosphate (cAMP), and their stimulation by AVP induces CAMP to mobilize aquaporins. Aquaporins are proteins that insert water channels into the membrane. V_2 stimulation causes one type of aquaporin (AQP_2) to insert water channels into the apical membrane of kidney cells in the collecting duct. This enables water to be reabsorbed directly back into the cells. Aquaporins are proteins that have a structure that forms a very narrow channel, only big enough for one molecule of H_2O to pass through at a time. With millions of aquaporins in the membrane, the passage of water in or out of the cell is very efficient. Different aquaporins at the basal membrane escort water out of the cell and back into the plasma. Studies in mice that have severe DI ($DI^{+/+}$) have low levels of cAMP, which is the second messenger for V_2 receptors, and low expression of AQP_2, which is the water channel protein in the apical cell membrane of the collecting duct cells. Thus, cAMP stimulates AQP_2 expression by activating cAMP-responsive elements on the AQP_2 gene.

B. Corticotropin-Releasing Hormone

Corticotropin-releasing hormone (CRH) provides another clear example of a hormone acting in different places with different clinical consequences. Through CRH receptors in the hypothalamus and pituitary gland, CRH controls release of ACTH, which itself has behavioral consequences but which primarily causes the elevation of stress hormones (adrenal corticosteroids) in the blood. In turn, ACTH itself has major effects on conditioned fear and avoidance. We were reminded of this in Chapter 2 with developments that go all the way back to the classic work of deWied and Greidanus in Utrecht, Netherlands. In addition, however, prolonged high levels of CRH expression in the amygdala are thought to be associated with anxiety and depression. In parallel, CRH receptors in the forebrain, analyzed by Holsboer in Munich by the use of conditional gene knockouts, are very important for anxiety. Finally, CRH-activating noradrenergic systems that act in the hindbrain through locus ceruleus neurons have effects on alertness and fear. CRH has widespread sites of actions, with different consequences.

C. Progestins

Progestins have obvious effects upon mood, acting through midbrain reticular system neurons. They depress the activity of neurons associated with arousal of the entire forebrain, perhaps because of their ability to potentiate the effectiveness of the inhibitory transmitter γ-aminobutyric acid (GABA). These cellular actions may be related to the severe and sometimes incapacitating mood changes that are suffered by some adult women taking sequential hormone replacement therapy. On the other hand, the main proven importance of progestins in the cerebral cortex, from the work of Donald Stein at Emory University, is to prevent swelling following injury. In experimental animals, progestins acting through ventromedial hypothalamic neurons enhance sex behavior; when they act through basal forebrain neurons, they facilitate courtship behaviors. Thus, for progestins, at least four different sites of action have four types of functional consequences. When peptide hormones produced in the nervous system have both endocrine effects (of behavioral importance by indirect routes) and straight behavioral actions, a question arises: Are the extrahypothalamic, behavioral, and autonomic effects of the pituitary hormone releasing and inhibiting factors coordinated with their effects on pituitary hormone release? As documented in previous chapters, many of the hypothalamic releasing and inhibiting factors for pituitary hormones also act as neurotransmitters and neuromodulators in other areas of the CNS. Following are some examples.

D. Thyrotropin-Releasing Hormone

Thyrotropin-releasing hormone (TRH) occurs in high concentrations in motor nuclei of the brainstem and spinal cord and in lower concentrations in the amygdala, mesencephalon, and cerebral cortex. It is co-localized with serotonin and substance P in brainstem raphe nuclei and the medulla. Behaviorally, TRH is activating and mildly euphorigenic, and it has a brief antidepressant effect when administered to humans.

E. Somatostatin

Somatostatin is widely distributed outside the hypothalamus, with fairly high concentrations in amygdala and certain brainstem regions and lower concentrations in limbic structures and cerebral cortex. It is co-localized with GABA in the hippocampus, thalamus, and cortex and with norepinephrine in the medulla and sympathetic ganglia. When administered into the cerebrospinal fluid (CSF) in humans, somatostatin ameliorates intractable pain, (*e.g.*, cluster headache).

Little is known about possible coordination between the extrahypothalamic, direct behavioral effects of these releasing and inhibiting factors and the behavioral effects of the hormones they regulate via their secretion into the pituitary portal circulation. If, for example, hypothalamo–pituitary–adrenal cortical (HPA) axis hyperactivity is a causative factor in major depression, as discussed in Chapter 2, then the extrahypothalamic and neuroendocrine effects of CRH would be consonant in promoting some of the clinical components of major depression. On the other hand, the majority of patients with major depression do not have increased HPA axis activity, although they still could have increased activity of extrahypothalamic CRH systems. The suggestive antianxiety or antidepressant effect of CRH receptor antagonists was shown in depressed patients who had, on average, normal HPA axis activity. As well, clinical improvement in these patients was not significantly related to their plasma ACTH or cortisol concentrations either prior to or during treatment. These findings support the concept of the independence of extrahypothalamic and neuroendocrine CRH systems, a hypothesis that needs to be verified experimentally.

III. SOME OUTSTANDING QUESTIONS

1. To what extent can it be shown that actions of a given hormone at different levels of the neuraxis are orchestrated to produce synergistic effects?
2. Are there positive feedback mechanisms in which different actions multiply each other? Or, in contrast, do different actions at different levels of the neuraxis counterbalance and check each other?
3. What other molecules are the size of water, and why do they not pass through the aquaporins?

Mechanisms: Molecular and Biophysical Mechanisms of Hormone Actions Give Clues to Future Therapeutic Strategies

CHAPTER 16

In Responsive Neurons, Rapid Hormone Effects Can Facilitate Later Genomic Actions

Mechanisms of steroid hormone action have experienced large swings in emphasis and are still being worked out. At first, because steroids are soluble in fat (lipid), it was assumed that major actions would occur directly at the cell membrane. Then, during the 1960s, the groundbreaking studies of Jensen and colleagues proved that classical target tissues exhibited a prolonged retention of isotopically labeled steroids. In fact, this retention was due to accumulation and prolonged action in the cell nucleus. Especially exciting was the discovery that the proteins that act as nuclear receptors for steroid, thyroid, and certain other hormones are also transcription factors. Thus, endocrine investigations could rapidly enter the era of eukaryotic molecular biology.

Neurobiologists and others interested in the causation of behavior, however, also realized that some hormone actions in the central nervous system are registered too rapidly to be dependent on nuclear, transcriptional mechanisms. Within the field of molecular endocrinology, interpretations of hormone action as being membrane-initiated were often considered antithetical to interpretations relying on transcriptional facilitation. This chapter illustrates rapid actions of hormones in the central nervous system (CNS) and shows how rapid, membrane-based effects actually can facilitate later genomic actions. Such data foster a unified view of steroid actions on neurons.

I. ESTROGENS

A breakthrough in analyzing steroid actions with fast kinetics was derived from Jack Gorski's discovery that long epochs of steady exposure to estrogens could be replaced by brief pulsatile administrations. For example, to trigger large-scale cell division in the uterus, constant exposure for 24 hours could be substituted by two 1-hour pulses properly timed. The same is true of behavior. Bruce Parsons, at Rockefeller, found that female-typical sex behavior as well as induction of the progesterone receptor in the hypothalamus could be stimulated by two 1-hour estrogen pulses almost as well as continuous exposure. (See also Chapter 12, Fig. 12.1.)

What is the relation of the first, early pulse to the later pulse? This question was answered by using Gorski's two-pulse approach during experiments on estrogen-facilitated gene transcription in neuroblastoma cells. An early pulse, limited by chemical linkage of estradiol to the membrane, actually amplified the transcription-facilitating action of the second pulse (Fig. 16.1). Amazingly, the behavioral mechanisms in the hypothalamus are more permissive. The membrane-limited actions of estradiol could be brought into play either as the first pulse or the second pulse. Signal transduction pathways such as those involving both protein kinase A (PKA) and protein kinase C (PKC) are both necessary and sufficient for the membrane-limited hormone effect to amplify the genomic effect of estradiol.

How and exactly where in the membrane does that early action take place? One possibility is that caveoli, specialized domains of membrane that are attached to a wide variety of signal transduction initiators, transduce the estrogen effect. Michael Mendelsohn and colleagues showed that, in endothelial cells, estradiol working through estrogen receptor-β in membrane caveoli could rapidly activate nitric oxide synthase (Fig. 16.2). A related possibility is that rapid hormone effects are organized by so-called *lipid rafts* on membranes. The main lesson is that hormone action at the membrane does not occur just any place. Relevant receptors are organized by membranous structures which, in turn, are thought to be linked physically to submembrane signal transduction molecules.

What then? Following cascades of protein phosphorylation by well-studied pathways such as PKC, PKA, mitogen-activated protein kinase K (MAP-K), and other routes, hormone receptors that will eventually act in the cell nucleus can be phosphorylated and thus activated. An early example from Kenneth Korach and colleagues at NIEHS demonstrated that prior

PULSATILE ADMINISTRATION		RESULTS	
Membrane-Limited Estrogen Admin.	Cell Nuclear Estrogen Access	Estrogen-Facilitated Transcription?	Estrogen-Facilitated Lordosis Behavior?
NO	NO	NO	NO
YES	YES	YES	YES
AS PULSE 1	NO	NO	NO
NO	AS PULSE 2	NO	NO
AS PULSE 1	**AS PULSE 2**	**YES**	**YES**
AS PULSE 2	AS PULSE 1	NO	YES

FIGURE 16.1. What schedules of neuronal cell membrane application and cell nuclear access allow estrogens to achieve the effects equal to unlimited estradiol administration (second row in figure)? For estrogen-facilitated transcription, membrane mechanisms in the first pulse and nuclear access in the second pulse are necessary and sufficient. For female sex behaviors, surprisingly, either order of pulsatile access (membrane and then nuclear *or* nuclear and then membrane) is sufficient. (Vasudevan *et al.*, 2001). For other hormones, as well, choreographing membrane mechanisms and nuclear mechanisms of cellular actions is certainly essential to understanding and obtaining the full panoply of CNS hormone effects. Their timing and details probably depend on the hormone administered and the cell involved.

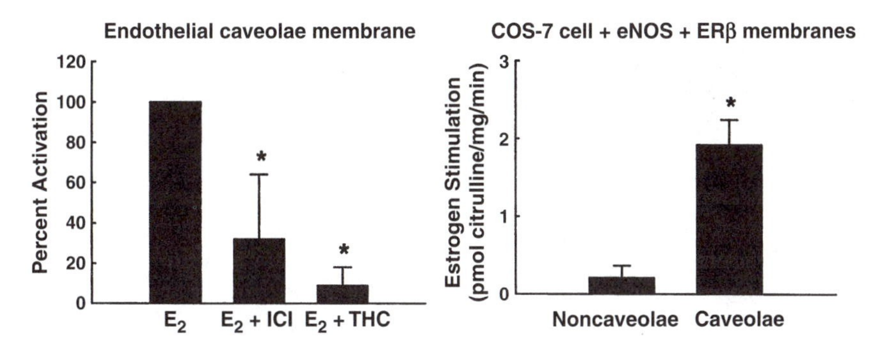

FIGURE 16.2. Rapid actions of steroid hormones. (Left panel) Enzymological activation of endothelial nitric oxide synthase (eNOS) by estradiol (E2) could be obtained by using endothelial cell membranes in the regions that contain specialized pits called *caveoli*. Surprisingly, estrogen receptor blockers such as ICI and THC inhibited the effect. (Right panel) The estrogenic stimulation worked much better using membranous regions with caveolae than without. In this effect, estradiol receptor ER-β is suspected to be the relevant molecule transducing the estrogenic effect. Such rapid actions are just beginning to be explored in the CNS with experimental designs relevant to behavioral studies. (From Chambliss, K.L. *et al.*, *Mol. Endocrinol.*, 16(5): 938–946, 2002. With permission.)

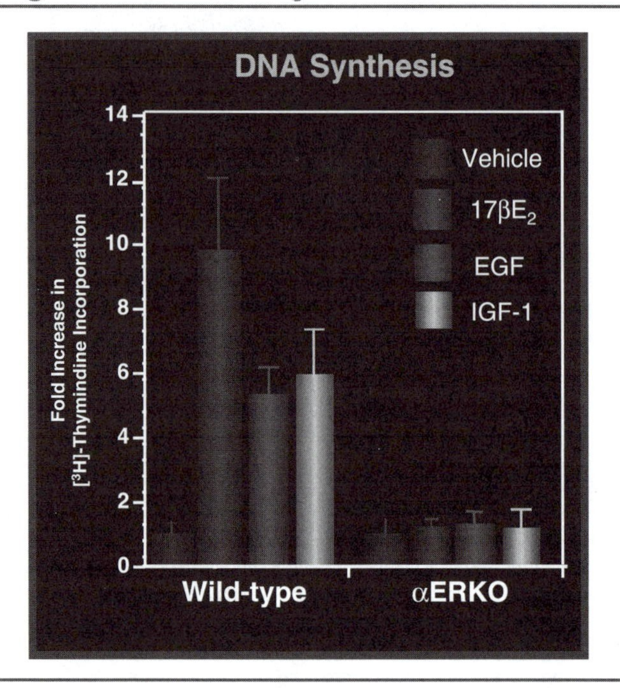

FIGURE 16.3. Estradiol treatment leads to a massive increase in the rate of cell division within the uterus of the wild-type female mouse. Suprisingly, growth factors can also have this effect. When the ER-α gene is disrupted (α-ERKO), all of these actions are abolished; therefore, even the growth factor effect relies upon the gene product for ER-α. (Courtesy of K. Korach *et al.*, NIEHS.)

delivery of a growth factor that acts at the membrane could activate the classical estrogen receptor (estrogen receptor-α) (Fig. 16.3).

Moreover, Benita Katzenellenbogen, at the University of Illinois, found that membrane stimuli, as could result from synaptic inputs, could activate estrogen receptors, thus leading to more efficient transcriptional facilitation. Power and O'Malley found the same for the ligand-free activation of progesterone receptors by dopamine-related agents. Not only second messenger systems but also membrane-crossing ion fluxes have been reported, over the years, to be sensitive to estrogens (reviewed by Segars and Driggers, *Hormones, Brain and Behavior*, 2002). All of these studies show how events in the environment of the cell interact with hormone administration, thus affecting the magnitude of the response of the cell to the hormone.

What do these molecular studies mean for behavior? Jill Becker and her students at the University of Michigan were pioneers in revealing rapid actions of estrogens on motor behaviors. Because these effects were likely mediated in the striatum, it was significant that parallel cellular studies by Paul Mermelstein reported rapid effects of estradiol treatment on calcium mobilization in striatal neurons. Coinciding with these reports were the electrophysiological recordings by Gu and Moss of PKA-dependent activation of hippocampal neurons immediately following estradiol administration. At several levels of analysis, therefore, these rapid actions are important.

II. THYROID HORMONES

Thyroid hormones also have rapid, physiologically important actions on a variety of cell types. Faith Davis and Paul Davis, at the State University of New York at Albany, have shown that thyroxine (T4) can induce phosphorylation and nuclear translocation (activation) of the enzyme MAP-K within 10 minutes. As a result, PAP kinase is able, within 10 to 30 minutes, to promote the phosphorylation (and consequent functional alteration) of an important transcription factor protein called p53. Another consequence of this nongenomic activation of MAP-K following thyroxine administration is the phosphorylation of nuclear receptors for thyroid hormones (TRs), presumably because of a physical association of MAP-K with TRs. Of potential direct relevance to nerve cells, thyroid hormones can promote the opening of sodium channels (causing action potential bursting) and, apparently because of changes in potassium channel currents, shorten action potential duration. For thyroid hormones as well, therefore, rapid nongenomic actions on signal transduction pathways play important roles and, under some circumstances, may potentiate later transcriptional effects (Fig. 16.4). There is every reason to expect that these or similar phenomena also occur in neurons governing T4-influenced behaviors, including emotional responses.

III. ANDROGENS

At a molecular level, androgens certainly exert rapid nongenomic effects, because they can activate kinase-signaling cascades with shorter latencies than transcriptional activation would require, and because androgen effects can occur in cell types that lack functional androgen receptors (ARs). Many of the male-typical mating behaviors do not rise in frequency and intensity until long after initiation of testosterone treatment. However, John Nyby, at Lehigh University, recently has reported that testosterone injections

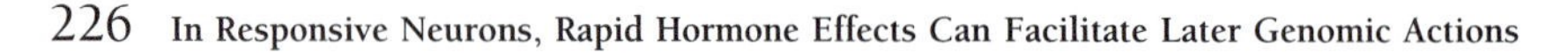

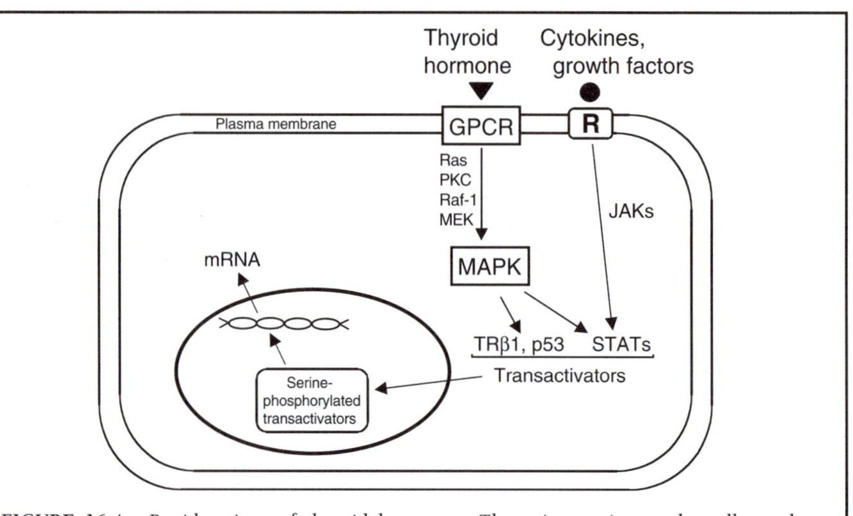

FIGURE 16.4. Rapid actions of thyroid hormones. Thyroxine, acting at the cell membrane and working through several signal transduction pathways, activates MAP-kinase. This kinase, in turn, phosphorylates transcription factors such as TR-β1, p53, and certain STATs, thus effecting changes in gene expression in the cell nucleus. In these rapid actions, thyroid hormones interact with cytokines and growth factors. (From Lin, H.-Y. *et al.*, *Biochem. J.*, 338: 427–432, 1999; Lin, H.-Y. *et al.*, *Am. J. Physiol.*, 276: C1014–C1024, 1999; Davis, P.J. *et al.*, *J. Biol. Chem.*, 275: 38032–38039, 2000; Shih, A. *et al.*, *Biochemistry*, 40: 2870–2878, 2001. With permission.)

can reduce anxiety-related behaviors within 30 minutes, too fast for most transcriptional effects (Fig. 16.5). Such effects must influence mating. In fact, Nyby's experiments showed that within 60 minutes of a subcutaneous injection of testosterone, latencies to mount females were significantly reduced. Early electrophysiological recordings from preoptic neurons support these behavioral data. Testosterone application elevated neuronal activity markedly within 40 minutes.

IV. PROGESTINS

The effects of progesterone on behavior clearly depend on transcriptional activation by its nuclear receptor. For example, antisense DNA against progesterone receptor (PR) mRNA, microinjected into the hypothalamus, can significantly reduce female reproductive behaviors. Further, the progesterone amplification of estrogenic effects on lordosis behavior disappears in female mice in which the gene for the progesterone receptor has been knocked out. But, two recent developments have uncovered rapid effects related to progestin action. First, membrane-initiated signals, as could

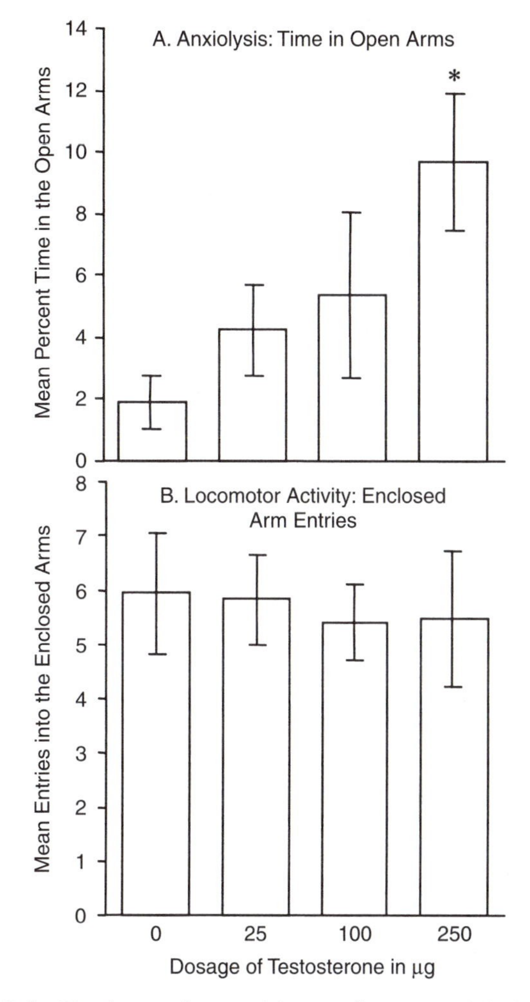

FIGURE 16.5. Only 30 minutes after receiving a subcutaneous injection of testosterone propionate (see *x*-axis for doses), male mice spent longer times in the open arms of an elevated plus maze (top of figure, part A), despite the fact that under these conditions general locomotor activity was not increased (bottom of figure, part B). (From Aikey, J.L. *et al.*, *Hormones Behav.*, 42(4): 448–460, 2002. With permission.)

result from synaptic excitation, activate the PR with short latency in a manner important for behavior. At the Baylor College of Medicine, scientists in the O'Malley laboratory have built a strong case for dopamine-initiated signals as an examplar of this principle. Second, in turn, the effects of injected progesterone sometimes are exerted by progesterone's metabolites (see Chapter 4). Reproductive behavior facilitation due to chemically reduced

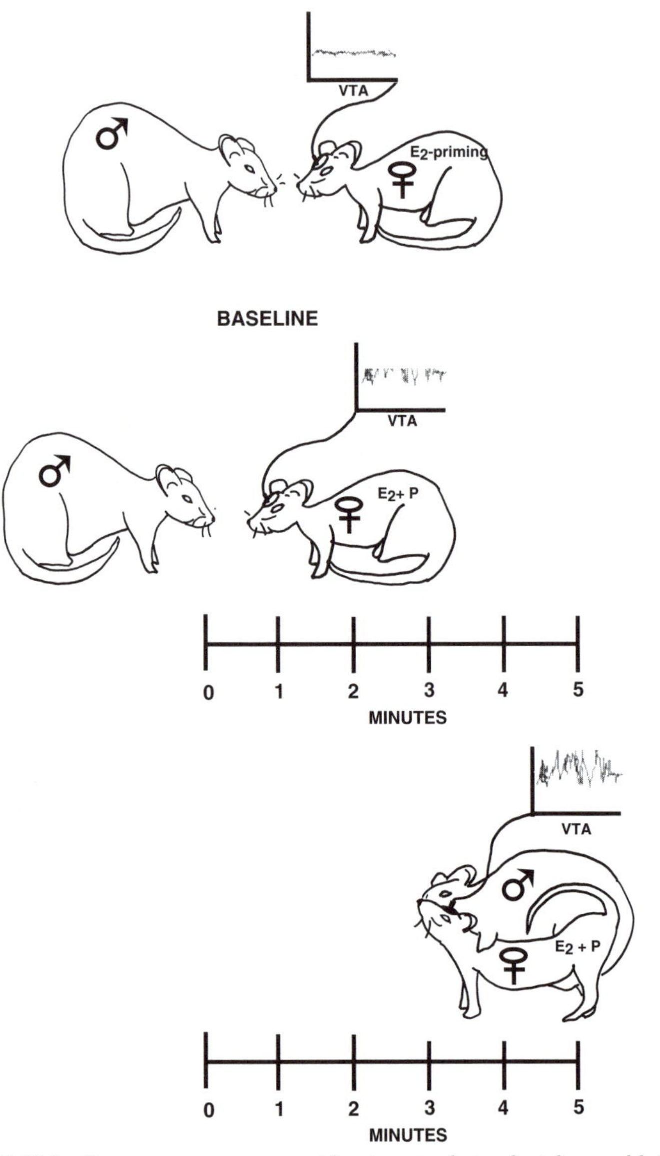

FIGURE 16.6. Progesterone can exert rapid actions on electrophysiology and behavior. Here, in the estrogen-primed female laboratory rodent, progestins given either intravenously or to the ventral tegmental area of the midbrain (VTA) have increased neuronal firing in the VTA within 2 minutes (middle panel). This is followed in another few minutes by increased lordosis behavior upon mounting by the male (bottom panel). (From Frye, C.A. *et al.*, *Psychobiology*, 28(1): 99–109, 2000; Frye, C.A. and Vongher, J.M., *J. Neuroendocrinol.*, 11(11): 829–837, 1999. With permission.)

progesterone metabolites in the midbrain, for example, occur too quickly to be accounted for by genomic changes (Fig. 16.6).

V. OUTSTANDING NEW BASIC OR CLINICAL QUESTIONS

1. Which novel cell-membrane steroid and thyroid hormone receptors will be unambiguously identified?
2. Even when question 1 has been answered, will there still be behaviorally important physicochemical modulation of the cell-membrane lipid bilayers by highly lipophilic steroids?
3. Where such modulations occur, do they make use of specialized membrane domains such as caveoli or lipid rafts?
4. Under what circumstances do the rapid signaling pathways lead to nuclear events, and when do they lead more directly to alteration of ion channels in nerve cells or glia?
5. What are the step-wise chemical mechanisms of the fastest effects? In particular, because several signal transduction pathways have been identified by the labs of McDonnell, Sjoberg, and others (*e.g.*, MAP-K, PKC, PKA, calcium mobilization), how do these signaling pathways relate to each other?
6. Are the rapid hormone actions, manifest in the first pulse of a two-pulse paradigm, the perfect site for cross-talk among biochemical systems of both (a) physiological and (b) medical importance? In regard to (a), physiologically, integration among hormonal actions or between hormonal actions and other extracellular changes could be exerted by a relaxation of requirements in the first pulse. In regard to (b), medically, if indeed the first pulse accepts many chemical substitutes for the hormone in question, a wide variety of environmental chemicals, including toxins, could intrude here. Especially during development, such contamination of the signaling system could have tremendous effects on public health.
7. In this chapter, several molecular phenomena have been cited, and rapid behavioral effects of hormones have been illustrated. How can we discover the exact routes from the former to the latter?

CHAPTER 17

Gene Duplication and Splicing Products for Hormone Receptors in the CNS Often Have Different Behavioral Effects

As a consequence of the operation of genetic recombination machinery during the passage of chromosomal information from mother and father to offspring, parts of chromosomes may be duplicated. Then, they may end up in tandem positions on the same chromosome, or, by the mechanisms of transposition, they may be moved to a different chromosome. These events happen with low probability, but they happen. In fact, it has been argued that such stochasticity represents a major driving force in evolution. That, in turn, would explain why mating behaviors are considered to be at the leading edge of evolutionary change. When one of the copies of the duplicated gene is altered subsequently, either of two consequences will occur. If the coding region is altered, then the properties of the encoded protein may be changed. If the promoter of the gene is altered, then the temporal and spatial patterns of expression and/or the physiological regulation of expression would be affected. The meaning of this train of events for hormone action is considered in this chapter.

The opportunities for small molecular change do not stop here. In the company of a cohort of RNA-binding proteins, nascent RNAs are spliced to make messenger RNAs, their half-lives are determined, their passage to the ribosome is controlled, the efficiency of their translation into protein is regulated, and elongation of the eventual chain of amino acids of the protein is completed. Then, the newly manufactured protein must be directed to the proper parts of the nerve cells or glial cells. Their rates of degradation constitute a subject of great research interest. All of the molecular steps (and accidents) mentioned here have implications for hormone/behavior relations. The principle is that closely related structures do not necessarily imply closely related functions.

I. BASIC EXPERIMENTAL EXAMPLES

A. Gene Duplication

1. Estrogen Receptor Genes

Following years of molecular endocrine work with the classical estrogen receptor (now called estrogen receptor-α, or ER-α), a new estrogen receptor, likely a gene duplication product, was cloned by Gustafsson and colleagues at the Karolinska Institute. It is called estrogen receptor-β (ER-β). They both bind ovarian estrogens with about the same affinity. Yet, their molecular properties in neurons and their behavioral effects are much different (Fig. 17.1). For example, on the promoter for the cyclin D1 gene (a major regulator of entry into the proliferative stage of the cell cycle), estrogen working through ER-α induces transcription, whereas ER-β inhibits the ER-α effect. In fact, in this investigation by Kushner's laboratory at the University of California, San Francisco, only with anti-estrogens could ER-β activate this promoter.

In neuroblastoma cell culture work by Nardini Vasudevan and her team, the relative abilities of ER-α and ER-β to manage the transcriptional facilitations upon estradiol treatment were dissimilar. Further, between the two estrogen receptors, molecular interactions with thyroid hormone effects frequently were in the opposite direction.

What about behavior? In many cases, primary reproductive behaviors simply depend upon ER-α. In fact, normal expression and regulation of the ER-α gene is required for normal sex behavior in both genders. On the other hand, social recognition and the suppression of aggression depend upon the ER-β gene working in concert with ER-α (refer back to Fig. 3.5). Conceptual scenarios for the relations between ER-α and ER-β have been tested extensively by comparison of normal wild-type mice with those in which either or both of the genes have been knocked out (Fig. 17.1). In the history

COMPARISONS of ER-α & ER-β FUNCTIONS in CNS				
Necessary?	**Sufficient?**		ASSAYS IN THE ♀	ASSAYS IN THE ♂
ER α & β	**Neither α nor β**	"ER α & β must synergize"	Social recognition	none
Neither α nor β	**Either α or β**	"ER α & β can subst for each other"	E induction of PR (ICC) E reduction of ER (ICC)	E induction of PR (ICC) E reduction of ER (ICC) Simple mounting
α, β each for its own	**α, β each for its own**	"ER α vs β contribs. not related"	Maternal behavior (α) Suppression of aggression (α) Reduction of food intake (α) Reduction of anxiety (α)	Ejaculation (α) Anxiety response (α)
ER α absent β	**ER α absent β**	"ER β can reduce α effect"	Lordosis behavior	Aggression
Optimal α/β balance	**Optimal α/β balance**	"ER α vs β always opposed"	none	none

FIGURE 17.1. Two likely gene duplication products, ER-*α* and EReceptor-*β*, have significantly different behavioral effects in the mouse CNS. Theoretical scenarios for combinations and patterns of action between the two ERs are charted on the left side of the matrix. On the right are data completed to date from the genetic female and the genetic male. Conclusions are derived from comparisons among wild-type mice, *α*-ERKOs, *β*-ERKOs, and double-ERKOs. (ERKO = Estrogen Receptor Knockout.) As a side point it is also clear that—well beyond the Beadle and Tatum "one gene/one enzyme" concept—different combinations of gene functions are required for different patterns of sociosexual behaviors in this mammal.

of biochemical genetics, the classic principle of "one gene, one enzyme" was derived from the pioneering work of Beadle and Tatum. Now, however, in light of the complexities of the mammalian central nervous system and human behavior, we must instead say: "Combinations and patterns of gene expression influence combinations and patterns of behavior." Two illustrations follow. *First*, from Fig. 17.1 it is evident that different patterns of male sociosexual behavior depend on different relations between ER-α and ER-β. Simple mounting flourishes from a synergy between the two. In aggressive behavior tests, the suppressive effect of ER-β actually counteracts the influence of ER-α. For aggression and for female lordosis behavior, the "yin/yang" principle of Gustafsson regarding the tendency of ER-β to oppose ER-α comes into play. Finally, for ejaculation, the behavioral function simply depends on ER-α. *Second*, looking across all behavioral functions tested, it appears that all possible combinations of requirements for ER-α and ER-β can be realized: dependence on ER-α, on ER-β, on both, on neither, and, finally, on ER-β actually opposing the influence of the ER-α gene product.

2. Thyroid Hormone Receptor Genes

What about nuclear receptors for a class of hormones that are not steroids? Great excitement attended the cloning of thyroid hormone receptors in the labs of Bjorn Vennstrom, at the Karolinska Institute, and Ron Evans at the Salk Institute. Two genes code for thyroid hormone receptor (TR)-α and TR-β. They have different patterns of expression and different ranges of physiological effects throughout the body. Importantly, animals with TR-β gene knockouts have greatly elevated circulating thyroid hormone levels, because the TR-β receptor gene is essential for negative feedback of thyroxine upon the pituitary. Surprisingly, these animals are also deaf. Mice with TR-α-1 knockouts are mildly hypothyroid and have heart defects and an abnormally low body temperature. If mice have both the TR-α-1 and TR-α-2 knocked out, they are severely hypothyroid, have marked retardation, and do not survive even until the age of puberty. In regard to the hormone binding capacity of the splice variants of each gene, both TR-β-1 and TR-β-2 bind triiodothyronine (T3), the most effective thyroid hormone ligand. TR-α-1 does also, but TR-α-2 does not.

Across a set of behavioral assays, there are significant differences among the effects of various types of thyroid hormone gene knockouts. Most striking are the results with lordosis behavior, in which—absolutely against predictions—the effects of an TR-α-1 knockout were the opposite of a TR-β gene knockout (Table 17.1). These effects remain to be explained. Also, against all expectations, the overall impact of these knockouts on tests of higher mental function, including learning and memory, have turned

TABLE 17.1 Summary of Behavioral Phenotypes of αTRKO and βTRKO Male Mice

	αTRKO	βTRKO
Sexual behavior	Increased	No change
Activity		
Open field test	No change	No change
Anxiety		
Elevated plus maze	Increased	Decreased
Dark/light transition	No change	Decreased
Arousal (startle reflexes)		
Baseline	Elevated	No change
Acoustic startle	Reduced	No change
Tactile startle	Reduced	No change
Passive avoidance	No change	No change

TR-α and TR-β gene deletion have differential behavioral effects. (From Vasudevan *et al.*, unpublished data.)

out so far to be mild or nonexistent, representing another problem to be solved.

3. Oxytocin Versus Vasopressin Genes

From an ancestral neuropeptide concerned with fluid distributions in various organs, there evolved two genes (look ahead to Fig. 21.3). Both code for neuropeptide hormones having nine amino acids, differing only in two of them, and both are shaped by an identically located disulfide bond. One gene is the oxytocin gene and the other is the gene coding for vasopressin. Both genes are expressed by neurons in magnocellular groups in the anterior hypothalamus, (*e.g.*, the paraventricular nucleus [PVN] and the supraoptic nucleus [SON]), although only an extremely small percentage of cells express both genes. Both are still concerned with fluid distributions, albeit not in the same organs of the body; however, their roles in a variety of social behaviors are markedly different.

Oxytocin fosters a wide range of behaviors that can only be described as "friendly." Many of them are in the service of reproduction: mating behaviors and parental behaviors. However, a still larger number simply go under the name *affiliative*. These behaviors may be species typical, such as side-by-side sitting by voles. In humans, the ability of oxytocin to cause the individual to *tend and befriend* has been reviewed.

In dramatic contrast, vasopressin frequently has behavioral effects that are almost opposite those of oxytocin. Not only can vasopressin administration increase the probability of frank aggressive behavior, but it can also intensify communicative behaviors that are preludes of aggression. For example, flank marking behaviors that deposit pheromones are enhanced by vasopression, and these are acts of territoriality which, if ignored by the recipient, can lead to a fight. Thus, gene duplication can lead to two products with strikingly different effects on behavior.

B. Immature RNA Splicing Variants

Messenger RNAs coding for progesterone receptors (PRs) can be spliced so that, in addition to the larger sequence (PR-B), there is a smaller variant (PR-A). Both are present in the hypothalamus and pituitary and are induced by estrogens, but not to the same extent. Their transcriptional actions in cell lines, however, are markedly different, as demonstrated by the startling results from the laboratories of Kate Horwitz at Colorado and Donald McDonnell at Duke (Fig. 17.2). Most prominent among the results is the finding that there are molecular chemical conditions under which the transcriptional facilitation operating through PR-B was unable to be mimicked by PR-A, and PR-A would have a transcriptional repressing effect.

How this striking difference plays out in neuroendocrine systems is not yet clear. Leung and colleagues in Vancouver give an early hint. Under conditions where expression of PR-B facilitates transcription through the promoter of the gonadotropin-releasing hormone (GnRH) receptor gene, PR-A expression actually decreases GnRH receptor transcription. If these data hold for neurons, the behavioral implications would be obvious. GnRH tends to facilitate certain aspects of courtship or mating responses in both males and females.

	Spliced form of Progesterone Receptor	
	PR-B	**PR-A**
Supports ligand-dependent transcription?	10 X	1 X
Activation by Anti-Progestin?	Minimal	Absent
Trans-repression of ER-dependent transcription?	No	Yes

FIGURE 17.2. Two splicing variants from the progesterone receptor (PR) gene have significantly different molecular effects. PR-β is the transcriptional facilitator *par excellence*. In contrast, PR-α can repress transcription. (Modified from Horwitz and McDonnell.)

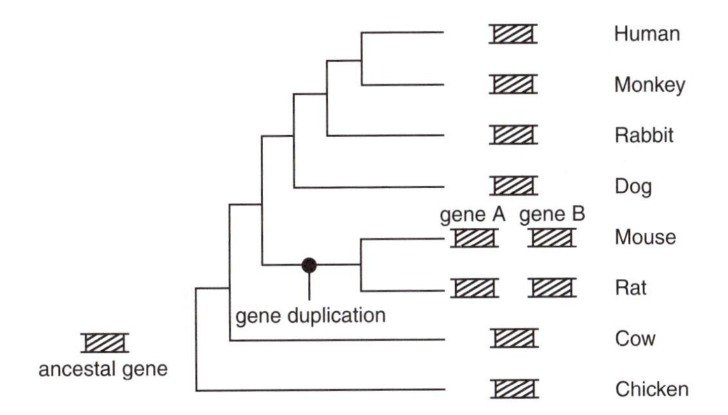

FIGURE 17.3. Evolution and gene duplication of the angiotensin II type T receptor gene produces a variety of receptors in various species. In mice and rats, but not in humans, the AT_1 receptor has two subtypes (AT_{1A} and AT_{1B}) due to gene duplication and subsequent mutation. (From Czelusniak, J. *et al.*, *Methods Enzymol.*, 183: 601–615, 1999. With permission.)

Vigorous transcription of the gene for its receptors in neurons would be a *sine qua non* of its behavioral influences.

C. Glucocorticoid Receptors

From the primary glucocorticoid receptor, now called GR-α, there is spliced a second form, GR-β. GR-β does not bind glucocorticoids such as corticosterone or cortisol, it does not activate transcription in a ligand-dependent manner, and, in fact, it can block the ability of GR-α to activate transcription (Fig. 17.4). From the work of Cidlowski and his colleagues at NIEHS, we see how a single gene coding for Grs can yield several proteins of differing functional capacities. This is just as important for the neuroscientist to consider as it is for the molecular endocrinologist.

D. Androgen Receptors

While only one androgen receptor has been cloned, genetic polymorphisms have been strongly associated with risk of prostate cancer. For example, there is a tract of glutamines the length of which is inversely correlated with tumor size in men. It seems highly likely that genetic variations in androgen receptors, especially in a transactivation domain, would affect the influences of androgens on behavior, especially aggressive behaviors; however, these have not been tested.

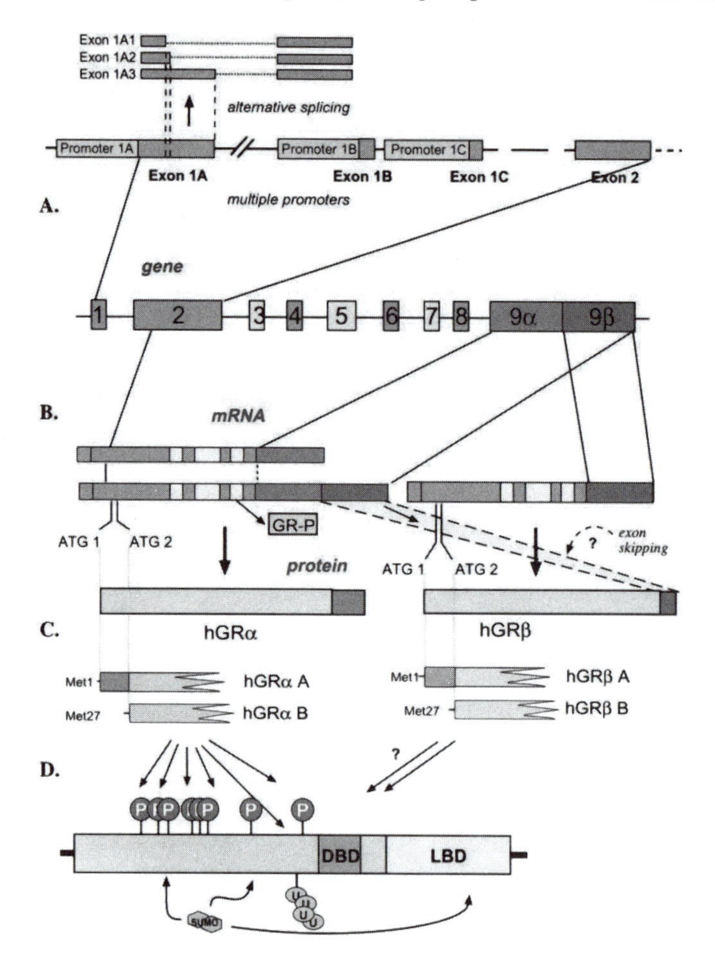

FIGURE 17.4. Several isoforms of the glucocorticoid receptor (GR; binds cortisol, corticosterone, etc.) come from the products of a single gene. The primary GR gene is composed of nine exons; there are three different promoter start sites (A). Alternative splicing of exon 9 at its 5′ end give rise to the GR-α and GR-β families of proteins (B). Putting together the variations in parts (A) and (B), we see at least four proteins in C. Then, posttranslational modifications by phosphorylation (P), ubiquitination (U), etc., of each of these proteins lead to different functional capacities (D). (From Yudt, M.R. and Cidlowski, J.A., *Mol. Endocrinol.*, 16(8): 1719–1726, 2002. With permission.)

E. Receptors for Angiotensin

Angiotensin has evolved two types of receptors: type 1 (AT_1R) and type 2 (AT_2R) (Fig. 17.3). Both are seven-transmembrane, G-protein-linked receptors, but the AT_2R is only 38% homologous with AT_1R. They have different distributions in the body and brain and different functions. AT_1R is the receptor involved in thirst and blood pressure; AT_2R is not. There are several reports that the actions of AT_2R activation have opposite effects from those resulting from stimulation of AT_1R. Yet, they are both highly sensitive to angiotensin II. The gene duplication has gone further in rodents, where two subtypes of AT_1R have been cloned: AT_1a and AT_1b. These are 95% homologous; yet, despite their similarity, recent evidence suggests that they have different functions. (See Chapter 21 for more discussion on gene duplication and multiple gene functions.)

II. CLINICAL EXAMPLES

A. Kallmann's Syndrome

Kallmann's syndrome is hypogonadotropic hypogonadism, often coupled with anosmia. The most striking behavioral problem with men suffering from this syndrome is that they have no libido. This syndrome illustrates not how two similar genes have opposite effects, but rather how disruption of two very different genes can have the same behavioral effect. The story starts with the discovery by Schwanzel-Fukuda and colleagues, in 1989, that the neurons that control mammalian reproduction—those coding for GnRH (also referred to as LHRH)—are not born in the brain, where other neurons are. Instead, they are born in the developing olfactory pit and migrate up the nose, along the bottom of the brain and into the basal forebrain. Strikingly, a human individual with the X-linked Kallmann's mutation at Xp-22.3 displayed a failure of migration. It was not that the GnRH gene was not being expressed normally. It was, but the GnRH neurons had never migrated out of the olfactory apparatus (refer back to Fig. 14.4).

Because the Kallmann's gene (now called anosmin) codes for an extracellular matrix protein evidently connected with GnRH neuronal migration, the entire universe of molecular and endocrine facts permits the following explanation of the behavioral disorder. Men suffering from X-linked Kallmann's syndrome have no libido because they have very low testosterone levels because they have no gonadotropic hormones coming from their pituitary to their testes because they have no GnRH coming from the basal forebrain to their pituitary because the GnRH neurons never make the migration into the brain during development because of a deletion in the Kallmann's gene. These facts, therefore, have yielded the first proof that damage to an individual gene can drastically reduce an important human social behavior. On the other hand, the long causal route discourages simple-minded thinking about such matters.

A second type of cause for Kallmann's syndrome comes from a completely different direction. In 1997, familial hypogonadotropic hypogonadism was reported as resulting from multiple mutations widely distributed across the gene for the receptor for GnRH. Clearly, loss of function of the GnRH receptor could lead to the same downstream endocrine consequences as listed above, including eventual loss of libido. The customary interpretation of GnRH receptor mutations would be that they reduce either the interaction of the receptor with its ligand or the ability of the receptor to trigger signal transduction pathways in the pituitary cell (or neuron). A novel interpretation has been presented by Michael Conn and colleagues in Oregon: Mutations leading to misfolding of the receptor protein and subsequent misrouting in the cell lead to the failure of GnRH signaling. Each of these interpretations, not mutually exclusive, provides an explanation of different forms of Kallmann's syndrome that is entirely different from each other.

B. Resistance to Thyroid Hormones

Another reminder of how small changes in receptor proteins can make large difference in endocrine and behavioral function comes from the work of Refetoff and colleagues, cited earlier (see, for example, Figs. 7.4 and 18.5). Those mutations that lead to a resistance of thyroid hormone effects should have clear behavioral consequences (*e.g.*, see Chapter 1).

III. OUTSTANDING NEW BASIC OR CLINICAL QUESTIONS

1. What can be done to distinguish the helpful effects of estrogen receptor ligands from the harmful effects associated with estrogen-dependent cancers? For the CNS, for example, can ER-β be activated in a way that does not activate ER-α, so that potential emotional and cognitive benefits of estrogenic ligands working through ER-β can be obtained without risk of breast cancer or endometrial cancer?
2. With respect to both PR-β (versus PR-α) and GR-α (versus GR-β), can we see, in neurons, how the transcription-facilitating properties of the main isoform work, as compared to the inhibiting properties of the splice variant? Are there subsequent behavioral effects?

3. It is puzzling that the behavioral phenotypes of thyroid hormone receptor knockouts are as minimal and mild as they are. If there is compensation during development, what is it and how does it work?
4. Failures of parental care are rife in segments of modern industrialized societies. Are there any ways in which an understanding of oxytocin (*vis á vis* vasopressin) actions can either (a) contribute to an understanding of how such lapses come about, or (b) make inroads into the problem?

CHAPTER 18

Hormone Receptors and Other Nuclear Proteins Influence Hormone Responsiveness

One of the outstanding strategic advantages for the neuroscientist studying hormone-influenced behaviors continues to be that nuclear hormone receptors comprise some of the best-studied transcriptional systems in eukaryotic molecular biology. Moreover, steroids are small lipophilic molecules that cross the blood–brain barrier easily and are widely distributed in the brain. Experiments so far have been divided into two parts: determining the molecular biology of the nuclear hormone receptors themselves and investigating other nuclear proteins that go under the names *co-activators* and *co-repressors*.

I. NUCLEAR RECEPTORS AS THEY INFLUENCE BEHAVIOR

The behavioral mechanisms best worked out are those that control the courtship and mating behaviors of female laboratory animals such as rats and mice. First, a significant number of genes have the following two properties: (1) they are turned on by estrogen administration, and (2) their products

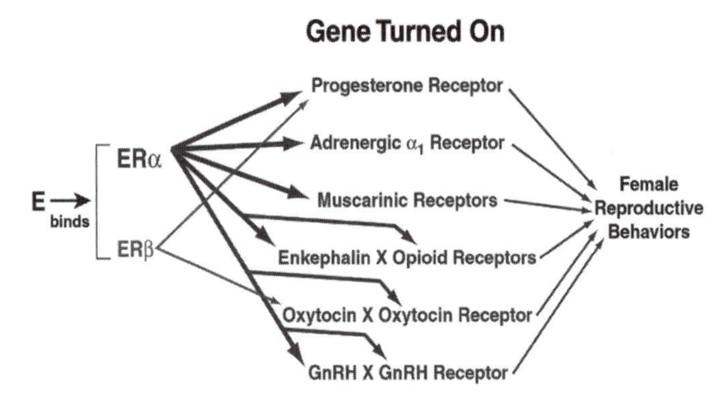

FIGURE 18.1. Estrogens (E), binding to ER-α or ER-β, turn on several genes which, in turn, have been implicated in female sex behaviors. Thinking in the form of a logical syllogism, if E turns a gene on and the gene's product helps drive the behavior, then that genomic effect is one causal route by which the hormone influences the behavior. Two special notes: (a) a biologically sensible formulation of this list of genes is presented in Fig. 18.2; (b) we continue to add to the list of relevant genes by the use of microarrays. (From Pfaff, D.W., *Drive: Neurobiological and continue to Molecular Mechanisms of Sexual Motivation*, MIT Press, Cambridge, MA, 1999. With permission.)

foster female reproductive behaviors (Fig. 18.1). A list of genes such as that shown in Fig. 18.1, however long it took to achieve, immediately leads to at least two new questions. First, is the list limited by the reagents available? Yes, it is, but this problem has now been solved by the use of microarray technology (see below). Second, can the genes be conceived in a biologically sensible fashion? Yes, they can. Even though the genes in Fig. 18.1 have very different biochemical roles, their routes of action in the organization of estrogen-dependent sex behaviors can be organized into modules that make clear biologic sense (Fig. 18.2). The several molecular steps they serve can be summarized as follows: Growth of ventromedial hypothalamic neurons, amplification of the estrogen effect by progesterone, preparation for reproductive behaviors, permissive actions on hypothalamic neurons at the top of the neural circuit for lordosis behavior, and synchronization with ovulation (these steps thus forming the acronym GAPPS).

Within this set of molecular mechanisms, how do ER-α and ER-β share the load? From studies with female mice in which the genes for either ER-α or ER-β, or both types of receptors, have been knocked out (see Chapter 17), the following seems clear: Estrogens working through ER-β foster social recognition and affiliative behaviors that help to bring reproductively competent females and males to the same place at the same time (refer back to Fig. 3.5). As well, ER-β is responsible for reducing aggression, at least by male mice, so that the animals will "make love, not war." ER-α mediates

MODULAR SYSTEMS DOWNSTREAM FROM HORMONE-FACILITATED TRANSCRIPTION RESPONSIBLE FOR A MAMMALIAN SOCIAL BEHAVIOR: "GAPPS."

- ***Growth*** (rRNA, cell body, synapses).
- ***Amplify*** (pgst/PR →→ downstream genes).
- ***Prepare*** (indirect behavioral means; analgesia (ENK gene) and anxiolysis (OT gene)).
- ***Permit*** (NE alpha-1b; muscarinic receptors).
- ***Synchronize*** (GnRH gene, GnRH Rcptr gene —synchronizes with ovulation).

FIGURE 18.2. The GAPPS systems for estrogen actions on sex behaviors. Estrogenic effects on neuronal growth (G) increase signaling capacity in the hypothalamic neurons that control lordosis. By facilitating transcription of progesterone receptors, estrogens arrange for their own amplification (A). Estrogens can prepare (P) the female for strong stimulation from the male by enhancing transcription of an opioid peptide gene (enkephalin, ENK) and can also prepare for the anxiety of courtship behaviors in an open environment by increasing transcription of oxytocin and its receptor. At the top of the lordosis behavior circuit, estrogenic effects on neurotransmitter receptors increase electrical activity in the relevant hypothalamic neurons, thus permitting (P) the rest of the circuit to operate. Finally, it is biologically adaptive in a small animal that is prey to synchronize (S) sex behavior with reproduction—estrogenic effects on gonadotrophin releasing hormone (GnRH) and its receptor has this effect. (From Mong, J.A. *et al.*, *Molecular Psychiatry*, 2004. With permission.)

the increased locomotor activity due to estrogenic action in the preoptic area. Such locomotion is an essential part of the courtship behaviors (*i.e.*, direct approaches by the female to the male) that in turn increase the likelihood of successful mating behaviors. Then, ER-α is absolutely essential for the primary copulatory behavior (lordosis) of the female. Later, ER-α subserves the positive effects of estrogens on maternal behavior. Thus, ER-α is closely tied to an entire chain of behavioral steps essential to the nitty gritty of reproductive physiology. This makes good biologic sense, because (1) the estradiol itself signals the readiness of the developing follicles to release eggs, and (2) ER-α also participates in the female's ovulation. Therefore, the same gene coordinates the endocrine and behavioral aspects of reproduction, helping to ensure that the female will not risk predation or waste energy in courtship and mating behaviors if fresh gametes (*i.e.*, eggs) would not be available for fertilization. Further, this set of roles for ER-α helps to satisfy the demand of the late, great British neurophysiologist, Sir Charles Sherrington, that the neurobiologist should account for the flow of behavioral responses through time.

Despite major advances in understanding how estrogen receptors, androgen receptors (ARs), glucocorticoid receptors (GRs), mineralocorticoid receptors (MRs), and thyroid hormone receptors (TRs) contribute to behavioral mechanisms, we have much more to learn. For example, the proper functioning of nuclear hormone receptors in cells depends upon proper folding of all parts of the protein, not just in the ligand-binding domain but also in the transactivation domains (proven important for hormone-dependent facilitation of transcription) at the N-terminal of the proteins and at the C-terminal of the proteins. Proper folding in many cases allows for high-affinity binding of the behaviorally relevant hormone, but, just as important, is essential for interactions with co-activator proteins in the nucleus.

A. Microarray Technology

While many advances revealing molecular mechanisms in the hormone/behavior field have come from the candidate gene approach, lists of genes as provided earlier may be limited by a lack of imagination or a lack of reagents. Microarray technology has the advantage that large numbers of transcripts can be screened at one time, without reference to the experimenter's theoretical biases (Mong *et al.*, 2002). For example, estrogen acting in the preoptic area was, surprisingly, discovered to downregulate mRNA levels for prostaglandin D-synthetase (PgDS), a transcript found mainly in glial cells with a product that promotes sleep (Mong *et al.*, 2003). This discovery would explain how estrogens, working through preoptic cells, as they are known to do, could enhance brain arousal and locomotor activity.

II. OTHER NUCLEAR PROTEINS: CO-ACTIVATORS AND CO-REPRESSORS

Nuclear hormone receptors do not sit naked upon the DNA of neurons and glia, thus facilitating or repressing transcription; instead, they participate in assemblies of substantial numbers of proteins that mediate their genomic effects. Such nuclear proteins either foster or block the formation of a complex bridge between the hormone-dependent enhancer sequence and the basal transcriptional machinery. Under circumstances where those nuclear proteins enhance hormone action, they are referred to as *co-activators*; when they block transcription, *co-repressors*. The nuclear protein assemblies surrounding hormone-influenced enhancers are so complex that it appears a combinatorial code among them eventually determines the sensitivity of a particular cell to that hormone at that time (Fig. 18.3). Co-activator functions have now been described for a number of major hormonal systems. In addition, they have been drawn into explanations of hormone actions on behavior.

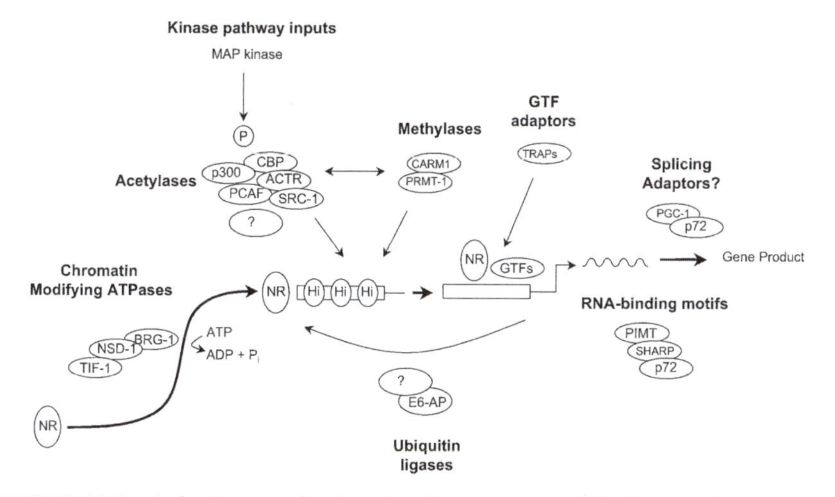

FIGURE 18.3. A dominant mode of nuclear hormone receptor facilitation of transcription is to bind to DNA response elements; however, a large number of nuclear proteins other than receptors are necessary to complete the reaction. The proteins that surround and shape the DNA (chromatin) can be modified by ATPases, acetylases, kinases, and methylases. Further, especially for hormone receptor response elements on gene promoters far from the transcription start site, adaptor proteins may be necessary to bridge the liganded hormone receptor to general transcription factors (GTFs). Because all of these complexities are presumed to be just as important in nerve cells as in other hormone-responsive cells, they all provide mechanisms for developmental and environmental events to alter brain tissue sensitivity to any given hormone. (From McKenna, N.J. and O'Malley, B.W., *J. Steroid Biochem. Mol. Biol.*, 74(5): 351–356, 2000. With permission.)

A. Estrogens

More than ten nuclear proteins have been shown to influence the efficacy of estrogens bound to ER-α in facilitating transcription in a wide variety of cell types. It is exciting to see that different estrogens impose different requirements upon the interactions between the estrogen receptor and the gene's promoter for co-activators to do their job. The precise nucleotide sequence on a particular gene's promoter influences the recruitment of co-activators to the estrogen receptor and thus the ability for a particular estrogen to be effective. Moreover, the protein structural bases for these phenomena are partly understood. That is, pure estrogens cause a conformational change in estrogen receptors such that a specific part of this protein, helix 12, is available for co-activators to bind. An estrogen antagonist, of the sort used to retard the growth of breast cancer, causes a different conformational shift such that the essential co-activator binding surface (helix 12) is blocked.

The nuclear protein story is not limited to positive interactions. A negative regulatory surface within ER-α has been identified by McDonnell and colleagues at Duke University. ER is known to interact with at least two types of co-repressors: NcoR and SMRT. For example, mutation of a single DNA codon, changing a leucine into an arginine within the estrogen receptor, abolished the ability of the estrogen receptor to interact with NcoR and dramatically increased the transcriptional activity following estradiol administration. Thus, molecular endocrinologists now believe that estrogenic signaling in cells includes the regulated relief from repression.

The importance of co-activator and co-repressor proteins for hormone action is clear. One example is breast cancer cells. Antisense sequences, composed of DNA sequences designed to bind specific messenger RNAs and thus block the function of specific genes, were constructed to target each of three co-activator gene products. Treatment of the breast cancer cells with two of them (against SRC-1 and TIF2) inhibited estrogen-dependent DNA synthesis and cell proliferation; yet, treatment with antisense DNA against the third, SRA, did not (Smith and O'Malley, 1999), thus demonstrating that cellular effects were co-activator-specific.

B. Androgens

Androgens affect transcriptional activity of specific genes in a large number of tissues, including the brain. Androgen receptor effectiveness depends on a large number of nuclear proteins. This set of proteins overlaps with, but is not identical to, the set of proteins important for ER action. Some of these proteins are thought to modify the conformation of the androgen receptor itself, while others work indirectly by acting on the protein that coats the DNA, chromatin. Both the amino terminus and the carboxyl terminus of androgen receptors are required for the receptor to interact with co-activators. Likewise, specific sequences with the co-activators have been identified as crucial for their interactions with nuclear receptors. For example, Wilson and French, at the University of North Carolina, deleted portions of the co-activator ARA-70 and found a 145-amino-acid domain effective for assisting transcriptional activation of androgen receptors. In fact, one of the mechanisms for these interactions has been revealed using three-dimensional imaging methods. The co-activator CBP helps an estrogen receptor to form specific spatial foci within the cell nucleus, a physical distribution of nuclear receptor shown by Gordon Hager and colleagues at NIH to be closely associated with transcriptional effectiveness.

Co-repressors play a part in the regulation of androgenic hormonal signaling, just as surely as they do with estrogens. Treuter *et al.* at the Karolinska

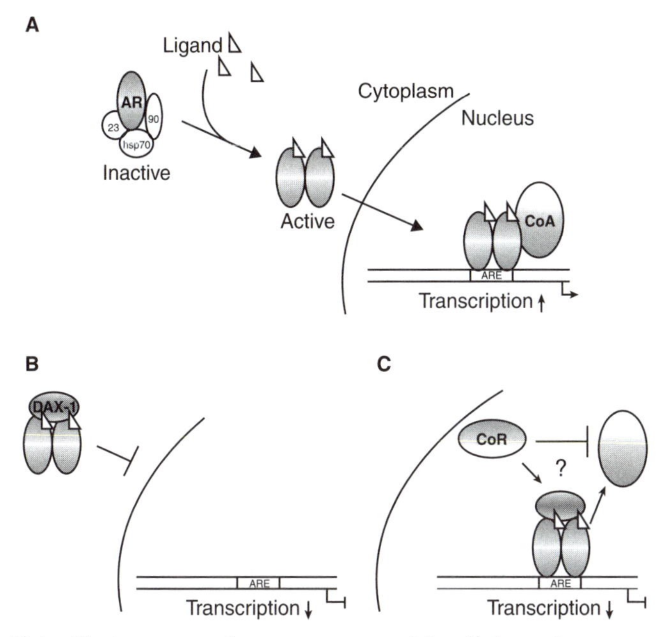

FIGURE 18.4. The importance of co-repressors; a model applied to androgen receptors (ARs) based on current molecular data. (A) Normally, the inactive AR surrounded by heat-shock proteins in the cytoplasm will be activated by the hormonal ligand (dropping off the heat-shock proteins) and will be translocated to the nucleus. There, AR not only binds to AREs on DNA but also achieves the required bonding to nuclear proteins called co-activators, thus facilitating transcription. (B) How does the protein DAX-1 antagonize androgenic activation of transcription? One possible way is to repress binding to AREs on DNA. (C) A second way in which the protein DAX-1 could work is to call in co-repressors in the nucleus, which will act either on AR itself or on the co-activators. (From Holter, E. *et al.*, *Mol. Endocrinol.*, 16(3): 515–528, 2002. With permission.)

Institute identified the target on androgen receptors for the co-repressor DAX-1 and then, correspondingly, mapped the DAX-1 domain involved in the interactions with androgen receptors. It was interesting and novel that initiation of this particular interaction does not occur exclusively in the cell nucleus (Fig. 18.4).

C. Other Hormonal Systems

While a tremendous amount of information has accumulated about the roles of nuclear co-activators and repressors in sex hormone signaling, the same principles seem to apply to other hormones with nuclear receptors. Most

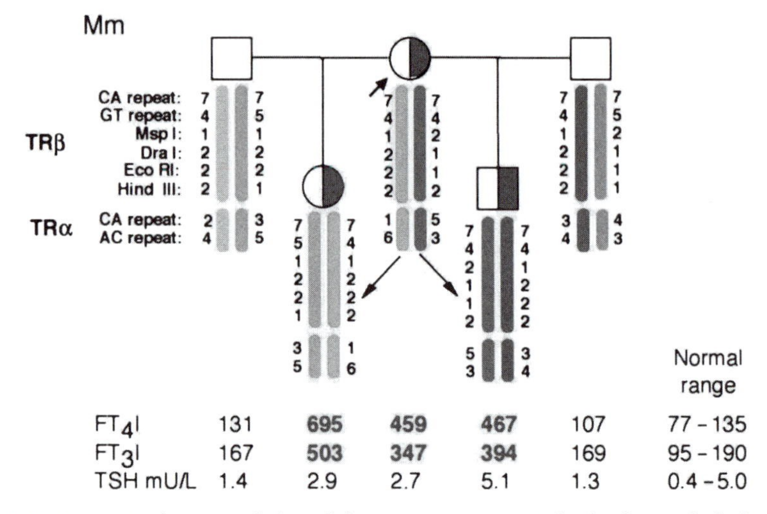

FIGURE 18.5. Haplotypes and thyroid function tests in part of a family in which there is familial resistance to thyroid hormones, but no mutation in the coding region of the thyroid hormone-beta receptor gene. The high levels of thyroid-stimulating hormone (TSH) and thyroxine and triiodothyronine (FT4I and FT3I) (for three affected individuals, see the middle three columns) are due to a lack of neuroendocrine negative feedback, which normally is mediated through the TR-β gene product. Because this gene has no apparent mutation in its coding region, either there may be a mutation elsewhere in the gene, or some of the problem may be due to mutations in other nuclear proteins, or both. Because of the effects of thyroid hormones on mood and cognition, these nuclear protein chemistries are of likely importance to hormone/behavior relations in the affected patients. (From Pohlenz, J. *et al.*, *J. Clin. Endocrinol. Metab.*, 84: 3919–3928, 1999. With permission.)

impressive are the data on thyroid hormones. On the one hand, TRAP-220 (thyroid-receptor-associated protein with an apparent molecular weight of about 220 Da) and SRC-1 (steroid receptor co-activator-1) can enhance the transcriptional activating ability of thyroid hormones. On the other hand, N-CoR (nuclear co-repressor) causes the unliganded TR-β to repress basal transcriptional activity of target genes. Thus, relief from this repression upon thyroxine binding to TR-β would elevate the transcriptional rate.

Two special points must be made about these thyroid hormone receptor co-regulators. First, vitamin D also works through a nuclear receptor. An amazing overlap of the TRAPs and nuclear proteins associated with vitamin D has emerged from the work of Roeder at Rockefeller University and Freedman at Memorial Sloan Kettering Cancer Center. That is, some of the vitamin D receptor interacting proteins (DRIPs) are identical to some of the TRAPs. While these similarities may derive from the near-identity of their DNA response elements, many other explanations are also possible. Second, in regard to the clinical importance of these co-regulators, the syndrome of resistance to thyroid hormone (elucidated by Samuel Refetoff at the University of Chicago) is associated with specific modifications in thyroid hormone receptor genes. These mutations might well have their pathogenicity in disturbed relationships between thyroid hormone receptors and nuclear co-activators and repressors. In fact, over a period of several years, Refetoff has found a substantial number of families with inherited resistance to thyroid hormones but who do not have mutations in TR-α or TR-β genes (Fig. 18.5). An ongoing search for abnormalities in nuclear co-repressors or co-activators associated with thyroid hormone receptors may uncover the mechanisms of hormone resistance in these patients.

D. Studies with Behavioral Measurements

Most exciting are those studies that approach the question of whether nuclear receptor co-activators are actually important for behavior. First among these was a series of experiments from the lab of Margaret McCarthy, at the University of Maryland Medical School. She and her colleagues worked with CBP (cAMP-response-element-binding protein). Early in postnatal life, the male hypothalamus has more of this protein than does the female hypothalamus. Does this protein participate in defeminization of the brain? That is, might it participate in the causal route by which testicular androgens, impacting the brain neonatally, affect later behavior by the male? Antisense DNA oligodeoxynucleotides directed against the mRNA for CBP were infused directly into the hypothalamus of neonatal female rats, which also received testosterone injections. Blocking the gene function of CBP interfered specifically with the defeminizing actions of testosterone on adult sex behavior (Fig. 18.6).

A powerful co-activator is steroid receptor coactivator-1 (SRC-1), which influences the transcriptional effectiveness of many nuclear hormone receptors. Apostolakis *et al.*, at Baylor College of Medicine, used the same type of antisense DNA technique as the McCarthy laboratory, but they did their infusion in the third ventricle of the adult female rat brain. Acute disruption of the function of the SRC-1 gene, but not other similar gene products, inhibited the ability of estrogens working through ER-α to turn on lordosis behavior (Fig. 18.7).

These pioneering studies have established beyond doubt that nuclear co-activator proteins participate within hypothalamic neurons to influence reproductive behaviors.

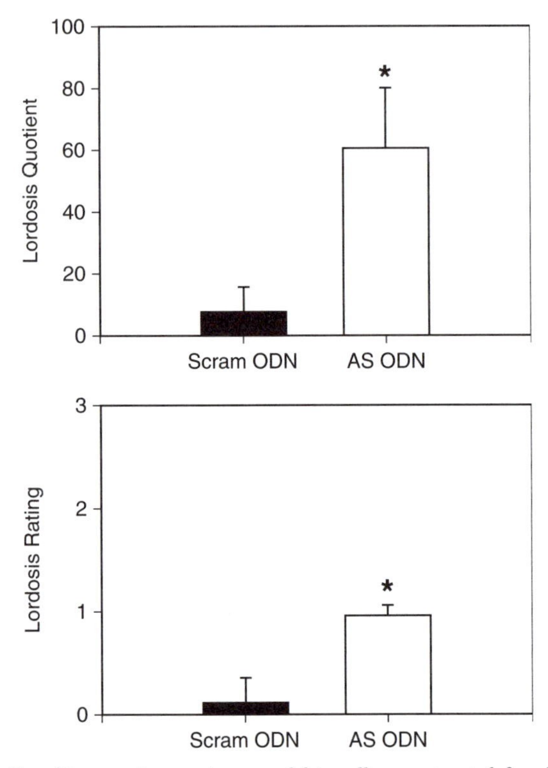

FIGURE 18.6. For this experiment, Auger and his colleagues treated female rats neonatally with testosterone so that they would not show the female-typical sex behavior, lordosis, when they were adults. Rats treated intracerebrally with antisense DNA oligomers interfering with mRNA for the nuclear protein CBP (CREB-binding protein) had higher lordosis quotients and ratings than control rats treated intracerebrally with scrambled DNA sequences; therefore, the nuclear protein CBP must be essential for the suppressive effects of testosterone on female sexual behavior. AS ODN = Antisense oligodeoxynucleotides; Scram = scrambled sequence controls. (From Auger, A.P. *et al.*, *Endocrinology*, 143(8): 3009–3016, 2002. With permission.)

III. SOME OUTSTANDING QUESTIONS

1. Under what circumstances do the nuclear receptors, even without hormones bound, influence behavior?
2. Microarray technology is revealing that large numbers of new genes are hormone-responsive in the forebrain. Do their transcript levels actually influence nerve cell activity and hormone-dependent behaviors?

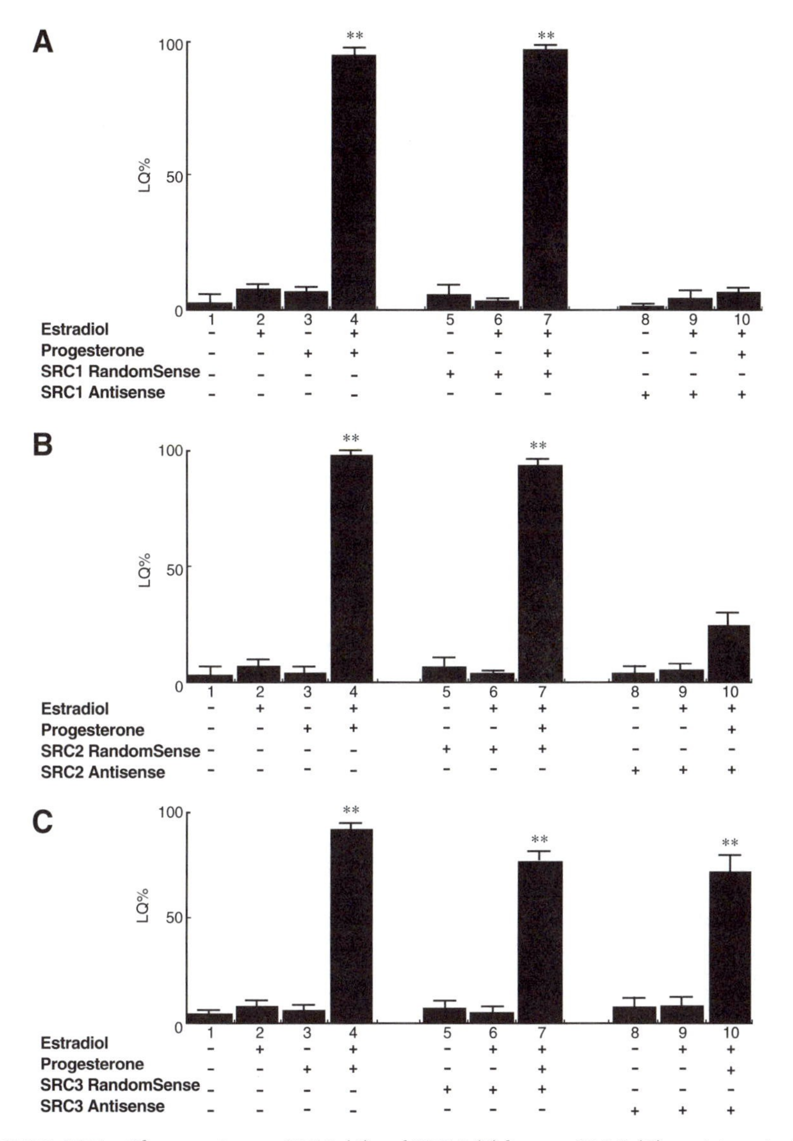

FIGURE 18.7. The co-activators SRC-1 (A) and SRC-2 (B) but not SRC-3 (C) participate in the facilitating effects of estradiol and progesterone on lordosis behavior (measured by the *lordosis quotient*, LQ%). Their participation is shown by the ability of antisense oligomers against SRC-1 or SRC-2 mRNA to block E+P-facilitated lordosis behavior. (From Apostolakis, E.M. *et al.*, *Mol. Endocrinol.*, 16(7): 1511–1523, 2002. With permission.)

3. How do rapid hormone effects (see Chapter 16) operate together with (or against) nuclear, genomic effects to influence behavior?
4. For each hormone that has effects on behavior, what are the crucial co-activator assemblies that allow the cognate nuclear receptor to be effective?
5. In the CNS, in cells contributing to hormonal influences on behavior, how do alterations in receptor protein folding affect neuronal (or glial) sensitivity to hormones?
6. At what point do co-repressors come into the CNS story and play an important role in the behavioral effects of hormones? That is, in the history of behavioral endocrinology, two major influences were studied: hormone levels and "tissue responsiveness" to those hormones. Perhaps co-repressor levels help to determine the latter.
7. The discovery of hormone effects on PgDS mentioned in this chapter raised the possibility of hormone-dependent neuronal/glial cooperation. Can this idea be proven?
8. How can changes in genomic structure exerted by hormones be brought into the analysis of CNS mechanisms? Over the long term, perhaps developmentally, nucleotide methylation; over the short term, histone acetylation or phosphorylation. Indeed, these molecular states may influence the tissue responsiveness to the hormones mentioned above.

SECTION VI

Environment: Environmental Variables Influence Hormone/Behavior Relations

Hormone Effects on Behavior Depend Upon Context

I. BASIC EXPERIMENTAL EXAMPLES

Even in lower animals, for which the richness of experience and the appreciation of environment should be much diminished compared to humans, the magnitude of hormone effects on particular behaviors can be affected by the context in which the hormone action is tested. In a fish that can be collected in Canadian rivers, the black goby, an androgenic hormone can induce male-typical ventilatory behaviors, but the hormone effect is strong only in the presence of certain pheromones thought to emanate from females (Fig. 19.1). In a different kind of fish, the brown ghost knife fish, communications among animals include weak electric discharges that signal aggression. Dunlap and colleagues found that male–male social interactions among these fish cause a rise in the blood levels of the stress hormone cortisol. In turn, that stress hormone is correlated with higher rates of electric discharges during aggressive bouts.

In mammals accessible for laboratory study, it is widely recognized that the promptness and avidity with which the male initiates mating behaviors

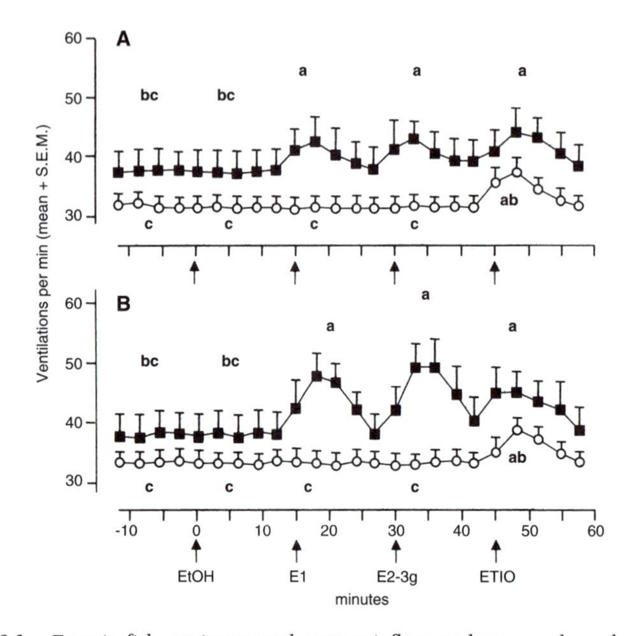

FIGURE 19.1. Even in fish, environmental context influences hormone-dependent behaviors. These experiments measured the ability of methyltestosterone to induce male-like behavior in female round gobies. Experimenters counted the ventilation response by females treated subcutaneously with methyltestosterone (black squares) or blank capsules (open circles). (A) Measurements 8 to 10 days after implantation. (B) Measurements 16 to 18 days after implantation. While the animals did not respond to the control (ethanol, EtOH), they did respond to estrogens in their water environment—E1 (estrone), E2 (estradiol-glucoronide), and, to a lesser extent, etiocholanolone (ETIO). (From Murphy, C.A. and Stacey, N.E., *Horm. Behav.*, 42(2): 109–115, 2002. With permission.)

with the female depend on how familiar the male is with his environment. If a receptive female is placed in the male's home cage, mating happens quickly. If both animals are put in a strange environment, the male will spend a lot of time exploring before attempting to mount the female. One way of explaining this is that there exists an optimal level of arousal for reproductive responses (Fig. 19.2). Obviously, a soporific animal is too little aroused for sex, but an animal too upset by frightening surroundings may be too aroused for optimal mating behavior. How would generalized arousal impact sexual arousal? A potential mechanism for this has been discovered by L.-M. Kow, at the Rockefeller University. Histamine is a neurotransmitter that drives arousal of brain and behavior. The highest levels of electrical activity in hypothalamic neurons that govern sex behavior in female rodents were achieved by histamine in the presence of estrogenic hormones. Conversely, an anesthetic

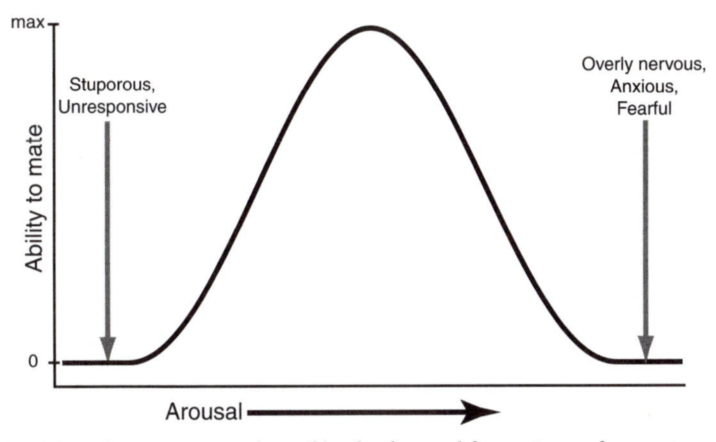

FIGURE 19.2. There is an optimal set of levels of arousal for mating, as for a variety of other emotional and cognitive behaviors. Many causes of changes in arousal are environmental or contextual. A person could be not aroused enough (stuporous) or could be too aroused, even fearful. This figure constitutes a version of the classical Yerkes–Dodson law, applied to the physiological and molecular analysis of natural behaviors.

suppressed the estrogen effect, even as it suppresses sex behavior. The combination of sex hormone and arousal-related neurotransmitter appears to turn on hypothalamic neuron electrical activity and to turn on female sex behavior.

Likewise, the degree of mating experience influences the intensity of sex behavior in many male lab animals. A highly experienced stud male with normal testosterone levels will mate rapidly and vigorously, whereas a less experienced male rat, for example, will be correspondingly more tentative even if his testosterone levels are also in the normal range. Roger Gorski and coworkers at the University of California, Los Angeles, have even found that sex partner preferences displayed by female rats are influenced by previous sexual experience.

Social environment is important. Stephen Glickman and co-workers, at the University of California, Berkeley, studied the frequencies of playful physical interactions in female and male hyenas as a function of grouping by gender. In all female groups, interactions were high. In all male groups, interactions were low. After putting the females together with the males, they found that the new social context reduced frequencies in the females, but increased them in the males. As a result, the sex difference disappeared.

Hormones associated with reproduction have behavioral effects far beyond sex. For example, Catherine Woolley, at Northwestern University, discovered that estrogens could influence the growth and architecture of dendrites in the hippocampus, a brain region associated with memory. Importantly, the effects of estrogens on hippocampal morphology depend not only on glutamatergic neurotransmission, but also upon inputs from basal forebrain cholinergic neurons (Fig. 19.3).

Other factors besides hormones are involved in the interactions between reproductive state and stress. James Pfaus, in Montreal, imposed an inescapable stress on female rats in particular hormonal states. As a result, they subsequently displayed anticipatory anxiety and also had greater expression of an early response gene, *fos*, in the amygdala, a brain region associated with fear and anxiety. These responses occurred in greater magnitude in the absence of ovarian sex hormones. The hormonal state interacted with the expectancy of stress in order to determine the level of stress reactivity.

In regard to simple locomotor behaviors, the effects of estrogens depend on context. Maria Morgan, at Rockefeller, found that, under safe conditions, estrogens increased running by female rats. But, in conditions associated with fear or anxiety, estrogens had the opposite effect. She interpreted the results as follows: Estrogens increase arousal, which (1) supports motor activity in the absence of danger, but also (2) heightens fear or anxiety when clear environmental signals for fear are present. George Wade's laboratory at the University of Massachusetts has uncovered a potential mechanism for suppression of behaviors essential for reproduction, which could include courtship-related locomotor behaviors as well as frank sex behavior. Newly discovered neuropeptides for stress responses, the urocortins, bind to receptors for corticotropin-releasing hormone (CRH), also essential to the stress response. These urocortins inhibit sexual receptivity in female hamsters, as they provide a specific neurochemical route by which frightening contexts can block the effects of sex hormones on mating behaviors.

Analyses of genetic influences on behavior have revealed clear specificities of gene/hormone/behavior relations that illustrate the effects of context. The increased aggressiveness of female mice with a knockout of the gene coding for oxytocin depends on the level of stress in the environment; the context of a stressful environment brings out the full oxytocin phenotype (see Chapters 1, 2, and 14). Likewise, the effect of the gene for an estrogen receptor (ER-β) on aggression is a function of the situation in which aggression is being assayed. Hormone–gene interactions depend exquisitely upon details of the experimental protocol. Sonoko Ogawa has shown that while testosterone-fueled fighting is reduced by the loss of a functional ER-β gene, many aspects of maternal aggression (*e.g.*, defending the nest) give the opposite result (Fig. 19.4), namely, increased maternal aggression.

Even as environment influences gene/hormone/behavior relations, there are strong environmental dependencies also evident in the control of hormone-dependent genes. Consider the conditions that control oxytocin gene expression. As noted previously, physiological conditions controlling oxytocin gene expression include development, puberty, plasmid osmolality, and circadian rhythms; therefore, the developmental context, the amount of water and salt in the environment, and the time of day all can influence gene expression for this neuropeptide, which is of such behavioral importance. Likewise, for expression of the opioid peptide gene encoding enkephalin, gender and hormone state interact with stress level (dependent on the environment, of course) to determine the amount of enkephalin transcript in hypothalamic neurons.

A. Seasons of the Year

An entirely different approach to how context can affect hormone/behavior relations is to think of all the environmental changes accompanying the different seasons of the year and offering contextual changes. Introduced in Chapter 13, these seasonal rhythms obviously depend on light, temperature, and melatonin signals. Ferkin and Zucker, at the University of California at Berkeley, showed the power of seasonal context by testing female meadow voles for their odor preferences. During the spring/summer breeding season, the voles prefer odors from males over those of females, but not during the winter. Experimentally, Ferkin and Zucker showed that in long photoperiods the females preferred male odors, just like during the summer. In contrast, during short photoperiods (only 10 hours of light per day), that preference was reversed. During these short photoperiods, not even estrogenic supplementation could reinstate the preferences for male odor stimuli, so the environmental signals truly had altered hormone/behavior relations.

II. CLINICAL EXAMPLES

David Rubinow and colleagues (*Hormones, Brain and Behavior*, 2002) found that the relations between hormones and mood in human subjects depend very much on context (Fig. 19.5). Their conclusions were drawn from studies in a clinical setting. In a more general social context, sex hormones, as they affect aggression in teenage boys, interact with such social factors as socioeconomic status, school size, and social environment (Devine *et al.*, 2004) (see also Chapter 10). Huge discrepancies in socioeconomic status (thus increasing the chance of personal humiliation of an adolescent boy), huge schools (thus increasing the sense of anonymity), and an absence of formal initiation rites (thus failing to provide a positive view of a boy's role in adult society) all conspire to raise the chance of violent behavior.

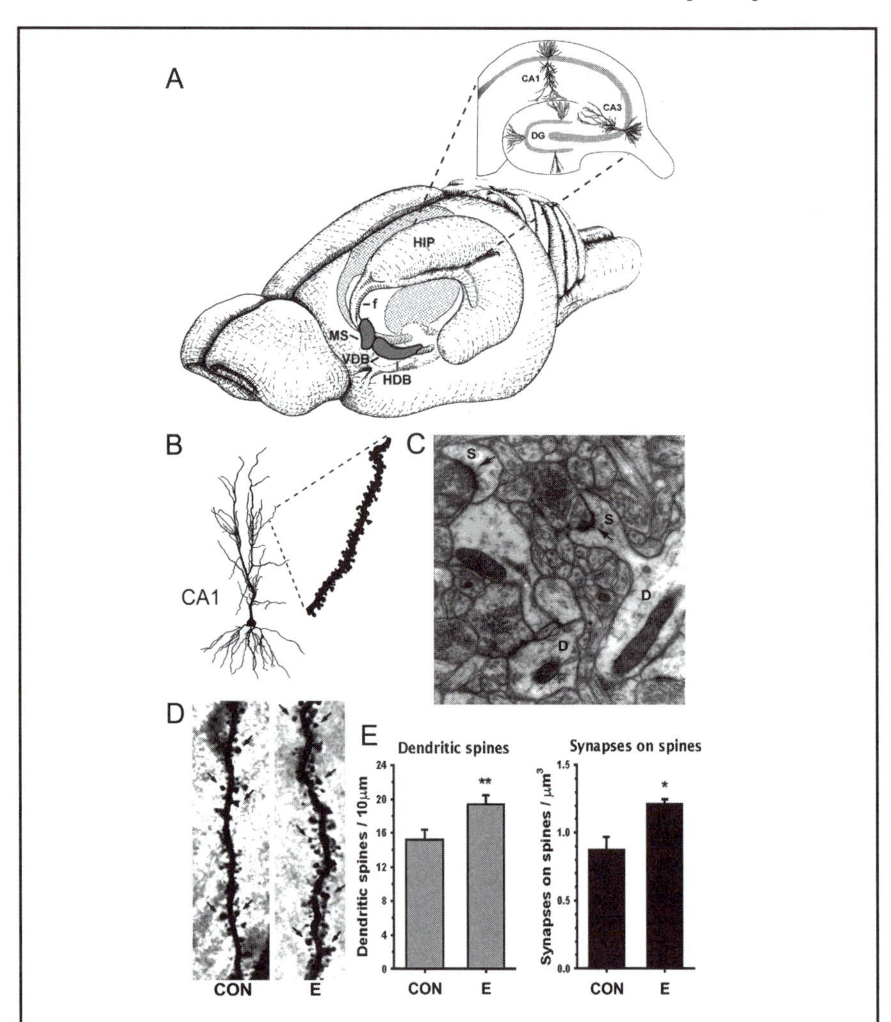

FIGURE 19.3. Hormonal state regulates excitatory synaptic connections in the hippocampus of adult female rats. (A) The hippocampus in the rat is a large, curved structure that is oriented septo-temporally just beneath the corpus callosum. In cross section, three types of principal cells are apparent: pyramidal cells of the CA1 and CA3 regions and granule cells of the dentate gyrus. Subcortical brain regions, such as the medial septum (MS) and the vertical and horizontal limbs of the diagonal band of Broca (VDB, and HDB, respectively) in the basal forebrain, project to the hippocampus via the fornix (f) and the fimbria, which extends as a sheet of fibers along the surface of the hippocampus. (B) The dendrites of each of the principal cell types in the hippocampus are densely studded with small protrusions termed *dendritic spines*. A CA1 pyramidal cell is shown with its dendrites covered by spines. Each CA1 pyramidal cell may have as many as 20,000 dendritic spines covering its dendrites. (C) Dendritic spines are the sites of the vast majority of excitatory synaptic input to CA1 pyramidal cells. An electron micrograph is shown in which the arrows indicate asymmetric (excitatory) synapses on dendritic spines (S) that arise from

	Wildtype Control Females	β-Estrogen Receptor Gene Knocked out
Postpartum Maternal Aggression	Normal	**Increased**
Testosterone-induced Aggression	Normal	**Reduced**

FIGURE 19.4. The effect of a gene disruption on a class of natural behaviors can depend on the context in which those behaviors are assayed. Here, knocking out the ER-β gene increases one form of aggression but reduces another form.

Figure 19.3. (*Continued.*)
the shafts of dendrites **(D)**. In adult female rats, removal of endogenous ovarian hormones by ovariectomy decreases the density of dendritic spines and synapses formed on spines on CA1 pyramidal cells; these decreases can be either prevented or reversed by systemic treatment with estradiol (Woolley and McEwen, 1992, 1993). The density of spines and synapses also fluctuates naturally as estradiol and progesterone levels fluctuate across the estrus cycle; spine/synapse density are greatest during proestrus, when estradiol and progesterone levels are high, and lowest during estrus, when estradiol and progesterone levels are low (Woolley *et al.*, 1990). (D) Light micrographs of CA1 pyramidal cell dendrites from a control (CON) rat that was ovariectomized and treated with oil vehicle and from an ovariectomized animal that was treated with systemic estradiol (E). Note that the density of dendritic spines is greater in the E-treated animal. **(E)** Quantitative analysis of dendritic spine and spine synapse density in CON and E-treated rats. Spine and synapse density each are ~25% greater in E-treated animals. Because each CA1 pyramidal cell has tens of thousands of dendritic spines, a 25% difference represents thousands of additional spine synapses per CA1 pyramidal cell in E-treated animals. Electrophysiological and behavioral studies reveal that the estrogen-induced increases in spine/synapse density on CA1 pyramidal cells are associated with increased sensitivity of CA1 pyramidal cells to excitatory synaptic input (Woolley *et al.*, 1997; Rudick and Woolley, 2001) and improved spatial working memory (Sandstrom and Williams, 2001; Daniel and Dohanich, 2001), a memory function known to depend on the hippocampus.

Estrogen-regulation of hippocampal synapses depends on the state of neural circuitry: Estrogen-regulation of synapses is activity dependent and involves both excitatory glutamate receptor activation (Woolley and McEwen, 1994) and suppression of inhibitory GABAergic neurotransmission (Murphy *et al.*, 1998). Additionally, estrogen regulation of synapses in the hippocampus requires input from subcortical brain regions (Leranth *et al.*, 2000) and can be blocked or mimicked by manipulations of the basal forebrain cholinergic system (Rudick *et al.*, 2002; Lam and Leranth, 2002). Thus, estrogen's effects in the hippocampus require coordination of multiple neural factors; that is, they depend on the state of neural circuitry. The state of neural circuitry, in turn, is influenced by experience.

Behavioral state interacts with hormonal state in regulation of hippocampal circuitry: Consistent with the concept that the state of neural circuitry plays a role in hormone dependent hippocampal plasticity, one recent study has demonstrated that stressful experience interacts with phase of the estrus cycle in regulating hippocampal dendritic spine density in female rats (Shors *et al.*, 2001). For example, a stressful experience preceding the proestrus phase of the cycle by 24 hours inhibits the normal rise in spine density that occurs at proestrus. Interestingly, stress in rats has been shown to influence neurons in the basal forebrain (*e.g.*, Krukoff and Khalili, 1997), which may provide a mechanistic link between this behavioral state and hormonal regulation of hippocampal synapses.

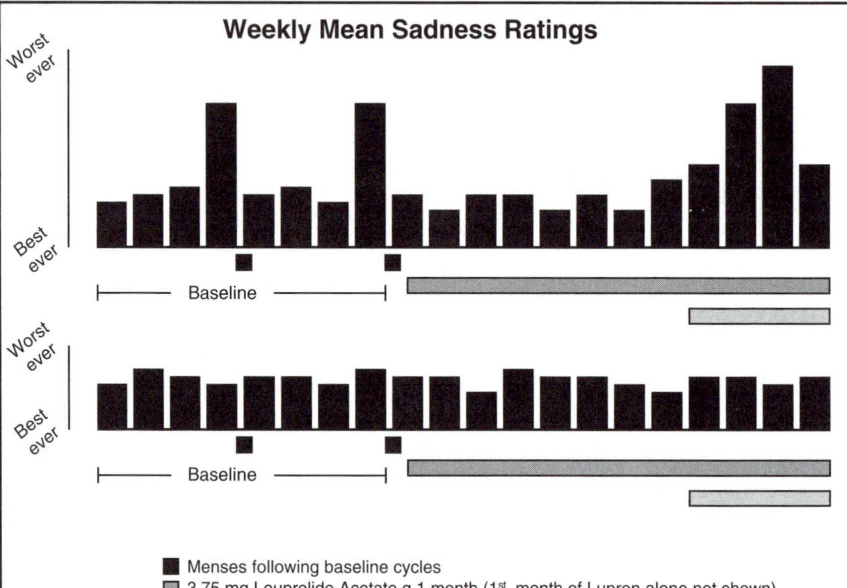

FIGURE 19.5. Ten women with premenstrual syndrome (see baseline cycles) and 15 controls had minimal mood and behavioral symptoms during Lupron administration. In contrast, women with premenstrual syndrome but not the controls, had a significant increase in sadness during either estradiol (E2) or progesterone (P4) administration. Histograms represent the mean of the seven daily scores on the Daily Rating Form sadness scale for each of the 8 baseline weeks, the 8 weeks preceding hormone replacement (Lupron alone), and the 4 weeks of Lupron plus P4 replacement. Menstrual cycle-related mood disorder (MRMD) is an affective disorder characterized by the appearance of negative mood symptoms (irritability, sadness, mood swings) in a menstrual cycle phase-specific fashion. Symptoms appear during the luteal phase and disappear at or soon after the onset of menstruation. Consequently, abnormalities of ovarian steroid (estradiol, progesterone) secretion were sought as the explanation for MRMD. Efforts to identify abnormal luteal phase levels of ovarian steroids or any of a myriad of hormones have been unsuccessful, leaving the etiology of the syndrome unknown. The role of ovarian steroids in MRMD was evaluated by first suppressing ovarian function (predicted to prevent symptoms) and then adding back physiologic levels of progesterone or estradiol (predicted to precipitate symptoms). In the top row of the figure, symptoms can be seen to appear during the premenstrual week in the two baseline cycles, to disappear (both symptom severity and cyclicity) during the second and third month of ovarian suppression (achieved with the gonadotropin-releasing hormone agonist, leuprolide acetate; first month of Lupron alone not shown), and to be precipitated during replacement with either progesterone or estradiol. Identical hormone manipulations are, however, without effect on mood in a group of comparison women without a history of MRMD (bottom row); that is, women with MRMD have a differential response to gonadal steroids such that levels or changes in hormones that have no impact on mood in normal women are capable of destabilizing mood in women with MRMD. The effects of the stimulus are, therefore, entirely contextual—presence or absence of a history of MRMD—with the physiologic underpinnings of the differential response being the focus of current investigations. (From Schmidt, P., Rubinow, D. *et al.*, *New Engl. J. Med.*, 338: 209–216, 1998. With permission.)

Further, Hancur *et al.*, suggest that under some conditions estrogens can actually increase responses to stress in post-menopausal women. This implies a context dependency of the hormone effect parallel to the suggestions of Morgan from her work with animals (mentioned above). To paraphrase this conclusion, estrogens heighten arousal. Under neutral or positive conditions, the hormone may heighten the positive affect; under stressful conditions, the same estrogen treatment may worsen mood.

A. Stress in the Environment

We are all aware that stress affects our behavior. Although it is difficult to prove quantitatively, some stress is useful, in that it drives us to meet deadlines and succeed in life, but high stress can be unhealthy and even debilitating. The concept of stress encompasses a variety of forms but does not differentiate the effects. Hemorrhage, for example, is a stress that elicits cortisol release and other hormones; yet, we tend to think of stress as having much more emotive power. Immobilization is a stress that activates high levels of adrenocorticotropic hormone (ACTH) and corticosterioids, but so does forced activity. Our autonomic system responds to stress by preparing us for flight or fight. Under a barrage of catecholamine release from autonomic nerves of the sympathetic nervous system, heart rate increases, blood pressure rises, bronchi of the lungs open for maximum ventilation, and blood flows more to the brain and muscles and away from the gut and skin—but it also does that when you fall in love! (See also previous discussions on ACTH and glucocorticoids.)

III. SOME OUTSTANDING NEW CLINICAL OR BASIC QUESTIONS

1. Why are hormone effects in humans so much more dependent on social context than in lower primates and mammals?
2. Hypertension is a word that conjures up the notion of stress, but what does that really mean? Can hypertension truly be induced by stress?
3. What are the neural and molecular steps involved when an environmental manipulation alters hormone effects on gene expression in a hypothalamic neuron?

CHAPTER 20

Behavioral/Environmental Context also Alters Hormone Release

Hormone/behavior relations are a two-way street. Throughout this text we have examined many of the effects of hormones on behavior. Now we explore the possibility that the environment—including the behavior of other organisms and even one's own past behaviors—alters hormone secretion. Some of the clearest examples involve stress. A stressful environment not only affects the hypothalamic–pituitary–adrenal cortical (HPA) axis, but also the hormones of reproduction.

I. BASIC EXPERIMENTAL EXAMPLES

One of the great general results of reproductive biology is that female mammals will not reproduce under inadequate environmental conditions. A clear example would be an adequate food supply allowing ovulatory hormones to be released from the anterior pituitary gland and its effect upon reproductive behavior. When a female is undernourished, luteinizing hormone (LH) will not be released from the pituitary to act upon the ovary

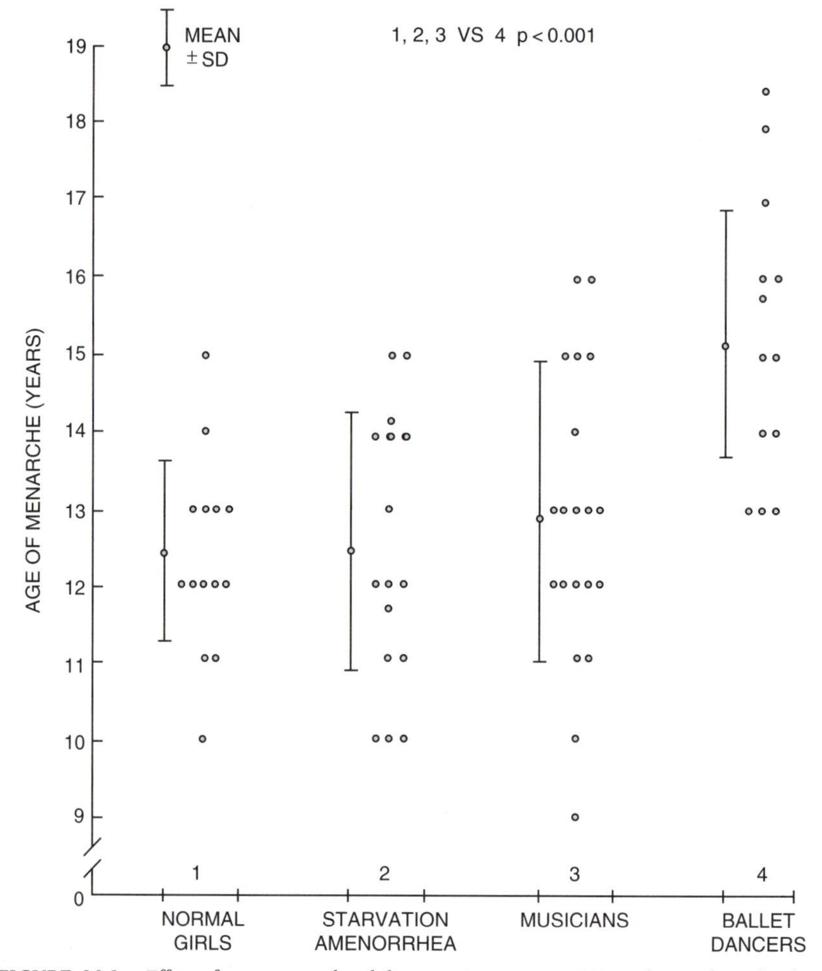

FIGURE 20.1. Effect of context on the ability to raise pituitary FSH and LH release levels so as to initiate menarche in young women. The combined physical stresses, social environment, and lean body mass of young ballet dancers lead to a later menarche (column 4) compared to "normal" girls (column 1) who do not have this history. The musical environment itself is not the key cause (see column 3) nor is the epochal lack of food intake (column 2). (From Warren, M.P. and Perlroth, N.E., *J. Endocrinol.*, 170(1): 3–11, 2001. With permission.)

in the huge surge necessary for ovulation (Fig. 20.1). From the work of Judy Cameron, in Oregon, we see that the same is true of males with respect to LH and the testes (Fig. 20.2). There are at least two causal routes by which this endocrine phenomenon becomes important for behavior in the female. First, during the normal estrus cycle (for animals) or menstrual cycle (for women),

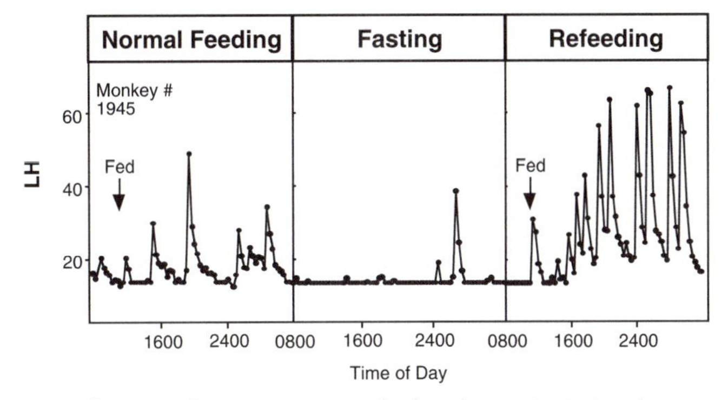

FIGURE 20.2. Changes in the environment can lead to changes in pituitary hormone secretion. Here, a lack of food in the experimental environment (fasting) causes a remarkable decline in pulses of luteinizing hormone (LH) secreted from the male monkey's anterior pituitary gland (Cameron *et al.*, 1993).

the ovulatory discharge causes the release of ovarian hormones, which circulate back to the brain and help to facilitate sex behavior. Second, even under circumstances where the ovarian hormones have been supplied in adequate quantities by the experimenter, it has been demonstrated by George Wade and colleagues, at the University of Massachusetts, that the underfed female animal will not mate. Their analysis of the neural systems responsible for the impact of nutrition on reproduction has shown that a very small region of the hindbrain, the area postrema, is important in the signaling that informs hypothalamic neurons as to whether enough food has been available. Testosterone levels, as well, are subject to environmental influence. Male monkeys defeated in a fight (as first shown by Robert Rose, at Boston University) have decreased levels of circulating testosterone, while the testosterone levels of the winners are elevated.

Seasonal breeding can reflect both food supply and stress in the environment. That is, the food supply issue (Fig. 20.2) noted above provides a mechanism for seasonality effects on hormone-controlled behavior in both the male and the female. Leptin levels reflect the availability of food in the environment leading to the deposition of fat, so, in turn, they provide a proximate mechanism for nutritional effects on sex behaviors.

Stress in the environment has multiple effects on hormone secretion (Figs. 20.3). Corticotropin-releasing hormone (CRH) is exquisitely responsive to a variety of environmental stressors. Of course, the whole point of CRH release in the median eminence is to act as the major driver for adrenocorticotropic hormone (ACTH) release. ACTH in turn can have behavioral effects on its own (for example, on certain forms of learning and memory) and, importantly, elevates corticosterone levels and thus alters

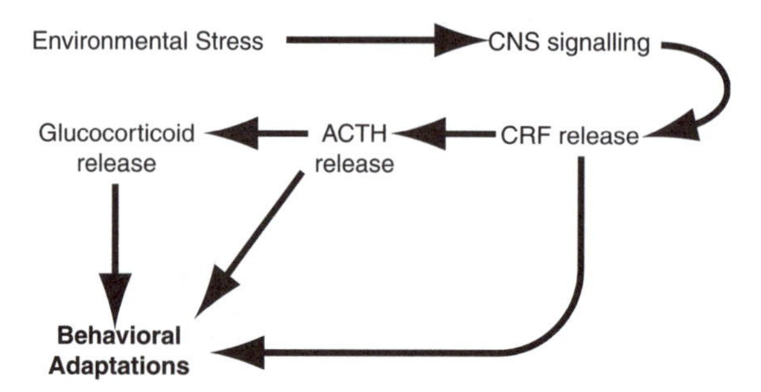

FIGURE 20.3. Classical examples of the effects of environmental context upon hormone secretion are those that have to do with stress. Illustrated here are three routes that comprise the minimum number of routes by which stress-related hormonal signaling could influence behavior.

behavior indirectly. We also note that ACTH is not the only stress hormone in experimental animals. Oxytocin and vasopressin levels in the blood are also sensitive to stressful stimuli.

In Chapters 1 and 3, we pointed out that hormones act in concert with one another to modulate metabolism and behavior and that they are produced and secreted in an orderly fashion over time in response to stressors. The example given was that of experimental stress (chair restraint and 72-hour sessions of foot-shock avoidance in monkeys), with measurement of multiple plasma and urinary hormones and their metabolites before, during, and after the stress sessions. As indicated in Fig. 20.4, this seminal set of experiments revealed an early secretion of catabolic hormones, including adrenal cortical glucocorticoids (17-OHCS) and adrenal medullary catecholamines (epinephrine, norepinephrine), that promote energy utilization by mobilizing energy stores in support of "fight or flight" behavior. This is followed by secretion of anabolic hormones, including female and male gonadal steroids (estrone, testosterone) and insulin, which help rebuild tissues that were catabolized for their energy-rich substrates. Figure 20.4 also shows that, during the immediate, catabolic hormone response, anabolic hormones are suppressed below baseline, and, conversely, during the later, anabolic hormone response, the earlier-secreted catabolic hormones are suppressed below baseline. These data highlight the finely orchestrated physiological reciprocity between these two groups of hormones in stress situations. This hormonal reciprocity parallels the reciprocity between the sympathetic and parasympathetic nervous systems, the former being activated in response to an immediate threat (*e.g.*, increasing heart rate and muscle activity) and the latter being more restorative in function.

Furthermore, repeated environmental stress of the same type results in adaptation of hormonal responses. In many instances, hormonal responses

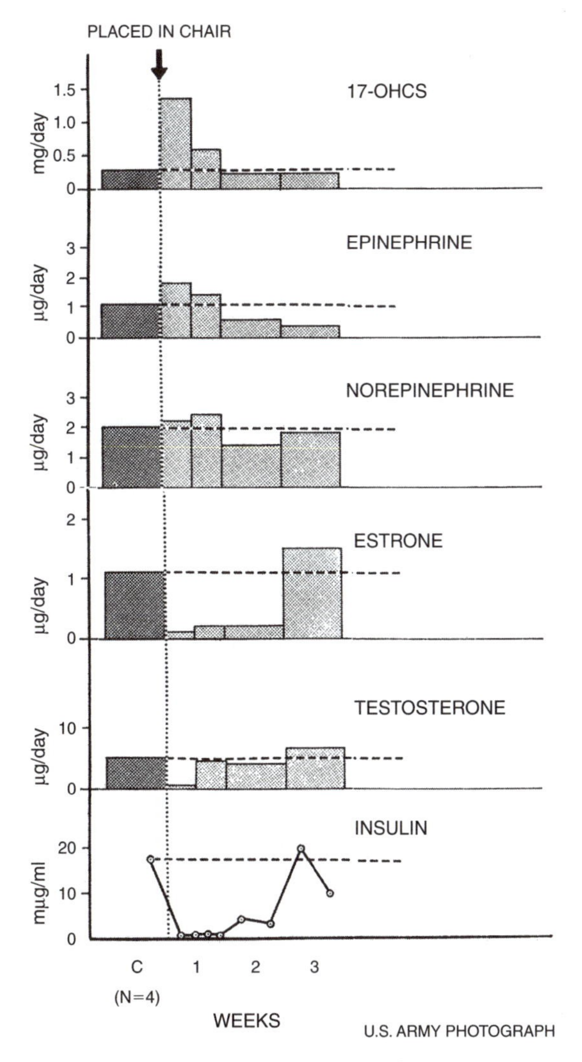

FIGURE 20.4. Multiple hormone responses during 3-week adaptation to restraining chair in a monkey. 17-Hydroxycorticosteroids (17-OHCS) is a measure of adrenal glucocorticoids. (Modified from Mason, J.W., *Psychosom. Med.*, 30: 774–790, 1968.)

can become habituated to repeated exposure to the same stressor. The chair-restrained monkey paradigm also illustrates this. Weekly sessions of 72-hour foot-shock avoidance initially resulted in clear HPA axis activation, as reflected by elevated 24-hour urinary 17-OHCS excretion. However, this response rapidly habituated, such that the second 72-hour session produced a much smaller response, and subsequent sessions produced little or no HPA axis activation. Were the stress-responsive HPA axis not to habituate, adrenal exhaustion could result from repeated stress activation, leading to metabolic collapse and ultimate death.

Are other organisms nearby? One of the features of the environment crucial for any socially relevant hormonal system is which other animals and humans, if any, are nearby. Availability of a mating partner can, in many species, facilitate release of reproductively important hormones. Especially in lower animals, pheromones can signal stress or sex, or even availability of food and water, thus influencing appropriate hormones in the test animal. In humans, complex social situations can influence stress hormones in the observer.

II. CLINICAL EXAMPLES

Family environment is important. One sensitive measure of social effects on the release of the reproductive hormones luteinizing hormone and follicle-stimulating hormone (FSH) from the pituitary is the age at which girls enter menarche—that is, the age at which LH is released in quantity and the menstrual cycles begin. Michelle Warren, at Columbia Medical School, has shown that familial factors, including the presence of a male, basic approval of the girl by the family, and the absence of conflict, all have significant effects. The earlier maturation of LH release patterns was associated with less positive family relations.

A second kind of influence on age of menarche in girls could be treated as an effect of family, an effect of stress, an effect of nutrition, or all three. Warren has found that trained ballet dancers enter menarche at a significantly later age than other girls. Of course, beginning and persisting with ballet could result in part from family pressures, and the dancing itself could represent a source of environmental stress. However, a dominating factor is that dancers are required to have a very low percentage of body fat, which is likely to delay reproductive hormonal maturation.

Oxytocin secretion is well known for its sensitivity to strong environmental signals that are relevant for a woman's emotional life. In a recent laboratory experiment by Turner and Amico, positive emotion induction (by recollection of intense happiness) lowered oxytocin levels, even as it increased the level of prolactin and failed to change ACTH. On the other hand, nursing mothers are familiar with distressing situations in which lactation is interrupted, perhaps because of μ opioid synaptic inputs inhibiting electrical activity in oxytocinergic neurons.

A surprising example of contextual influences over hormone secretion came from the work of an international team led by Hirschenhauser in Lisbon, Portugal. They studied 27 volunteer males during young adulthood. First, intense sexual activity tended to be associated with testosterone peaks. Second, higher hormone levels tended to occur around weekends. Finally, some men who reported a current wish for children had a 28-day temporal pattern of testosterone levels. While these findings are complex and require further examination, they remind us that even in a simple hormone/behavior causal relationship, subtleties may crop up when humans are under study.

A. Stress

Some environmental circumstances alter the degree to which stressful influences govern CRH and ACTH release. On the one hand, individuals may be encouraged to forget about it to recover quickly from the stress. In other cases, glucocorticoid levels remain high and the eventual effects of chronic stress are quite damaging. As McEwen (2002) points out, this difference is medically important because chronic stress without adequate recovery can lead to depression.

More generally, there have been many studies in human subjects that illustrate the importance of central nervous system (CNS) control of hormone systems. The frontal lobes of the cerebral cortex are the cortical area for higher executive function, and the demand for decision making, mastery, and control over novel situations (*i.e.*, learning and competence) is a powerful influence on the HPA axis.

The military setting in many respects is ideal for human stress studies. Personnel are for the most part young and healthy, they are available for repeat studies and follow up, they are frequently willing to be volunteer subjects, and they often experience extreme stressors in "naturalistic" settings (*i.e.*, as part of their training and in combat). In the U.S. Navy, underwater demolition team (SEAL) training represents such a situation; trainees who enter this program may have little knowledge of swimming and may be in poor physical condition. The training program requires the development of expert swimming skills, outstanding physical endurance, and expertise in handling potentially dangerous situations (*e.g.*, underwater explosives). Figure 20.5 portrays mean early-morning serum cortisol concentrations in 20 young, healthy Navy men during the first two months of underwater demolition team (UDT) training. The first week's training was primarily long-distance running, and the trainees felt confident of their abilities. A peak in cortisol occurred coincident with the start of swimming practice (January 9); at this time, most of the men were better runners than swimmers.

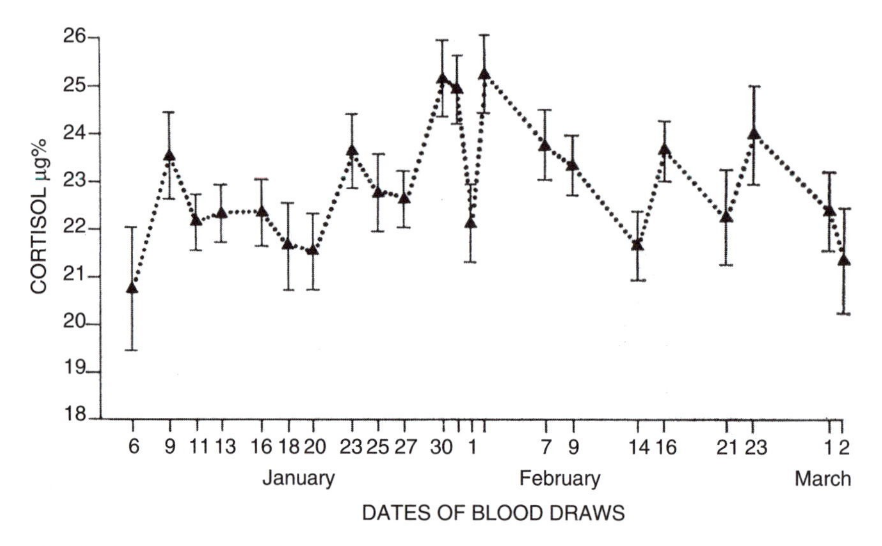

FIGURE 20.5. Mean (±SEM) serum cortisol concentrations for 20 U.S. Navy underwater demolition team (SEAL) candidates during their first 2 months of training. (From Rubin, R.T. *et al.*, *Psychosom. Med.*, 31: 557, 1969. With permission.)

Over the next two weeks, swimming increased in distance and shifted from a heated pool to the ocean (water temperature 56°F); nevertheless, cortisol values declined as the men adapted to their swimming training. A significant increase occurred coincident with the introduction of the face mask back in the heated pool (January 23); most of the trainees had little underwater diving experience, and this phase was met with overt anxiety. Over the next week, long ocean swims with the face mask were conducted, and the mean cortisol level again trended lower.

In the fifth week of training ("hell week"), the men were kept constantly on the move for 5 days and nights, with little or no sleep. The objective was to teach them that, no matter how tired they felt, they could always manage further effort. Mean serum cortisol peaked significantly on the first day of hell week (January 30) and fluctuated considerably over the 5 days. During the week, when the men realized they could not keep up such a pace, their *esprit de corps* quickly diminished, and by the end of the week their tolerance for the grueling schedule had been depleted. During this week, cortisol values were the highest at any time during the study.

For the two weeks following hell week, the schedule was relatively easy, classroom reconnaissance work began, and mean cortisol values decreased significantly. Cortisol then peaked significantly, coincident with the introduction of diving fins and UDT weapons (February 16). During the next week, cortisol levels declined somewhat as ocean swimming with fins was practiced. An increase in serum cortisol again occurred on the first day of ocean drops and pickups by helicopter (February 23). Values trended lower during the next week of helicopter maneuvers and with the beginning of night ocean swims.

At the end of these two months, the group was divided into two for the demolition phase of training. Cortisol concentrations for the two new groups followed similar patterns of increase in relation to new activities, such as the introduction of the self-contained underwater breathing apparatus (SCUBA), mine searching, and long night compass swims.

In this UDT training study, mean serum cortisol levels increased coincident with novel experiences that evoked anticipatory anxiety. As each new technique was practiced and became familiar, cortisol levels trended lower, even though increasing demands for the use of the new technique were being made by instructors. The determinant of the transient increases in HPA axis activity, therefore, appeared to be the anticipation of an unknown situation more than the inherent difficulty of the situation itself. There was a fairly rapid adaptation of the HPA axis once the novelty had passed and mastery of the technique had begun, even though the need for the newly learned technique was increasing. These findings highlight not only the stress reactivity but also the adaptability of the HPA axis to real-life situations. Many additional studies in both military and civilian circumstances have corroborated these principles.

Another example illustrates activation of the HPA axis in response to decision making. If a pair of chair-restrained monkeys is subjected to a noxious stimulus in an avoidance situation, the monkey permitted to press a lever to avoid the noxious stimulus to both animals develops gastrointestinal stress lesions, whereas the other monkey, which has no control over the stimulus, does not. This led to the concept of the "executive" monkey: An animal in a situation requiring decision making undergoes more physiological stress than does an animal experiencing the same adversity but having little or no control over its fate.

Several studies have tested this concept in humans, including another investigation of Navy personnel. Naval aviators in training during the Vietnam conflict had blood drawn for serum cortisol determinations shortly after they exited their aircraft following a series of landing practices of increasing difficulty. These included night mirror landing practice (MLP) on shore, daytime aircraft carrier landing practice (DAYQUALS), and nighttime carrier landing practice (NITEQUALS). Serum cortisol also was measured on a control, non-flying day.

The aircraft in use during these landing practices was the F-4B Phantom, which carries two aviators: the pilot in the front seat, who flies the aircraft,

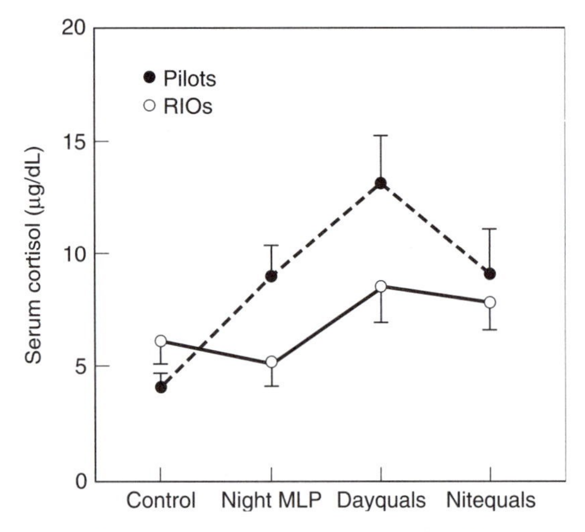

FIGURE 20.6. Mean (±SEM) serum cortisol concentrations for 9 U.S. Navy pilots and 10 radar intercept officers (RIOs) on a non-flying control day compared to land-based (night MLP) and aircraft-carrier-based (DAYQUALS, NITEQUALS) landing practice. (Modified from Miller, R.G. *et al.*, *Psychosom. Med.*, 32: 585, 1970.)

and the radar intercept officer (RIO) in the back seat, who has no control over the aircraft. During landing, the RIO calls out the airspeed to the pilot but in all other respects is a passive partner. Thus, the pilot is the "executive" naval aviator, whose split-second decisions during the most hazardous aspect of naval aviation, carrier landings, affect the fate of both him and his partner. In an emergency, both individuals could eject from the aircraft, although not necessarily safely during a final landing approach.

Figure 20.6 shows mean serum cortisol concentrations for 9 pilots and 10 RIOs on the 4 test days. Mean cortisol on the control, non-flying day was not different between the two groups. During all 3 landing practices, the pilots had highly significantly increased cortisol concentrations compared to their control day, whereas the RIOs had no significant increases in cortisol. Similar to the observations in monkeys, these findings indicate that a person in a situation requiring decision making undergoes more physiological stress than does a person experiencing the same potential adversity but having little or no control over his fate, thus extending the concept of "executive" stress to humans. Other studies in military personnel (*e.g.*, B-52 aircraft commanders versus their crews and Special Forces officers versus men under their command) support this concept.

Modern warfare is increasingly being conducted at a distance, through computers and remotely guided weapons. Military pilots are now being trained more frequently on flight simulators and in the art of computer-guided drones (*e.g.*, the Predator, a pilotless reconnaissance aircraft that can be armed). One concern that has been expressed by "old-school," combat-hardened senior military personnel is that this new generation of simulator-trained aviators, if needed in an actual combat situation, might have "inexperienced" physiological anxiety reactions, including stress-hormone responses, that could be deleterious to their optimal functioning when faced with a real threat of bodily harm.

Next, consider the activation and adaptation of the HPA axis in hospitalized patients. An important clinical example of HPA axis adaptation is in patients subjected to the stress of hospitalization, particularly psychiatric patients who may have HPA activation as part of their illness, such as major depression. As indicated in Chapter 2, 30 to 50% of patients with major depression have increased HPA axis activity, as indicated by increased circulating ACTH and cortisol concentrations and resistance of the HPA axis to suppression by the synthetic glucocorticoid, dexamethasone. The more severe the depression, the more likely it is that there will be increased activity of this endocrine axis. Successful treatment results in return of HPA activity to normal, as well as relief of depressive symptoms.

Patients with major depression remain responsive to environmental stressors that can contribute to their increased HPA axis activity. One indicator of this increase is the dexamethasone suppression test (DST), which involves administration of low-dose dexamethasone about midnight and measurement of circulating cortisol at intervals thereafter. Dexamethasone, acting primarily at the pituitary, normally strongly suppresses ACTH and cortisol secretion for the next 24 hours. If CRH driving of the HPA axis is increased, ACTH and cortisol escape from this suppression, and their circulating levels increase.

Figure 20.7 shows plasma cortisol before and during repeated DSTs in a hospitalized, depressed woman before and during treatment. All of the midnight blood samples were taken immediately prior to dexamethasone administration (2 mg orally), and all but the last sample (May 15, after 5 weeks of antidepressant treatment) show elevated cortisol levels. On April 2, just after hospital admission, the DST was strongly abnormal, with almost no suppression of post-dexamethasone cortisol. One week later, on April 9, there was normal suppression of cortisol at 8 a.m. but clear escape thereafter.

Of relevance to this lesson about HPA axis activation and adaptation, during the week between April 2 and April 9 the patient was treated with placebo medication, so that the lessening of the abnormal DST on April 9 compared to April 2 likely reflects the patient's increasing familiarity with and adaptation to the hospital environment. After active treatment was instituted, the DST levels gradually improved, becoming fully normal (cortisol suppression for a full 24 hours) on May 15. These data indicate that both

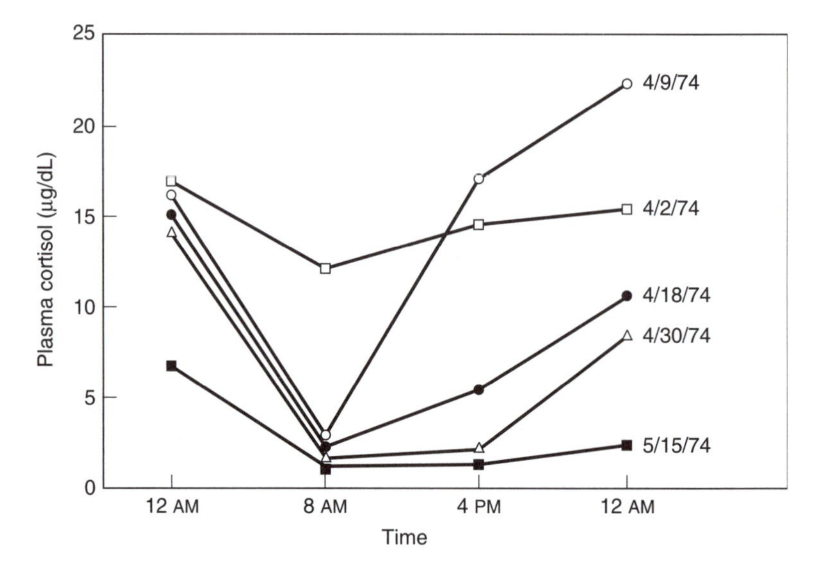

FIGURE 20.7. Plasma cortisol concentrations prior to (12 a.m.) and following administration of dexamethasone (2 mg) at 12 a.m. on the first test day to a hospitalized patient with major depression. April 2 and 9 are prior to antidepressant treatment; April 18, 30, and May 15 are during treatment. (From Carroll, B.J. *et al.*, *Arch. Gen. Psychiatry*, 33: 1041, 1976.)

internal factors (putative neurotransmitter abnormalities related to the disease process) and external factors (environmental stressors) may be simultaneous activators of a stress-responsive hormone axis.

III. SOME OUTSTANDING QUESTIONS

1. How plastic are hormonal responses to stressors; that is, which ones are reversible when an animal is removed from a stressful environment? Are stressful influences on hormone secretion ameliorated when a subject is placed in a hygienic/therapeutic environment? How does this vary with age (*e.g.*, pre- versus post-pubertal comparisons and studies of the aged)?
2. What severity and duration of a given stressor are required to produce irreversible changes in endocrine function and associated metabolic effects?

   a. What are the characteristics of stressors in early life that lead to increased HPA axis activity in adulthood?

   b. Which factors can mitigate the effects of stressors in early life, such that irreversible changes in endocrine function are not produced?

3. How do competing forces from the environment become integrated to form a net force of defined direction, amplitude, and time course on the release of a given hormone?
4. To what extent have certain environmental/endocrine relations remained intact during evolution from lower mammals to nonhuman primates to humans? To what extent have they changed in an understandable and lawful manner?
5. Which ascending neural pathways to the basomedial hypothalamus from the brainstem subserve any given environmental/endocrine causal sequence? Likewise, which pathways descending from the cerebral cortex and limbic system do the same?

SECTION VII

Evolution

CHAPTER 21

Neuroendocrine Mechanisms Have Been Conserved to Provide Biologically Adaptive Body/Brain/Behavior Coordination

Hormone-controlled behaviors, especially social, courtship, sex, and parental behaviors, constitute primary means by which individuals compete to pass on their genes. Evolutionary theorists, from Darwin to Ernst Mayr, have seen such behaviors as the "leading edge of evolutionary change." It has been argued (Pfaff, 1999, ch. 8) that a large fraction of neuroendocrine mechanisms have been conserved from lower mammalian brains into the human brain. Thus, studies of hormone/brain/behavior mechanisms in laboratory animals are likely to contribute to a medical understanding of disorders involving the same neurons and same biochemical reactions in humans. More generally, the principle of hormones evolving for specific functions can be illustrated by several hormones and their receptors.

Throughout this book, we have seen many examples of how hormones affect specific behaviors. *Pari passu*, the material has illustrated (1) how hormones have evolved to provide essential physiological functions for survival and (2) that behavioral mechanisms, in fact, comprise one aspect of that evolution. The most extensive examples quoted in this short text include steroid and peptide hormones for feeding, drinking, stress, and sex—clearly basic for any organism. Therefore, it is not surprising that several important families of hormones have been conserved for millions of years to serve a specific function. Neuropeptide Y (NPY), angiotensin II, arginine–vasopressin (AVP), oxytocin (OT), insulin, and other hormone peptides have been found not only in all mammalian species but also in the most ancient vertebrates. Further, they have been identified in invertebrates such as flies, worms, and mollusks, which have been in existence for hundreds of millions of years. This prolonged conservation is Nature's way of saying, "If it ain't broke, don't fix it."

Hormones are synthesized from DNA encoding specific sequences of a gene. Over time, mutations will occur in individual nucleotide bases within these sequences and different species may develop different variants of primordial hormones. But, as a principle, the biologically active part of the hormone is the most highly conserved sequence. Thus, we now explore examples where the general physiological function often remains the same over time.

I. FEEDING

As we have seen, feeding is a very complex hormonally controlled behavioral event, but the involvement of many hormones (recounted in Chapters 1 and 3) relates to a single function. The function of eating is to maintain energy. Energy is required to drive the machine—the body. We need energy to move around, to forage and hunt for food, to grow, to fight, to mate, and to survive. Energy also generates heat, and in poikilothermic animals, such as ourselves, feeding and eating are necessary for thermogenesis. Leptin, insulin, and NPY, for example, are not only key hormones in the control of feeding, but also key hormones in the regulation of energy balance and thermogenesis.

The vital importance of feeding to any organism is reflected in the continuity found throughout evolution of the NPY receptor family. Even in a primitive worm, *Caenorhabditis elegans*, which has only 900 cells, a seven-transmembrane protein has been identified as the key receptor in feeding behavior and is remarkably homogeneous with the NPY receptor family. From peptide sequencing , molecular cloning, and comparative genomics, the NPY family appears to have a common ancestral gene (NYY), and by gene duplications has produced NPY, PYY, pancreatic polypeptide (PP), and others. By the time the first vertebrates such as lamprey had evolved, they had NPY and PYY.

Insulin has also been shown in the fruit fly (*Drosophila*) to play a role in carbohydrate homeostasis and growth. Even *Drosophila* juvenile hormone is a member of the insulin family. Relaxin is also found in the fly in the form of an insulin/relaxin-like peptide. The preservation of function is shown by the fact that flies with mutations that reduce insulin signaling grow up to be small and infertile, just as do mice with insulin deficiency.

II. ANGIOTENSIN AND THIRST

Another essential biological function is thirst. We evolved out of the sea, and a major problem with becoming terrestrial is maintaining the sea within us. The levels of sodium in blood represent our past environment and must be maintained to achieve the delicate balance for life. Neurons and other cells depend on the precise balance of high sodium concentration outside cells versus a low level of sodium inside cells in order for neurons to communicate by action potentials. The balance of sodium must also be matched with a balance of high intracellular and low extracellular potassium in active cells. Therefore, hormones associated with fluid balance, including angiotensin, aldosterone, vasopressin, and atrial natriuretic peptide have evolved to maintain fluid balance as well as sodium and potassium balance. When water is lost it has to be replaced and, as much as possible, retained.

The renin–angiotensin system has evolved over millions of years to maintain fluid balance, even in fish in the sea. Because the salinity of our contemporary seas is higher than the sodium levels of the ancient seas from which we evolved, there is a constant battle to maintain a sodium level of around 144 mEq/L. Saltwater fish use angiotensin to stimulate thirst and drink, and gill cells to excrete sodium. Freshwater fish switch off their angiotensin levels, and sodium is not excreted. Thus, the renin–angiotensin system that can be found in sharks and crabs has evolved over millions of years as the hormone for controlling water volume. It does this by inducing thirst and drinking and by stimulating the release of antidiuretic hormone (ADH), prolactin (PRL), and aldosterone to reabsorb sodium. Prolactin is very important in fish to control sodium excretion from the gills and is a major water balance hormone. In mammals, it has evolved a unique fluid balance role, accumulating fluid in milk for lactation.

A clear example of a function of angiotensin is its role in water conservation during hemorrhage. In hemorrhage (Fig. 21.1), when blood is lost, there is a reflexive release of renin from the kidneys in response to the drop in pressure. Renin forms angiotensin II, which releases vasopressin. When the volume loss exceeds 6%, vasopressin is released to conserve water. Angiotensin released in the brain activates circuits that lead to thirst. This is a complex process because it also requires memory and motor function. In the vascular system, a further complexity of angiotensin's role is its activation of components of the thrombotic pathway to initiate clotting to prevent blood loss. Angiotensin releases plasminogen activator inhibitor type 1 (PAI-1).

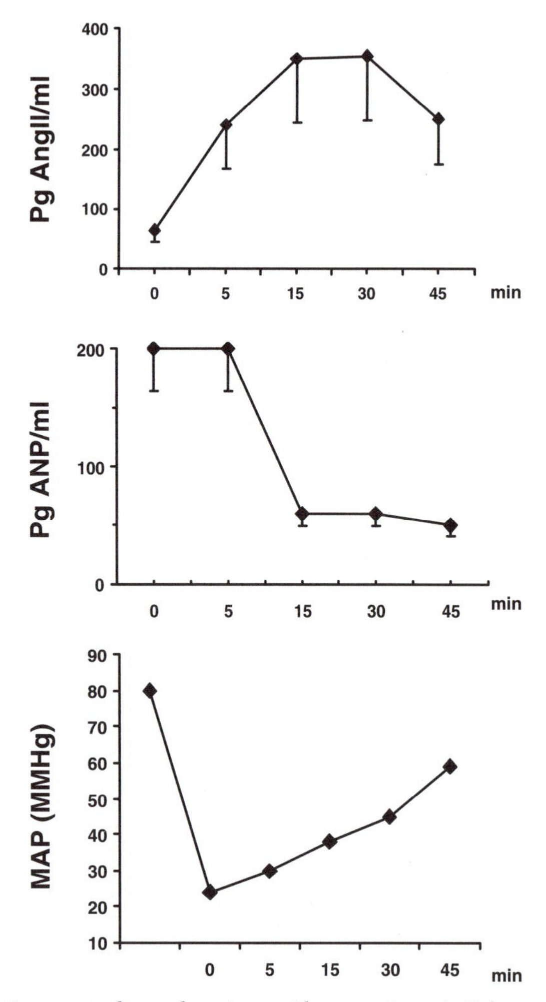

FIGURE 21.1. Response to hemorrhage in rats. Plasma angiotensin II (measured in picograms [pg]) rose markedly (top panel) within the first 5 minutes after hemorrhage, and plasma atrial natriuretic peptide (ANP) fell significantly (middle panel). Mean arterial blood pressure (MAP) fell (bottom panel). This rise in pressure after 5 minutes reflects the vasoconstriction due to angiotensin II. (From Galli, S.M. and Phillips, M.I., *Proc. Soc. Exp. Biol.*, 213: 128–137, 1996. With permission.)

Inhibiting plasminogen activation allows the fibrinogen clotting pathway to form clots, which seal the blood vessels and stop the hemorrhage.

If the hemorrhage is due to a wound, then wound healing becomes another vital function. Again, angiotensin has evolved to function in wound healing. Levels of angiotensin rise in wounds, which stimulates transforming growth factor-beta (TGF-β) to increase the growth rate and activates the immune system to prevent infections.

Another action of angiotensin that comes into play during volume loss and hemorrhage is vasoconstriction. Angiotensin is a powerful vasoconstrictor which it achieves by acting on smooth muscles which are the cells that surround blood vessels. Angiotensin type 1 receptors on these smooth muscles respond to angiotensin and activate a calcium influx which causes contraction of the muscle cell. This contraction narrows the lumen of the blood vessel, increasing the blood pressure. In hemorrhage and low volume, this is a very useful function for keeping up pressure. In normal volemia, when everything is in balance, excessive angiotensin has deleterious effects. It increases blood pressure and leads to hypertension; it also increases inflammatory responses and leads to atherosclerosis, and it stimulates growth factors which lead to left ventricular hypertrophy. Thus, from a fairly simple principle, that angiotensin can stimulate thirst, we end up with a complex set of functions, all related to each other and all dependent on a peptide that was created at least as far back as the beginnings of anthropoids.

III. VASOPRESSIN

Consider the conserved role of the nine-amino-acid peptide produced in certain magnocellular neuronal groups in the anterior hypothalamus. When animals moved out of the sea and onto land, they had to use hormones that already existed to maintain their water balance. The amount of water ingested has to be balanced by the quantity and concentration of urine produced in order to regulate fluid volume and osmolarity within a normal range. As we have seen in Chapter 5, the principal hormone for controlling urine output is vasopressin. Vasopressin and oxytocin are nonapeptides evolved from the same gene. As early as the *Hydra*, conserved sequences of vasopressin have been found. In fish, isotocin is the equivalent of oxytocin, while vasotocin is the equivalent of vasopressin. Even though fish and humans have had separate lineages for 400 million years, the hormones are similarly involved in fluid and electrolyte balance (see Fig. 21.3).

IV. ATRIAL NATRIURETIC PEPTIDE

Atrial natriuretic peptide (ANP) plays a key role in euryhaline fish as they move from seawater to freshwater. A physiological challenge for fish going from seawater to freshwater is the danger of exploding under excess osmotic pressure due to high sodium in the environment, versus drowning in sodium-free water. However, salmon obviously manage this as part of their lifecycle, as they move between freshwater rivers and the deep blue sea. ANP is high in fish in the seawater condition (Galli and Phillips, 1996). ANP increases cyclic adenosine monophosphate (cAMP) in the gills to secrete large amounts of sodium. In freshwater fish, ANP blood levels are reduced, and angiotensin effects are the reverse. In seawater fish, angiotensin keeps the fish constantly drinking, and they excrete excess sodium to maintain an osmotic balance. In freshwater fish, however, angiotensin II levels are reduced in the brain and drinking is inhibited (see Fig. 21.2).

V. MASTER HORMONES

Building on the examples above, we can address a more general concept: As we have evolved from simpler organisms to more complex organisms and face the challenges of living in different environments, Nature has retained families of hormones and given some of them new but related roles. As a result, many of these hormones have assumed the status of what we might call "*master hormones*". Master hormone families are those that control a number of functions that have been physiologically and systematically related throughout evolution. Even as "ontogeny recapitulates phylogeny," such hormones have major functions in the same systems in adults that they affected during development. "Master hormones" include, but are not limited to, gonadotropin-releasing hormone (GnRH), corticosteroids, proopiomelanocortin (POMC), insulin, angiotensin, oxytocin/vasopressin, estrogens, growth hormone (GH)/prolactin, somatostatins, and leptin. Each of these hormone families has multiple effects at different stages in life, but the sets of functions are in some way related internally. Here are several examples, to make the point.

Angiotensin, as we have noted, is primarily involved in thirst and maintaining the water balance, which involves the control of sodium. But, angiotensin is also involved in apoptosis in the late development of the fetus and in development of both the heart and blood vessels. Apoptosis, the orderly death of cells, is a way of sculpting tissues into shape. Current theory is that angiotensin, through its type 2 receptor (AT_2R), sculpts the shape of the heart and blood vessel beds, particularly in the kidney. Estradiol, of course, is the primary female hormone, but it also plays a role in bone density. A lack of estradiol is a major cause of osteoporosis. Insulin is the primary hormone in the uptake of glucose, but, through insulin-like growth receptors, it also affects growth and development, including brain function. Thyroid hormone is primarily involved in metabolism, but because it affects every cell, it is critical in growth and brain development. Ghrelin is involved in

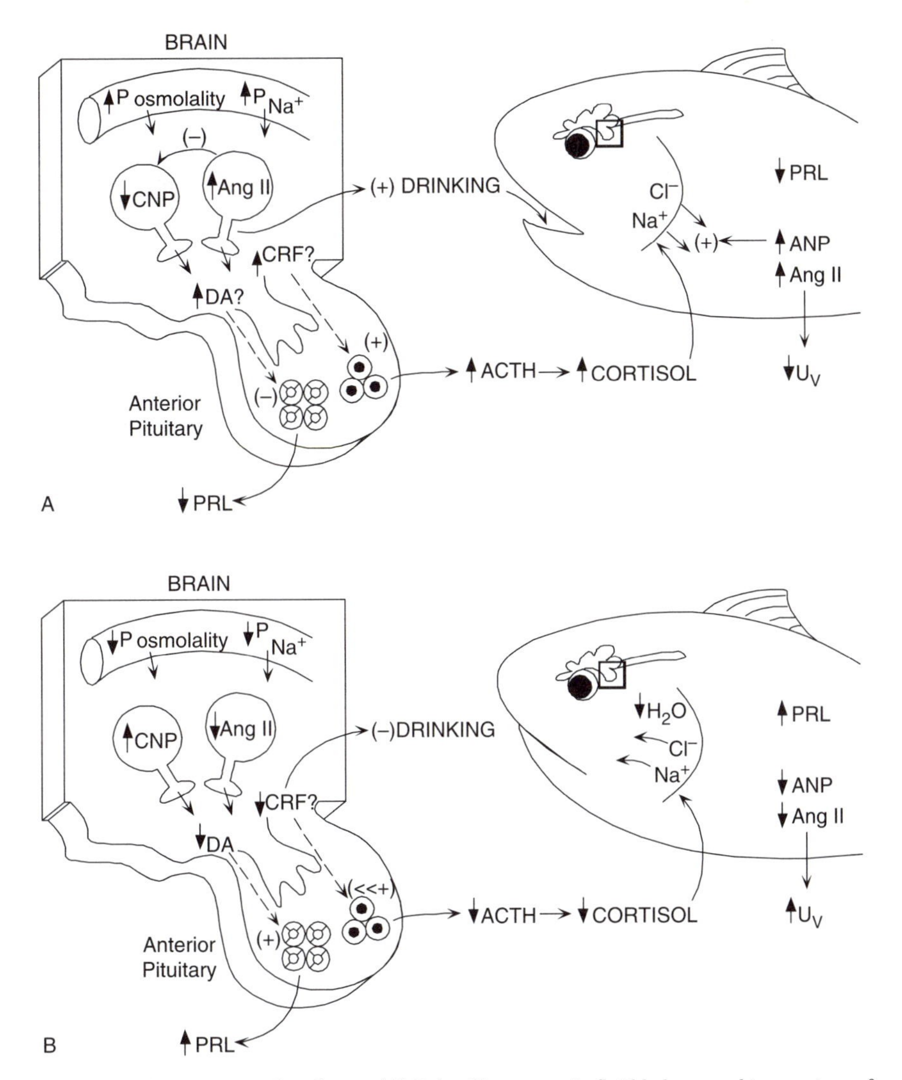

FIGURE 21.2. Functional evolution. **(A)** Role of hormones in fluid balance and interactions of brain CNP and angiotensin II in fish in seawater. The angiotensin II increases drinking and inhibits CNP. The reduction in CNP increases dopamine (DA), which inhibits prolactin (PRL) and increases adrenocortical hormone (ACTH) and cortisol. The effects are increased loss of sodium (Na) and conservation of water. **(B)** Interactions of brain ANP and angiotensin II in fish in freshwater. The ANP levels increase and inhibit DA, which releases prolactin from the pituitary gland into the plasma and acts on gills to reduce permeability to water. The decrease in angiotensin II inhibits drinking and cortisol release. The net effect is conservation of sodium (Na) and protection from water lowering the osmotic pressure. (From Galli, S.M. and Phillips, M.I., *Proc. Soc. Exp. Biol.*, 213: 128–137, 1996. With permission.)

stimulating feeding behavior, but it is also a ligand for growth-hormone-related receptors (hence the Ghr in the name), and growth hormone has multiple effects, through insulin-like growth factor (IGF) secretion, on liver, muscle, bone, cartilage, kidney, and skin. Leptin is primarily involved in feeding, energy balance, and long-term weight control, but it is also important in reproduction (for example, relating body fat mass to the initiation of puberty) and maintaining bone mass (Ducy, 2000). Leptin-deficient mice have a very high bone mass. Obese people resistant to leptin have protection from osteoporosis (Ahimer and Flyer, 2000). Leptin also plays a role in vascular permeability in adipose tissue, where it acts like a paracrine growth factor. Leptin stimulates the Jak/STAT pathway, which causes endothelial cells to lose the adhesion molecules that keep them together, thus allowing molecules to flow from the blood vessels into the adipose tissue.

Now, trying to relate these seemingly disparate functions of leptin, let us consider how growth effects and bone mass are related to leptin's effect on controlling body weight. Puberty is marked by the adolescent growth spurt, a surge of linear growth that is affected by nutrition as well as genetics. Normal growth is a sure indicator of health and well-being in a child. At puberty, dynamic developments in body size, shape, and composition are sexually dimorphic. The distribution of fat and bone growth occur at this time. The hormonal regulation of the growth spurt and alterations in body composition depend on the release of gonadotropins, sex steroids, growth hormone, and leptin.

Gonadotropin-releasing hormone is the master neuropeptide for reproduction in every vertebrate that has been studied. All of the 16 different forms of GnRH are involved in reproduction, including mating behavior (Temple, 2003). In mammals, the two forms in the brain are GnRH I and GnRH II. The latter is the most conserved during evolution and has been shown to coordinate reproductive behavior, but only when the nutritional state is appropriate.

Thus, we can make the case that certain master hormones have evolved not only for maintaining a vital function in the adult organism, but also for processes related to reproduction, fetal development, and adolescence—in good times and in bad.

VI. THE VALUE OF COMPARATIVE STUDIES

Comparative studies of lower order animals have revealed that these master hormones have similar functions among all mammals and, in most cases, among all vertebrates, including the earliest vertebrates such as elasmobranchs (*e.g.*, sharks) and the simplest vertebrates (agnathia), such as amphioxus. The history of master hormones can be traced back even further when we look at invertebrate hormonal systems. Those hormones are basically neurosecretory, and neurosecretory cells are present in all invertebrates. They are involved in growth, metabolism, water balance, reproduction, and other activities. The neurosecretory systems of invertebrates such as insects are very similar to the processes and genes involved in the hormonal systems of the hypothalamus in mammals. As far back as the most primitive multicellular animals, the *Cnidarians* (including sea anemones, hydra, jelly fish, and corals) contain neuropeptides that are found in the mammalian brain. The best example is FMRF-amide. The FMRF gene consists of five exons that are alternatively spliced to form different peptides. They are related to the enkephalin family in mammals. Vasopressin or vasopressin-like peptides have been shown in *Hydra*. In flat worms, an NPY-related peptide demonstrates the consistency of hormone families. Enkephalins have been found in parasitic worms. Neuropeptide F, which is found in all invertebrates, is very similar to NPY, and these peptides are involved in feeding behavior. This means that, over many millions of years, despite literally Earth-shaking changes in the environment, few adjustments have occurred in what Nature has found to be a successful peptide hormone. There are many examples of longevity of a chemical structure, from primitive animals to complex, but what is so compelling is that they are often associated with the same or similar physiological and behavioral functions throughout these millions of years of evolution. The coelacanth fish, *Latimeria*, which is considered to be a living fossil from the Devonian period that represents the beginnings of four-legged tetrapods, has pituitary hormones including proopiomelanocortin (POMC), alpha-melanocyte-stimulating hormone (α-MSH), and corticotropin-like intermediate lobe peptide (CLIP). The fish MSH is identical to mammalian MSH.

Further removed on the tree of life, leeches have a precursor molecular of angiotensin that is similar to human angiotensinogen with a 78% homology. The leech, which belongs to an ancient group of metazoans, has an angiotensin-I-like molecule that is highly homologous with the human angiotensin I molecule. The angiotensin-I-like molecules even have a role in water diuresis in the leech; therefore, the function of angiotensin in fluid homeostasis has long been preserved.

Even insulin is found in mollusks, in the cerebral ganglia of *Lymnaea*. It is found in giant neuron cells that are associated with growth, metabolism and reproduction. Insulin in the molluscan sea slug, *Aplysia*, is very similar to vertebrate insulin. Molluscan insulin is transcribed from four genes of precursor peptides that are structurally the same as the human insulin gene. The insulin receptor isolated in *Aplysia* is also similar to the vertebrate insulin receptor. Insulin in invertebrates has been shown to alter nerve action potentials. A similar study on insulin in the rat brain showed that insulin inhibits nerve potentials in the hippocampus; this inhibition is probably part of a negative feedback loop in response to high insulin levels. The ancient role of insulin on neuronal activity has never been lost. In the fruit fly, *Drosophila*, insulin acts like a growth factor, and mutations that reduce insulin cause a growth deficiency.

The POMC-derived peptides, ACTH, α-MSH, and β-endorphin are widely distributed in leeches, schistosomes, and other lower vertebrates, where their role is antiinflammatory. Again, there seems to be a principle of conservation in evolution, and opiates have evolved from being immunoregulatory for these lower forms. The analgesic effect may be a byproduct for survival, such as dulling pain in the host it infects, thus altering the behavior of the host.

VII. CORRESPONDING EVOLUTIONARY ROLES OF HORMONE RECEPTORS

One of the ways in which a single molecule can be conserved and yet serve many related but diverse functions is by the evolution of its receptor into numerous receptor subtypes. Although it is not clear how receptors evolved, we recognize simple types from a single transmembrane protein connected to inside the cell, such as tyrosine kinase (TK), to complex 7- and 12-transmembrane receptors. An example is neuropeptide Y, which is highly conserved in all vertebrates but has at least five receptor subtypes (Y1, Y2, . . . , Y5) that have probably allowed NPY to expand its functions. Another example is vasopressin. Vasopressin, with 9 amino acids, has its origins in mollusks, where it affects the firing rate of cells in the snail (*Helix*). Vasopressin is identical in the human and the rat, but the principal receptor subtypes V_1 or V_2 are approximately 80% homologous with the rat V_1 or V_2 receptors. Within these species, the V_1 receptor is less than 50% homologous with the V_2 receptor. V_1 receptors are in blood vessels and cause smooth muscle cells to contract. V_2 receptors are found in the kidney collecting duct, where they mediate the antidiuretic effect of vasopressin.

Oxytocin is very similar to vasopressin (also a 9-amino-acid peptide), and both peptides are derived from a common ancestral gene that underwent duplication to form isotocin and vasotocin, which became oxytocin and vasopressin, respectively (Fig. 21.3). Species as distant as octopus and squid exhibit a 78% sequence homology with mammalian oxytocin. Such families of peptides (*e.g.*, the oxytocin/vasopressin family) are further examples of the high degree of gene conservation related to function, yet in many species the oxytocin receptor is only 30 to 50% homologous with vasopressin receptors.

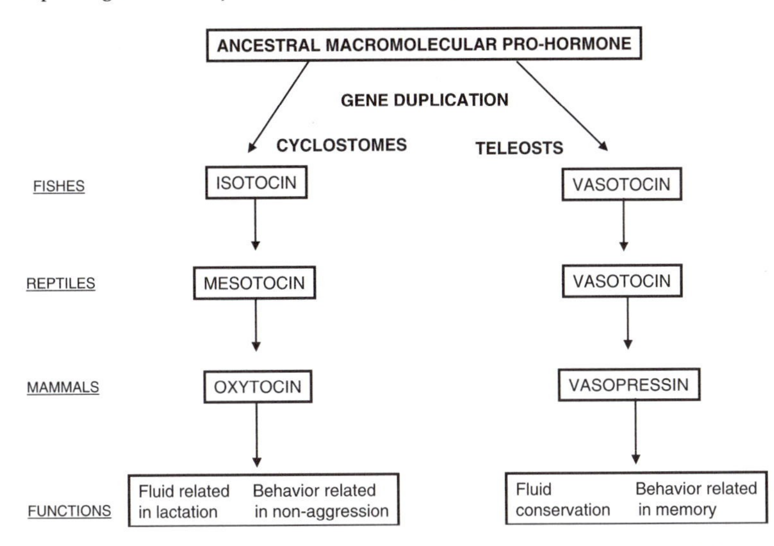

FIGURE 21.3. An example of the evolution of neuropeptides. Conservation and change across vertebrate evolution are illustrated by gene duplication of an ancient large molecule containing a nine-amino-acid sequence in fish. The duplication occurred as fish evolved from cyclostomes to teleosts. Subsequent mutations led to two evolutionary lines, resulting in oxytocin and vasopressin. Both peptides are composed of nine amino acids and have similar but specific behavioral and physiological functions.

Yet another example is α-MSH, which has five receptor subtypes in mammals and six subtypes in fish.

Thus, throughout the course of evolution, receptors have given a variety of functions to their conserved ligand. In humans, only two types of angiotensin receptors are known: AT_1R and AT_2R. In rodents, however, two additional subtypes of the AT_1R receptors have been identified: AT_{1a} and AT_{1b}. Their distributions in the brain are different; based on transgenic experiments, Davisson *et al.* (2000), at the University of Iowa, has proposed that AT_{1a} mediates thirst and AT_{1b} mediates blood pressure.

These examples, and the view of hormonal evolution provided by comparative studies, show that once Nature finds a functional peptide, it sticks with it, but, in order to make this peptide function in different roles, multiple receptor subtypes have evolved. Multiple receptor subtypes enable different functions to be served by the conserved peptide and expand the actions of the peptide to a variety of tissues. The reason why receptors have more variation than the ligand is probably because the peptide hormones are relatively short sequences of amino-acids, whereas receptors, particularly the 7-transmembrane receptors, have long amino acid sequences. The greater the number of amino acids in the sequence, the greater the chance of mutations to form a different subtype. This process is always going on, but at a rate we cannot experience in a single lifetime. Thus, while receptor mutations lead to many endocrine and behavioral malfunctions observed in clinical medicine, over a longer time course the biologically adaptive receptor subtypes should survive. As Darwin said, "It is not the strongest that survive, but those most adaptive to change."

FURTHER READING

This is not a comprehensive bibliography; instead, we have tried to provide enough literature citations to support the figures and tables in this text and to lead the reader to some of the primary data in our field. Note that a comprehensive reference source, Hormones, Brain and Behavior, *has been published recently in five volumes by Academic Press (2002). As well, two other excellent texts treat neuroendocrinology from different points of view:* Behavioral Endocrinology *by J. Becker, M. Breedlove, D. Crews, and M. McCarthy (MIT Press, 2002), and* An Introduction to Behavioral Endocrinology *by R. Nelson (Sinauer Associates, 2000).*

Introduction

Clark JH, Schrader WT, O'Malley BW. 1992. Mechanism of action of steroid hormones. In: Wilson JD, Foster DW, Eds., *Williams Textbook of Endocrinology*, 8th ed. Philadelphia: Saunders, pp. 35–90.

Deutch AY, Roth RH. 2003. Neurotransmitters. In: Squire LR, Bloom FE, McConnell SK, Roberts JL, Spitzer NC, Zigmond MJ, Eds., *Fundamental Neuroscience*, 2nd ed. Amsterdam: Academic Press, pp. 163–196.

Kahn CR, Smith RJ, Chin WW. 1992. Mechanism of action of hormones that act at the cell surface. In: Wilson JD, Foster DW, Eds., *Williams Textbook of Endocrinology*, 8th ed. Philadelphia: Saunders, pp. 91–134.

Kandel ER, Schwartz JH, Jessel TM, Kandel ER, Schwartz JH, Jessel TM, Kandel ER, Schwartz JH, Jessel TM. 2000. *Principles of Neural Science*, 4th ed. New York: McGraw-Hill.

Krieger DT. 1980. The hypothalamus and neuroendocrinology. In: Krieger DT, Hughes JC, Eds., *Neuroendocrinology*. Sunderland, MA: Sinauer Associates, pp. 3–12.

Taylor SE, Klein LC *et al.* 2000. Female responses to stress: tend and befriend; not fight or flight. *Psychol. Rev.* 107:411–429.

Weitzman ED. 1980. Biologic rhythms and hormone secretion patterns. In: Krieger DT, Hughes JC, Eds., *Neuroendocrinology*. Sunderland, MA: Sinauer Associates, pp. 85–92.

Chapter 1: Hormones Can Both Facilitate and Repress Behavioral Responses

Barbaccia ML, Serra M, Purdy RH, Biggio G. 2001. Stress and neuroactive steroids. *Int. Rev. Neurobiol.* 46:243–272.

Barbarich N. 2002. Is there a common mechanism of serotonin dysregulation in anorexia nervosa and obsessive compulsive disorder? *Eat Weight Disord.* 7(3):221–231.

Barsh GS, Schwartz MW. 2002. Genetic approaches to studying energy balance: perception and integration. *Nat. Rev. Genet.* 3(8):589–600.

Baskin DG, Blevins JE, Schwartz MW. 2001. How the brain regulates food intake and body weight: the role of leptin. *J. Pediatr. Endocrinol. Metab.* 14(Suppl. 6): 1417–1429.

Blevins JE, Schwartz MW, Baskin DG. 2002. Peptide signals regulating food intake and energy homeostasis. *Can. J. Physiol. Pharmacol.* 80(5):396–406.

Branson R, Potoczna N, Kral JG, Lentes KU, Hoehe MR, Horber FF. 2003. Binge eating as a major phenotype of melanocortin 4 receptor gene mutations. *N. Engl. J. Med.* 348(12):1096–103.

Carter CS, Keverne EB. 2002. The neurobiology of social affiliation and pair bonding. In: Pfaff DW, Arnold AP, Etgen AM, Fahrbach SE, Rubin RT, Eds., *Hormones, Brain and Behavior*, Vol. 1. San Diego, CA: Academic Press, pp. 299–337.

Clark JT, Kalra PS, Kalra SP. 1984. Neuropeptide Y stimulates feeding but inhibits sexual behavior in rats. *Endocrinology* 117(6):2435–2442.

Cummings DE, Clement K, Purnell JQ, Vaisse C, Foster KE, Frayo RS, Schwartz MW, Basdevant A, Weigle DS. 2002. Elevated plasma ghrelin levels in Prader–Willi syndrome. *Nat. Med.* 7:643–644.

Cummings DE, Schwartz MW. 2002. Genetics and pathophysiology of human obesity. *Annu. Rev. Med.*

DelParigi A, Tschop M, Heiman ML, Salbe AD, Vozarova B., Sell SM, Bunt JC, Tataranni PA. 2002. High circulating ghrelin: a potential cause of hyperphagia and obesity in Prader–Willi syndrome. *J. Clin. Endocrinol. Metab.* 87(12):5461–5464.

Engel SR, Grant KA. 2001. Neurosteroids and behavior. *Int. Rev. Neurobiol.* 46:321–348.

Grill HJ, Schwartz MW, Kaplan JM, Foxhall JS, Breininger J, Baskin DG. 2002. Evidence that the caudal brainstem is a target for the inhibitory effect of leptin on food intake. *Endocrinology* 143(1):239–246.

Halaas JL, Gajiwala KS, Maffei M, Cohen SL, Chait BT, Rabinowitz D, Lallone RL, Burley SK, Friedman JM. 1995. Weight-reducing effects of the plasma protein encoded by the obese gene. *Science* 269(5223):543–546.

Hewson AK, Tung LY, Connell DW, Tookman L, Dickson SL. 2002. The rat arcuate nucleus integrates peripheral signals provided by leptin, insulin, and a ghrelin mimetic. *Diabetes* 51(12):3412–3419.

Insel TR, Young LE. 2000. Neuropeptides and the evolution of social behavior. *Curr. Opin. Neurobiol.* 10:784–789.

Jacoangeli F, Zoli A, Taranto A, Staar-Mezzasalma F, Ficoneri C, Pierangeli S, Menzinger G, Bollea MR. 2002. Osteoporosis and anorexia nervosa: relative role of endocrine alterations and malnutrition. *Eat Weight Disord.* 7(3):190–195.

Kaufman J, Plotsky PM, Nemeroff CB, Charney DS. 2000. Effects of early adverse experiences on brain structure and function: clinical implications. *Biol. Psychiatry* 48:778–790.

Lal S, Kirkup AJ, Brunsden AM, Thompson DG, Grundy D. 2001. Vagal afferent responses to fatty acids of different chain length in the rat. *Am. J. Physiol. Gastrointest. Liver Physiol.* 281(4):G907–G915.

MacDougald OA, Hwang CS, Fan H, Lane MD. 1995. Regulated expression of the obese gene product (leptin) in white adipose tissue and 3T3-L1 adipocytes. *Proc. Natl. Acad. Sci. USA* 92(20):9034–9037.

Maffei M, Fei H, Lee GH, Dani C, Leroy P, Zhang Y, Proenca R, Negrel R, Ailhaud G, Friedman JM. 1995. Increased expression in adipocytes of *ob* RNA in mice with lesions of the hypothalamus and with mutations at the *db* locus. *Proc. Natl. Acad. Sci. USA* 92(15):6957–6960.

Margetic S, Gazzola C, Pregg GG, Hill RA. 2002. Leptin: a review of its peripheral actions and interactions. *Int. J. Obes. Relat. Metab. Disord.* 26(11):1407–1433.

Mason JW. 1968. Organization of the multiple endocrine responses to avoidance in the monkey. *Psychosom. Med.* 30:774–790.

McMinn JE, Baskin DG, Schwartz MW. 2002. Neuroendocrine mechanisms regulation food intake and body weight. *Obes. Rev.* 26(11):1407–1433.

Morton GJ, Niswender KD, Rhodes CJ, Myers MG, Jr, Blevins JE, Baskin DG, Schwartz MW. 2003. Arcuate nucleus-specific leptin receptor gene therapy attenuates the obesity phenotype of koletsky (fa(k)/fa(k)) rats. *Endocrinology* 144(5):2016–2024.

Morton GJ, Schwartz MW. 2001. The NPY/AgrP neuron and energy homeostasis. *Int. J. Obes. Relat. Metab. Disord.* 5(Suppl.):S56–S62.

Nijenhusiu WA, Garner KM, VanRozen RJ, Adan RA. 2003. Poor cell surface expression of human melanocortin-4 receptor mutations associated with obesity. *J. Biol. Chem.*

Pacak K, Palkovits M. 2001. Stressor specificity of central neuroendocrine responses: implications for stress-related disorders. *Endocrine Rev.* 22:502–548.

Parry BL, Berga SL. 2002. Premenstrual dysphoric disorder. In: Pfaff DW, Arnold AP, Etgen AM, Fahrbach SE, Rubin RT, Eds., *Hormones, Brain and Behavior*, Vol. 5. San Diego, CA: Academic Press, pp. 531–552.

Phillips MI, Heininger F, Toffolo S. 1996. The role of brain angiotensin in thirst and AVP release induced by hemorrhage. *Regul. Peptides* 66(1–2):3–11.

Phillips MI, Olds J. 1969. Unit activity: motivation-dependent responses from midbrain neurons. *Science* 165(899):1269–1271.

Reame NE. 2001. Premenstrual syndrome. In: DeGroot LJ, Jameson JL, Eds., *Endocrinology*, 4th ed. Philadelphia, PA: Saunders, pp. 2147–2152.

Rubin RT. 2004. Anorexia nervosa, bulimia nervosa, and other eating disorders. In DeGroot W, Jameson JL, Eds., *Endocrinology, Fifth Edition*. Philadelphia: W.B. Saunders, in press.

Rubinow DR, Schmidt PJ, Roca CA, Daly RC. 2002. Gonadal hormones and behavior in women: concentrations versus context. In: Pfaff DW, Arnold AP, Etgen AM, Fahrbach SE, Rubin RT, Eds., *Hormones, Brain and Behavior*, Vol. 5. San Diego, CA: Academic Press, pp. 37–73.

Schwartz MW. 2001. Brain pathways controlling food intake and body weight. *Exp. Biol. Med.* 226(11):978–981.

Schwartz MW, Morton GJ. 2002. Obesity: keeping hunger at bay. *Nature* 418:595–597.

Schwartz MW, Woods SC, Seeley RJ, Barsh GS, Baskin DG, Leibel RL. 2003. Is the energy homeostasis system inherently biased toward weight gain? *Diabetes* 52(2): 227–231.

Shibuya I, Utsunomiya K, Toyohira Y, Ueno S, Tsutsui M, Cheah TB, Ueta Y, Izuma F, Yanagihara N. 2002. Regulation of catecholamine synthesis by leptin. *Ann. N.Y. Acad. Sci.* 971:522–527.

Smith SS. 2002. Novel effects of neuroactive steroids in the central nervous system. In: Pfaff DW, Arnold AP, Etgen AM, Fahrbach SE, Rubin RT, Eds., *Hormones, Brain and Behavior*, Vol. 3. San Diego, CA: Academic Press, pp. 747–778.

Swart I, Jahng JW, Overton JM, Houpt TA. 2002. Hypothalamic NPY, AgrP, and POMC mRNA responses to leptin and refeeding mice. *Am. J. Physiol. Regul. Integ. Comp. Physiol.* 283(5):R1020–R1026.

Wisse BE, Frayo RS, Schwartz MW, Cummings DE. 2001. Reversal of cancer anorexia by blockade of central melanocortin receptors in rats. *Endocrinology* 142(8):3292–3301.

Wisse BE, Schwartz MW. 2001. Role of melanocortins in control of obesity. *Lancet* 358:857–859.

Zhou QY, Palmiter RD. 1995. Dopamine-deficient mice are severely hypoactive, adipsic, and aphagic. *Cell* 83(7):1197–1209.

Chapter 2: One Hormone Can have Many Effects: A Single Hormone Can Affect Complex Behaviors

Blaustein JD, Erskine MS. 2002. Feminine sexual behavior: cellular integration of hormonal and afferent information in the rodent forebrain. In: Pfaff DW, Arnold AP, Etgen AM, Fahrbach SE, Rubin RT, Eds., *Hormones, Brain and Behavior*, Vol. 1. San Diego, CA: Academic Press, pp. 139–214.

Buggy J, Jonklaas J. 1984. Sodium appetite decreased by central angiotensin blockade. *Physiol. Behav.* 32(5):737–742.

de Bold AJ. 1985. Atrial natriuretic factor: a hormone produced by the heart, *Science* 15;230(4727):767–770, 1985.

De Wied D, Jolles J. 1982. Neuropeptides derived from pro-opiocortin: behavioral, physiological, and neurochemical effects. *Physiol. Rev.* 62:976–1059.

Dhingra H, Roongsritong C, Kurtzman NA. 2002. Brain natriuretic peptide: role in cardiovascular and volume homeostasis, *Semin. Nephrol.* 22(5):423–437.

Ehrlich KJ, Fitts DA. 1990. Atrial natriuretic peptide in the subfornical organ reduces drinking induced by angiotensin or in response to water deprivation. *Behav. Neurosci.* 104(2):365–372.

Epstein AN, Fitzsimons JT, Rolls BJ. 1970. Drinking induced by injection of angiotensin into the brain of the rat. *J. Physiol.* 210(2):457–474.

Fluharty SJ, Epstein AN. 1983. Sodium appetite elicited by intracerebroventricular infusion of angiotensin in the rat. II. Synergistic interaction with systemic mineralocorticoids. *Behav. Neurosci.* 97(5):746–758.

Ganesan R, Sumners C. 1989. Glucocorticoids potentiate the dipsogenic action of angiotensin II. *Brain Res.* 499(1):121–130.

Gonzalez-Mariscal G, Poindron P. 2002. Parental care in mammals: immediate internal and sensory factors of control. In: Pfaff DW, Arnold AP, Etgen AM, Fahrbach SE, Rubin RT, Eds., *Hormones, Brain and Behavior*, Vol. 1. San Diego, CA: Academic Press, pp. 215–298.

Holsboer F. 1999. The rationale for corticotropin-releasing hormone receptor (CRH-R) antagonists to treat depression and anxiety. *J. Psychiatric Res.* 33:181–214.

Insel TR, Gingrich BS, Young LJ. 2001. Oxytocin: who needs it? *Prog. Brain Res.* 133:59–66.

Kaufman J, Plotsky PM, Nemeroff CB, Charney DS. 2000. Effects of early adverse experiences on brain structure and function: clinical implications. *Biol. Psychiatry* 48:778–790.

Lewko B, Stepinski J. 2002. Cyclic GMP signaling in podocytes. *Microsc. Res. Tech.* 15:574(4):232–235.

Li Z, Ferguson AV. 1993. Subfornical organ efferents to paraventricular nucleus utilize angiotensin as a neurotransmitter. *Am. J. Physiol.* 265(2, Pt. 2):R302–R309.

Ma LY, McEwen BS, Sakai RR, Schulkin J. 1993. Glucocorticoids facilitate mineralocorticoid-induced sodium intake in the rat. *Hormones Behav.* 27(2):240–250.

Melis MR, Arigolas A. 1995. Nitric oxide donors in penile erection and yawing when infected in the central nervous system of male rats. *Eur. J. Pharm.* 294(1):9.

Phillips MI. 1987. Functions of angiotensin in the central nervous system. *Annu. Rev. Physiol.* 49:413–435.

Phillips MI, Gyurko R. 1995. *In vivo* applications of antisense oligonucleotides for peptide research. *Regul. Pept.* 59(2):131–141.

Phillips MI, Heininger F, Toffolo S. 1996. The role of brain angiotensin in thirst and AVP release induced by hemorrhage. *Regul. Pept.* 66(1–2):3–11.

Rubin RT, Reinisch JM, Haskett RF. 1981. Postnatal gonadal steroid effects on human behavior. *Science* 211:1318–1324.

Sakai RR, McEwen BS, Fluharty SJ, Ma LY. 2000. The amygdala: site of genomic and nongenomic arousal of aldosterone-induced sodium intake. *Kidney Int.* 57(4): 1337–1345.

Schulkin J. 2002. Hormonal modulation of central motivational states. In: Pfaff DW, Arnold AP, Etgen AM, Fahrbach SE, Rubin RT, Eds., *Hormones, Brain and Behavior*, Vol. 1. San Diego, CA: Academic Press, pp. 633–657.

Schulkin J. 2003. *Rethinking Homeostasis: Allostatic Regulation in Physiology and Pathophysiology*. Cambridge, MA: MIT Press.

Simon NG. 2002. Hormonal processes in the development and expression of aggressive behavior. In: Pfaff DW, Arnold AP, Etgen AM, Fahrbach SE, Rubin RT, Eds., *Hormones, Brain and Behavior*, Vol. 1. San Diego, CA: Academic Press, pp. 339–392.

Stewart WK, Fleming LW. 1973. Features of a successful therapeutic fast of 382 days' duration. *Postgrad. Med. J.* 49(569):203–209.

Sumners C, Fregly MJ. 1989. Modulation of angiotensin II binding sites in neuronal cultures by mineralocorticoids. *Am. J. Physiol.* 256(1, Pt. 1):C121–C129.

Sumners C, Myers LM. 1991. Angiotensin II decreases cGMP levels in neuronal cultures from rat brain. *Am. J. Physiol.* 260(1, Pt. 1):C79–C87.

Wilson WL, Starbuck EM, Fitts DA. 2002. Salt appetite of adrenalectomized rats after a lesion of the SFO. *Behav. Rain Res.* 136(2):449–453.

Wingfield JC, Romero LM. 2001. Adrenocortical responses to stress and their modulation in free-living vertebrates. In: McEwew, BS, Ed., *Coping with the Environment: Neural and Endocrine Mechanisms.* Oxford: Oxford University Press.

Chapter 3: Hormone Combinations Can be Important for Influencing an Individual Behavior

Abiko T, Kimura Y. 2001. Syntheses and effect of bombesin-fragment 6–14 and its four analogues on food intake in rats. *Curr. Pharm. Biotechnol.* 2(2): 201–207.

Blair-West JR, Carey KD, Denton DA, Weisinger RS, Shade RE. 1998. Evidence that brain angiotensin II is involved in both thirst and sodium appetite in baboons. *Am. J. Physiol.* 275(5, Pt. 2):R1639–R1646.

Brown M, Allen R, Villarreal J, Rivier J, Vale W. 1978. Bombesin-like activity: radioimmunologic assessment in biological tissues. *Life Sci.* 23(27–28):2721–2728.

Choleris E, Gustafsson JA *et al.* 2003. An estrogen-dependent four-gene micronet regulating social recognition: a study with oxytocin and estrogen receptor-alpha and -beta knockout mice. *Proc. Natl. Acad. Sci. USA* 100(10):6192–6197.

Dellovade TL, Zhu Y-S, Krey L, Pfaff DW. 1996. Thyroid hormone and estrogen interact to regulate behavior. *Proc. Nat. Acad. Sci. USA*, 93:12581–12586.

Dohanich G. 2002. Gonadal steroids, learning, and memory. In: Pfaff DW, Arnold AP, Etgen AM, Fahrbach SE, Rubin RT, Eds., *Hormones, Brain and Behavior*, Vol. 2. San Diego, CA: Academic Press, pp. 265–327.

Figlewicz DP, Nadzan AM, Sipols AJ, Green PK, Liddle RA, Porte D, Jr, Woods SC. 1992. Intraventricular CCK-8 reduces single meal size in the baboon by interaction with type-A CCK receptors. *Am. J. Physiol.* 263(4, Pt. 2):R863–R867.

Fitts DA, Thurnhorst RL, Simpson JB. 1985. Diuresis and reduction of salt appetite by lateral ventricular infusions of atriopeptin II. *Brain Res.* 348:118–124.

Fitzsimons JT, Stricker EM. 1971. Sodium appetite and the renin–angiotensin system. *Nat. New Biol.* 231(19):58–60.

Fletcher SW, Colditz GA. 2002. Failure of estrogen plus progestin therapy for prevention [editorial]. *JAMA* 288:366–367.

Gibbs J, Fauser DJ, Row EA, Rolls BJ, Rolls ET, Maddison SP. 1979. Bombesin suppresses feeding in rats. *Nature* 282(5735):208–210.

Gibbs J, Young RC, Smith GP. 1997. Cholecystokinin decreases food intake in rats. *Obes. Res.* 5(3):284–290.

Hamilton RB, Norgren R. 1984. Central projections of gustatory nerves in the rat. *J. Comp. Neurol.* 222(4):560–577.

Johnson AK, Thunhorst RL. 1997. The neuroendocrinology of thirst and salt appetite: visceral sensory signals and mechanisms of central integration. *Front. Neuroendocrinol.* 18(3):292–353.

Kojima M, Hosoda H, Date Y, Nakazato M, Matsuo H, Kangawa K. 1999. Ghrelin is a growth-hormone-releasing acylated peptide from stomach. *Nature* 402(6762): 656–660.

Mohon MA. 2001. An airman with Schmidt's syndrome. *Fed. Air Surgeon's Med. Bull.* 1–3.

Moran TH, Sawyer TK, Seeb DH, Arriglio PJ, Lombard MA, McHugh PR, 1992. Potent and sustained satiety actions of a cholecystokinin octapeptid analogue. *Am. J. Clin. Nutr.* 2865–2905.

Morgan M, Dellovade TL, Pfaff DW. 2000. Effect of thyroid hormone administration on estrogen-induced sex behavior in female mice. *Hormones Behav.* 37:15–22.

Niswender KD, Schwartz MW. 2003. Insulin and leptin revisited: adiposity with overlapping physiological and intracellular signaling capabilities. *Front. Neuroendcrinol.* 24(1):1–10.

Odell WD, Burger HG. 2001. Menopause and hormone replacement. In: DeGroot LJ, Jameson JL, Eds., *Endocrinology*, 4th ed. Philadelphia, PA: Saunders, pp. 2153–2162.

Parry BL, Berga SL. 2002. Premenstrual dysphoric disorder. In: Pfaff DW, Arnold AP, Etgen AM, Fahrbach SE, Rubin RT, Eds., *Hormones, Brain and Behavior*, Vol. 5. San Diego, CA: Academic Press, pp. 531–552.

Pfaff DW. 1999. *Drive: Neurobiological and Molecular Mechanisms of Sexual Motivation.* Cambridge, MA: MIT Press.

Porte D, Jr, Baskin DG, Schwartz MW. 2002. Leptin and insulin action in the central nervous system. *Nutr. Rev.* 60(10, Pt. 2):S20–S29.

Reame NE. 2001. Premenstrual syndrome. In: DeGroot LJ, Jameson JL, Eds., *Endocrinology*, 4th ed. Philadelphia, PA: Saunders, pp. 2147–2152.

Sakai RR, Ma LY, Zhang DM, McEwen BS, Fluharty SJ. 1996. Intracerebral administration of mineralocorticoid receptor antisense oligonucleotides attenuate adrenal steroid-induced salt appetite in rats. *Neuroendocrinology* 64(6):425–429.

Schumacher M, Robert F. 2002. Progesterone: synthesis, metabolism, mechanisms of action, and effects in the nervous system. In: Pfaff DW, Arnold AP, Etgen AM, Fahrbach SE, Rubin RT, Eds., *Hormones, Brain and Behavior*, Vol. 3. San Diego, CA: Academic Press, pp. 683–745.

Seibel MM. 2001. Ovulation induction and assisted reproduction. In: DeGroot LJ, Jameson JL, Eds., *Endocrinology*, 4th ed. Philadelphia, PA: Saunders, pp. 2138–2145.

Takeda R, Takayama Y, Tagawa S, Komel L. 1999. Schmidt's syndrome: autoimmune polyglandular disease of the adrenal and thyroid glands. *Israel Med. Assoc. J.* 1:285–286.

Vasudevan N, Davidkova G, Zhu YS, Koibuchi N, Chin WW, Pfaff DW. 2001. Differential interactions of estrogen receptor and thyroid hormone receptor isoforms on the rat oxytocin receptor promoter leads to differences in transcriptional regulation. *Neuroendocrinology* 74(5):309–324.

Vasudevan N, Koibuchi N, Chin WW, Pfaff DW. 2001. Differential crosstalk between estrogen receptor (ER)α and ERβ and the thyroid hormone receptor isoforms results in flexible regulation of the consensus ERE. *Brain Res. Mol. Brain Res.* 95(1–2):9–17.

Vasudevan N, Zhu YS, Daniels S, Koibuchi N, Chin WW, Pfaff DW. 2001. Crosstalk between oestrogen receptors and thyroid hormone receptor isoforms results in differential regulation of the preproenkephalin gene. *J. Neuroendocrinol.*

Wilson WL, Starbuck EM, Fitts DA. 2002. Salt appetite of adrenalectomized rats after a lesion of the SFO. *Behav. Brain Res.* 136(2):449–453.

Chapter 4: Hormone Metabolites Can be the Behaviorally Active Compounds

Baulieu E-E, Robel P. 1996. Dehydroepiandrosterone (DHEA) and dehydroepiandrosterone sulfate (DHEAS) as neuroactive steroids. *J. Endocrinol.* 150:5221–5239.

Baulieu E-E, Robel P. 1998. Dehydroepiandrosterone (DHEA) and dehydroepiandrosterone sulfate (DHEAS) as neuroactive steroids [commentary]. *Proc. Natl. Acad. Sci. USA* 95:4089–4091.

Brown NJ, Kumar S, Painter CA, Vaughan DE. 2002. ACE inhibition versus angiotensin type 1 receptor antagonism: differential effects on PAI-1 over time. *Hypertension* 40(6):859–865.

De Wied D, Jolles J. 1982. Neuropeptides derived from pro-opiocortin: behavioral, physiological, and neurochemical effects. *Physiol. Rev.* 62:976–1059.

Ferrario CM. 2003. Contribution of angiotensin-(1–7) to cardiovascular physiology and pathology. *Curr. Hypertens. Rep.* 5(2):129–134.

Forest MG. 2001. Diagnosis and treatment of disorders of sexual development. In: DeGroot LJ, Jameson JL, Eds., *Endocrinology*, 4th ed. Philadelphia: Saunders, pp. 1974–2010.

Frye CA. 2001a. The role of neurosteroids and non-genomic effects of progestins and androgens in mediating sexual receptivity of rodents. *Brain Res. Brain Res. Rev.* 37(1–3):201–222.

Frye CA. 2001b. The role of neurosteroids and nongenomic effects of progestins in the ventral tegmental area in mediating sexual receptivity of rodents. *Hormones Behav.* 40(2):226–233.

Gooren LJG. 2001. Gender identity and sexual behavior. In: DeGroot LJ, Jameson JL, Eds., *Endocrinology*, 4th ed. Philadelphia, PA: Saunders, pp. 2033–2042.

Gorski RA. 2000. Sexual differentiation of the nervous system. In: Kandel ER, Schwartz JH, Jessel TM, Eds., *Principles of Neural Science*, 4th ed. New York: McGraw-Hill, pp. 1131–1148.

Green R. 2002. Sexual identity and sexual orientation. In: Pfaff DW, Arnold AP, Etgen AM, Fahrbach SE, Rubin RT, Eds., *Hormones, Brain and Behavior*, Vol. 4. San Diego, CA: Academic Press, pp. 463–485.

Kasckow J, Geracioti JTD. 2002. Neuroregulatory peptides of central nervous system origin: from bench to bedside. In: Pfaff DW, Arnold AP, Etgen AM, Fahrbach SE, Rubin RT, Eds., *Hormones, Brain and Behavior*, Vol. 5. San Diego, CA: Academic Press, pp. 153–208.

Pavlides C, Kimura, A *et al.* 1995. Hippocampal homosynaptic long-term depression/depotentiation induced by adrenal steroids. *Neuroscience* 68(2):379–385.
Pavlides C, McEwen BS. 1999. Effects of mineralocorticoid and glucocorticoid receptors on long-term potentiation in the CA3 hippocampal field. *Brain Res.* 851(1–2):204–214.
Pavlides C, Watanabe Y *et al.* 1995. Opposing roles of type I and type II adrenal steroid receptors in hippocampal long-term potentiation. *Neuroscience* 68(2): 387–394.
Rubin RT, Reinisch JM, Haskett RF. 1981. Postnatal gonadal steroid effects on human behavior. *Science* 211:1318–1324.
Styne DM, Grumbach MM. 2002. Puberty in boys and girls. In: Pfaff DW, Arnold AP, Etgen AM, Fahrbach SE, Rubin RT, Eds., *Hormones, Brain and Behavior*, Vol. 4. San Diego, CA: Academic Press, pp. 661–716.
White A, Ray DW. 2001. Adrenocorticotropic hormone. In: DeGroot LJ, Jameson JL, Eds., *Endocrinology*, 4th ed. Philadelphia, PA: Saunders, pp. 221–233.
Wright JW, Tamura-Myers E, Wilson WL, Roques BP, Llorens-Cortes C, Speth RC, Harding JW. 2003. Conversion of brain angiotensin II to angiotensin III is critical for pressor response in rats. *Am. J. Physiol. Regul. Integr. Comp. Physiol.* 284(3): R725–R733.

Chapter 5: There are Optimal Hormone Concentrations: Too Much or Too Little Can be Damaging

Bauer M, Whybrow PC. 2000. Thyroid hormone, brain, and behavior. In: Fink G, Ed., *Encyclopedia of Stress*, Vol. 2. San Diego, CA: Academic Press, pp. 239–264.
Bornstein SR, Stratakis CA, Chrousos GP. 2000. Cushing's syndrome: medical aspects. In: Fink G, Ed., *Encyclopedia of Stress*, Vol. 1. San Diego, CA: Academic Press, pp. 615–621.
Chiovato L, Barbesino G, Pinchera A. 2001. Graves' disease. In: DeGroot LJ, Jameson JL, Eds., *Endocrinology*, 4th ed. Philadelphia, PA: Saunders, pp. 1422–1449.
Joffe RT. 2000. Hypothyroidism. In: Fink G, Ed., *Encyclopedia of Stress*, Vol. 2. San Diego, CA: Academic Press, pp. 496–499.
Laron Z. 2000. Growth hormone and insulin-like growth factor I: effects on the brain. In: Fink G, Ed., *Encyclopedia of Stress*, Vol. 5. San Diego, CA: Academic Press, pp. 75–96.
Lesser IM. 2000. Hyperthyroidism. In: Fink G, Ed., *Encyclopedia of Stress*, Vol. 2. San Diego, CA: Academic Press, pp. 439–440.
Loriaux DL, McDonald WJ. 2001. Adrenal insufficiency. In: DeGroot LJ, Jameson JL, Eds., *Endocrinology*, 4th ed. Philadelphia, PA: Saunders, pp. 1683–1690.
McEwen BS. 2002. *The End of Stress As We Know It*. Washington D.C.: Joseph Henry Press, p. 64.
Melmed S. 2001. Acromegaly. In: DeGroot LJ, Jameson JL, Eds., *Endocrinology*, 4th ed. Philadelphia, PA: Saunders, pp. 300–312.

Netter FH. 1965a. The suprarenal glands (adrenal glands). In: Forsham PH, Ed., *Endocrine System and Selected Metabolic Diseases: The Ciba Collection of Medical Illustrations*, Vol. 4. New York: Ciba Pharmaceutical Co., pp. 75–108.

Netter FH. 1965b. The thyroid gland; the parathyroid glands. In: Forsham PH, Ed., *Endocrine System and Selected Metabolic Diseases: The Ciba Collection of Medical Illustrations*, Vol. 4. New York: Ciba Pharmaceutical Co., pp. 39–74.

Niemann LK. 2001. Cushing's syndrome. In: DeGroot LJ, Jameson JL, Eds., *Endocrinology*, 4th ed. Philadelphia, PA: Saunders, pp. 1691–1715.

Rosenfeld RG. 2001. Growth hormone deficiency in children. In: DeGroot LJ, Jameson JL, Eds., *Endocrinology*, 4th ed. Philadelphia, PA: Saunders, pp. 503–519.

Rosenfield RL, Cuttler L. 2001. Somatic growth and maturation. In: DeGroot LJ, Jameson JL, Eds., *Endocrinology*, 4th ed. Philadelphia, PA: Saunders, pp. 477–502.

Starkman M. 2000. Cushing's syndrome, neuropsychiatric aspects. In: Fink G, Ed., *Encyclopedia of Stress*, Vol. 1. San Diego, CA: Academic Press, pp. 621–625.

Stumvoll M, Haring H. 2002. Glitazones: clinical effects and molecular mechanisms. *Am. Med.* 34(3):217–224.

Visser TJ, Fliers E. 2000. Thyroid hormones. In: Fink G, Ed., *Encyclopedia of Stress*, Vol. 3. San Diego, CA: Academic Press, pp. 605–612.

Walker C-D, Welberg LAM, Plotsky PM. 2002. Glucocorticoids, stress, and development. In: Pfaff DW, Arnold AP, Etgen AM, Fahrbach SE, Rubin RT, Eds., *Hormones, Brain and Behavior*, Vol. 4. San Diego, CA: Academic Press, pp. 487–534.

White A, Ray DW. 2001. Adrenocorticotropic hormone. In: DeGroot LJ, Jameson JL, Eds., *Endocrinology*, 4th ed. Philadelphia, PA: Saunders, pp. 221–233.

Wiersinga WM. 2001. Hypothyroidism and myxedema coma. In: DeGroot LJ, Jameson JL, Eds., *Endocrinology*, 4th ed. Philadelphia, PA: Saunders, pp. 1491–1506.

Willenberg HS, Bornstein SR, Chrousos GP. 2000. Adrenal insufficiency. In: Fink G, Ed., *Encyclopedia of Stress*, Vol. 1. San Diego, CA: Academic Press, pp. 58–63.

Chapter 6: Hormones Do Not Cause Behavior; They Alter Probabilities of Responses to Given Stimuli

Abeck DS, McKittrick CR, Blanchard DC, Blanchard RJ, Nikulina J, McEwen BS, Sakai RS. 1997. Chronic social stress alters expression of corticotrophin-releasing factor and arginine vasopressin mRNA expression in rat brain. *J. Neurosci.* 12:4895–4903.

American Psychiatric Association. 1994. *Diagnostic and Statistical Manual of Mental Disorders*, 4th ed. Washington, D.C.: American Psychiatric Press.

Bauer M, Whybrow PC. 2000. Thyroid hormone, brain, and behavior. In: Fink G, Ed., *Encyclopedia of Stress*, Vol. 2. San Diego, CA: Academic Press, pp. 239–264.

Blanchard DC, McKittrick CR, Hardy MP, Blanchard RJ. 2002. Social stress effects on hormones, brain and behavior. In: Pfaff DW, Arnold AP, Etgen AM, Fahrbach SE, Rubin RT, Eds., *Hormones, Brain and Behavior*, Vol. 1. San Diego, CA: Academic Press, pp. 735–772.

Blanchard RJ, Blanchard DC. 1989. Anti-predator defensive behaviors in a visible burrow system. *J. Comp. Psychol.* 103:70–82.

Carroll BJ, Feinberg M, Greden JF, Tarika J, Albala AA, Haskett RF, James NM, Kronfol Z, Lohr N, Steiner M, de Vigne JP, Young E. 1981. A specific laboratory test for the diagnosis of melancholia: standardization, validation, and clinical utility. *Arch. Gen. Psychiatry* 38:15–22.

Chao HM, Blanchard DC, Blanchard RJ, McEwan BS, Sakai RM. 1993. The effect of social stress on hippocampal gene expression. *Mol. Cell. Neurosci.* 4:543–548.

Chiovato L, Barbesino G, Pinchera A. 2001. Graves' disease. In: DeGroot LJ, Jameson JL, Eds., *Endocrinology*, 4th ed. Philadelphia, PA: Saunders, pp. 1422–1449.

Figlewicz DP. 2003. Adiposity signals and food reward: expanding the CNS roles of insulin and leptin. *Am. J. Physiol. Regul. Integr. Comp. Physiol.* 284(4): R882–R892.

Holsboer F. 1999. The rationale for corticotropin-releasing hormone receptor (CRH-R) antagonists to treat depression and anxiety. *J. Psychiatric Res.* 33:181–214.

Kow LM, Montgomery MO, Pfaff DW. 1979. Triggering of lordosis reflex in female rats with somatosensory stimulation: quantitative determination of stimulus parameters. *J. Neurophysiol.* 42:195–202.

Kow LM, Pfaff DW. 1979. Responses of single units in sixth lumbar dorsal root ganglion of female rats to mechanostimulation relevant for lordosis reflex. *J. Neurophysiol.* 42:203–13.

Kow LM, Zemlan FP, Pfaff DW. 1980. Responses of lumbosacral spinal units to mechanical stimuli related to analysis of lordosis reflex in female rats. *J. Neurophysiol.* 43:27–45.

Lesser IM. 2000. Hyperthyroidism. In: Fink G, Ed., *Encyclopedia of Stress*, Vol. 2. San Diego, CA: Academic Press, pp. 439–440.

McKittrick CR, Blanchard DC, Blanchard RJ, McEwen BS, Sakai RR. 1995. Serotonin receptor binding in a colony model of chronic social stress. *Biol. Psychiatry* 37:383–393.

McKittrick CR, Magarinos AM, Blanchard DC, Blanchard RJ, McEwen BS, Sakai RR. 2000. Chronic social stress reduces dendritic arbors in CA3 of hippocampus and decreases binding to serotonin transporter sites. *Synapse* 36(2):85–94.

Rubin RT, Dinan TG, Scott LV. 2002. Affective disorders. In: Pfaff DW, Arnold AP, Etgen AM, Fahrbach SE, Rubin RT, Eds., *Hormones, Brain and Behavior*, Vol. 5. San Diego, CA: Academic Press, pp. 467–514.

Spencer RL, Miller AH, Moday H, McEwen BS, Blanchard RJ, Blanchard DC, Sakai RR. 1996. Chronic social stress produces reductions in available splenic type II corticosteroid receptor binding and plasma corticosteroid binding globulin levels. *Psychoneuroendocrinology* 21:95–109.

Starkman M. 2000. Cushing's syndrome, neuropsychiatric aspects. In: Fink G, Ed., *Encyclopedia of Stress*, Vol. 1. San Diego, CA: Academic Press, pp. 621–625.

Zobel AW, Nickel T, Künzel HE, Ackl N, Sonntag A, Ising M, Holsboer F. 2000. Effects of the high-affinity corticotropin-releasing hormone receptor 1 antagonist

R121919 in major depression: the first 20 patients treated. *J. Psychiatric Res.* 34:171–181.

Chapter 7: Familial/Genetic Dispositions to Hormone Responsiveness Can Influence Behavior

Anisman H, Zaharia MD, Meaney MJ, Merali Z. 1998. Do early-life events permanently alter behavioral and hormonal responses to stressors? *Int. J. Develop. Neurosci.* 16:149–164.

Bennis-Taleb N, Remacle C, Hoett JJ, Reusens B. 1999. A low-protein isocaloric diet during gestation affects brain development and alters permanently cerebral cortex blood vessels in rat offspring. *J. Nutr.* 129(8):1613–1619.

Forest MG. 2001. Diagnosis and treatment of disorders of sexual development. In: DeGroot LJ, Jameson JL, Eds., *Endocrinology*, 4th ed. Philadelphia: Saunders, pp. 1974–2010.

Francis DD, Young LJ, Meaney MJ, Insel TR. 2002. Naturally occurring differences in maternal care are associated with the expression of oxytocin and vasopressin (V1a). *J. Neuroendocrinol.* 14(5):349–353.

Glover V. 1999. Maternal stress or anxiety during pregnancy and development of the baby. *Pract. Midwife* 2(5):20–22.

Glushakov AV, Dennis DM, Morey TE, Sumners C, Cucchiara RF, Seubert CN, Martynuk AE. 2002. Specific inhibition of *N*-methyl-D-aspartate receptor function in rat hippocampal neurons by L-phenylalanine at concentrations observed during phenylketonuria. *Mol. Psychiatry* 7(4):359–367.

Gooren LJG. 2001. Gender identity and sexual behavior. In: DeGroot LJ, Jameson JL, Eds., *Endocrinology*, 4th ed. Philadelphia, PA: Saunders, pp. 2033–2042.

Green R. 2002. Sexual identity and sexual orientation. In: Pfaff DW, Arnold AP, Etgen AM, Fahrbach SE, Rubin RT, Eds., *Hormones, Brain and Behavior*, Vol. 4. San Diego, CA: Academic Press, pp. 463–485.

Greer JM, Capecchi MR. 2002. Hoxb8 is required for normal grooming behavior in mice. *Neuron* 33(1):23–43.

Hines M. 2002. Sexual differentiation of human brain and behavior. In: Pfaff DW, Arnold AP, Etgen AM, Fahrbach SE, Rubin RT, Eds., *Hormones, Brain and Behavior*, Vol. 4. San Diego, CA: Academic Press, pp. 425–462.

Kaufman J, Plotsky PM, Nemeroff CB, Charney DS. 2000. Effects of early adverse experiences on brain structure and function: clinical implications. *Biol. Psychiatry* 48:778–790.

Leo CP, Hsu SY, Hsueh AJW. 2002. Hormonal genomics. *Endocrine Rev.* 23: 369–381.

Leon DA, Johanson M, Rasmussen, F. 2000. Gestational age growth rate of fetal mass are inversely associated with systolic blood pressure in young adults: an epidemiologic study of 165,136 Swedish men aged 18 years. *Am. J. Epidemiol.* 152(7):597–604.

McCarton CM, Brooks-Gunn J, Wallace IF, Bauer CR, Bennett FC, Bernbaum JC, Broyles RS, Casey PH, McCormick MC, Scott DT, Tyson J, Tonascia J, Meinert CL. 1997. Results at age 8 years of early intervention for low-birth-weight premature infants: the Infant Health and Development Program. *JAMA* 277(2):126–132.

McKinney WT, Suomi SJ, Harlow HF. 1971. Depression in primates. *Am. J. Psychiatry* 127:1313–1320.

Müller MB, Keck ME, Steckler T, Holsboer F. 2002. Genetics of endocrine-behavior interactions. In: Pfaff DW, Arnold AP, Etgen AM, Fahrbach SE, Rubin RT, Eds., *Hormones, Brain and Behavior*, Vol. 5. San Diego, CA: Academic Press, pp. 263–301.

Pedersen CA, Boccia ML. 2002. Oxytocin links mothering received, mothering bestowed and adult stress responses. *Stress* 5(4):259–267.

Pohlenz J, Weiss RE *et al.* 1999. 5 new families with resistance to thyroid hormone not caused by mutations in the thyroid hormone receptor beta gene. *J. Clin. Endocrinol. Metab.* 84(11):3919–3928.

Refetoff S. 2000. Resistance to thyroid hormone. In: Braverman LE, Utiger RE, Eds., *Werner & Ingbar's The Thyroid: A Fundamental and Clinical Text*, 8th ed. Philadelphia, PA: Lippincott Williams & Wilkins, pp. 1028–1043.

Reutrakul S, Sadow PM *et al.* 2000. Search for abnormalities of nuclear corepressors, coactivators, and a coregulator in families with resistance to thyroid hormone without mutations in thyroid hormone receptor beta or alpha genes. *J. Clin. Endocrinol. Metab.* 85(10):3609–3617.

Sadow P, Reutrakul, Weiss, Refetoff. 2000. Resistance to thyroid hormone in the absence of mutations in the thyroid hormone receptor genes. *Curr. Opin. Endocrinol. Diabetes* 7:253–259.

Seminara SB, Crowley WF. 2002. Genetic approaches to unraveling reproductive disorders: examples of bedside to bench research in the genomic era. *Endocrine Rev.* 23:382–392.

Snoeck A, Remacle C, Reusens B, Hoett JJ. 1990. Effect of a low protein diet during pregnancy of the fetal rat endocrine pancreas. *Biol. Neonate* 57(2):107–118.

Styne DM, Grumbach MM. 2002. Puberty in boys and girls. In: Pfaff DW, Arnold AP, Etgen AM, Fahrbach SE, Rubin RT, Eds., *Hormones, Brain and Behavior*, Vol. 4. San Diego, CA: Academic Press, pp. 661–716.

Thompson EB. 2002. The impact of genomics and proteomics on endocrinology [editorial]. *Endocrine Rev.* 23:366–368.

Walker C-D, Welberg LAM, Plotsky PM. 2002. Glucocorticoids, stress, and development. In: Pfaff DW, Arnold AP, Etgen AM, Fahrbach SE, Rubin RT, Eds., *Hormones, Brain and Behavior*, Vol. 4. San Diego, CA: Academic Press, pp. 487–534.

Weaver SA, Aherne FX, Meaney MJ, Schaefer AL, Dixon WT. 2000. Neonatal handling permanently alters hypothalamic–pituitary–adrenal axis function, behavior, and body weight in boars. *J. Endocrinol.* 164:349–359.

Welberg LA, Seckl JR, Holmes MC. 2001. Prenatal glucocorticoid programming of brain corticosteroid receptors and corticotrophin-releasing hormone: possible implications for behaviour. *Neuroscience* 104(1):71–79.

Chapter 8: The Sex of the Recipient Can Influence the Behavioral Response

Chodorow N. 1999. *The Reproduction of Motherhood: Psychoanalysis and the Sociology of Gender*, Berkeley: University of California Press.

DeVries GJ, Simerly RB. 2002. Anatomy, development, and function of sexually diorphic neural circuits in the mammalian brain. In: Pfaff DW, Arnold AP, Etgen AM, Fahrbach SE, Rubin RT, Eds., *Hormones, Brain and Behavior*, Vol. 4. San Diego, CA: Academic Press, pp. 137–191.

Gonzalez-Mariscal 2002. In: Pfaff DW, Arnold AP, Etgen AM, Fahrbach SE, Rubin RT, Eds., *Hormones, Brain and Behavior*, Vol. San Diego, CA: Academic Press, pp.

Goy RW, McEwen BS *et al.* 1980. *Sexual Differentiation of the Brain: Based on a Work Session of the Neurosciences Research Program*. Cambridge, MA: MIT Press.

Handa RJ, McGivern RF. 2000. Gender and stress. In: Fink, G., Ed., *Encyclopedia of Stress*, Vol. 2. San Diego, CA: Academic Press, pp. 196–204.

Imperato-McGinley 2002. In: Pfaff DW, Arnold AP, Etgen AM, Fahrbach SE, Rubin RT, Eds., *Hormones, Brain and Behavior*, Vol. San Diego, CA: Academic Press, pp.

McCormick CM, Linkroum W *et al.* 2002. Peripheral and central sex steroids have differential effects on the HPA axis of male and female rats. *Stress* 5(4): 235–247.

Mogil JS, Wilson, SG *et al.* 2003. The melanocortin-1 receptor gene mediates female-specific mechanisms of analgesia in mice and humans. *Proc. Natl. Acad. Sci. USA* 100(8):4867–4872.

Ogawa S, Taylor J *et al.* 1996. Reversal of sex roles in genetic female mice by disruption of estrogen receptor gene. *Neuroendocrinology* 64:467–470.

Ogawa S, Washburn T *et al.* 1998b. Modifications of testosterone-dependent behaviors by estrogen receptor-alpha gene disruption in male mice. *Endocrinology* 139:5058–5069

Ogawa S, Eng V *et al.* 1998a. Roles of estrogen receptor-alpha gene expression in reproduction-related behaviors in female mice. *Endocrinology* 139:5070–5081.

Rubin RT, O'Toole SM, Rhodes ME, Sekula LK, Czambel RK. 1999. Hypothalamo–pituitary–adrenal cortical responses to low-dose physostigmine and arginine vasopressin administration: sex differences between major depressives and matched control subjects. *Psychiatry Res.* 89:1–20.

Rubin RT, O'Toole SM, Rhodes ME, Sekula LK, Czambel RK. 1999. Hypothalamo–pituitary–adrenal cortical responses to low-dose physostigmine and arginine vasopressin administration: sex differences between major depressives and matched control subjects. *Psychiatry Res.* 89:1–20.

Suzuki S, Lund TD, Price RH, Handa RJ. 2001. Sex differences in the hypothalamo–pituitary–adrenal axis: novel roles for androgen and estrogen receptors, *Recent Res. Develop. Endocrinol.* 69–86.

Simerly RB. 1990. Hormonal control of neuropeptide gene expression in sexually dimorphic olfactory pathways. *Trends Neurosci.* 13:104–110.

Simerly RB. 2002. Wired for reproduction: organization and development of sexually dimorphic circuits in the mammalian forebrain. *Ann. Rev. Neurosci.* 25:507–5036.

Simon N. 2002. In: Pfaff DW, Arnold AP, Etgen AM, Fahrbach SE, Rubin RT, Eds., *Hormones, Brain and Behavior*, Vol. San Diego, CA: Academic Press, pp.

Wilhelm K, Parker G, Dewhurst J. 1998. Examining sex differences in the impact of anticipated and actual life events. *J. Affect. Disord.* 48:37–45.

Chapter 9: Hormone Actions Early in Development Can Influence Hormone Responsiveness in the CNS During Adulthood

Barbazanges A, Piazza PV. *et al.* 1996. Maternal glucocorticoid secretion mediates long-term effects of prenatal stress. *J. Neurosci.* 16(12):3943–3949.

Berenbaum SA, Duck SC, Bryk K. 2000. Behavioral effects of prenatal versus postnatal androgen excess in children with 21-hydroxylase-deficient congenital adrenal hyperplasia. *J. Clin. Endocrinol. Metab.* 85(2):727–733.

Blaustein JD, Erskine MS. 2002. Feminine sexual behavior: cellular integration of hormonal and afferent information in the rodent forebrain. In: Pfaff DW, Arnold AP, Etgen AM, Fahrbach SE, Rubin RT, Eds., *Hormones, Brain and Behavior*, Vol. 1. San Diego, CA: Academic Press, pp. 139–214.

Canlon B, Erichsen S, Nemlander E, Chen M, Celsi G, Ceccatelli S. 2004. Alterations in the intrauterine environment by glucocorticoids modifies the developmental programme of the auditory system. *Eur. J. Neurosci.* In press.

Dowling ALS, Iannacone EA, Zoeller RT. 2001. Maternal hypothyroidism selectively affects the expression of neuroendocrine-specific protein-A messenger ribonucleic acid in the proliferative zone of the fetal rat brain cortex. *Endocrinology* 142:390–399.

Dowling ALS, Martz GU, Leonard JL, Zoeller RT. 2000. Acute changes in maternal thyroid hormone induce rapid and transient changes in specific gene expression in fetal rat brain. *J. Neurosci.* 20:2255–2265.

Dowling ALS, Zoeller RT. 2000. Thyroid hormone of maternal origin regulates the expression of RC3/Neurogranin mRNA in the fetal rat brain. *Brain Res.* 82:126–132.

Forest MG. 2001. Diagnosis and treatment of disorders of sexual development. In: DeGroot LJ, Jameson JL, Eds., *Endocrinology*, 4th ed. Philadelphia: Saunders, pp. 1974–2010.

Gooren LJG. 2001. Gender identity and sexual behavior. In: DeGroot LJ, Jameson JL, Eds., *Endocrinology*, 4th ed. Philadelphia, PA: Saunders, pp. 2033–2042.

Gorski RA. 2000. Sexual differentiation of the nervous system. In: Kandel ER, Schwartz JH, Jessel TM, Eds., *Principles of Neural Science*, 4th ed. New York: McGraw-Hill, pp. 1131–1148.

Green R. 2002. Sexual identity and sexual orientation. In: Pfaff DW, Arnold AP, Etgen AM, Fahrbach SE, Rubin RT, Eds., *Hormones, Brain and Behavior*, Vol. 4. San Diego, CA: Academic Press, pp. 463–485.

Hines M. 2002. Sexual differentiation of human brain and behavior. In: Pfaff DW, Arnold AP, Etgen AM, Fahrbach SE, Rubin RT, Eds., *Hormones, Brain and Behavior*, Vol. 4. San Diego, CA: Academic Press, pp. 425–462.

Kamphuis PJ, Bakker JM *et al.* 2002. Enhanced glucocorticoid feedback inhibition of hypothalamo–pituitary–adrenal responses to stress in adult rats neonatally treated with dexamethasone. *Neuroendocrinology* 76(3):158–69

Meaney MJ, Diorio J *et al.* 1994. Environmental regulation of the development of glucocorticoid receptor systems in the rat forebrain: the role of serotonin. *Ann. N.Y. Acad. Sci.* 746:260–273; discussion: 274, 289–293

Mohammed AH, Henriksson BG, Södeström S, Ebendal T, Olsson T, Seckl JR. 1993. Environmental influences on the central nervous system and their implications for the aging rat. *Behav. Brain Res.* 57:183–192.

Nathanielsz PW. 1999. *Life in the Womb: The Origin of Health and Disease*. Ithaca NY: Promethean Press.

Rubin RT. 1982. Testosterone and aggression in men. In: Beumont PJV, Burrows GD, Eds., *Handbook of Psychiatry and Endocrinology*. Amsterdam: Elsevier, pp. 355–366.

Simon NG. 2002. Hormonal processes in the development and expression of aggressive behavior. In: Pfaff DW, Arnold AP, Etgen AM, Fahrbach SE, Rubin RT, Eds., *Hormones, Brain and Behavior*, Vol. 1. San Diego, CA: Academic Press, pp. 339–392.

Styne DM, Grumbach MM. 2002. Puberty in boys and girls. In: Pfaff DW, Arnold AP, Etgen AM, Fahrbach SE, Rubin RT, Eds., *Hormones, Brain and Behavior*, Vol. 4. San Diego, CA: Academic Press, pp. 661–716.

Wallen K, Baum MJ. 2002. Masculinization and defeminization in altricial and precocial mammals: comparative aspects of steroid hormone action. In: Pfaff DW, Arnold AP, Etgen AM, Fahrbach SE, Rubin RT, Eds., *Hormones, Brain and Behavior*, Vol. 4. San Diego, CA: Academic Press, pp. 385–423.

Welberg LA, Seckl JR. 2001. Prenatal stress, glucocorticoids and the programming of the brain. *J. Neuroendocrinol.* 13(2):113–128.

Zoeller RT, Rovet J. 2004. *Timing of Thyroid Hormone Action in the Developing Brain: Clinical Observations and Experimental Findings*, in press.

Zoeller TR, Dowling AL *et al.* 2002. Thyroid hormone, brain development, and the environment. *Environ. Health Perspect.* 110(Suppl. 3):355–361.

Chapter 10: Puberty Alters Hormone Secretion and Hormone Responsivity and Heralds Sex Differences

Clark PA, Iranmanesh A *et al.* 1997. Comparison of pulsatile luteinizing hormone secretion between prepubertal children and young adults: evidence for a mass/

amplitude-dependent difference without gender or day/night contrasts. *J. Clin. Endocrinol. Metab.* 82(9):2950–2955.

Ducy P, Amling M, Takeda S, Priemel M, Schilling AF, Beil FT, Shen J, Vinson C, Rueger JM, Karsenty G. 2000. Leptin inhibits bone formation through a hypothalamic relay: a central control of bone mass. *Cell* 100(2):197–207.

Marshall WA, Tanner JM. 1999. Variations in pattern of pubertal changes in girls. *Arch. Dis. Child.* 44(235):291–303.

Marshall WA, Tanner JM. 1970. Variations in the pattern of pubertal changes in boys. *Arch. Dis. Child.* 45(239):13–23.

Mauras N, Blizzard RM *et al.* 1987. Augmentation of growth hormone secretion during puberty: evidence for a pulse amplitude-modulated phenomenon. *J. Clin. Endocrinol. Metab.* 64(3):596–601.

Olney JW. 1969. Brain lesions, obesity, and other disturbances in mice treated with monosodium glutamate. *Science* 164(880):719–721.

Rogol AD, Roemmich JN, Clark PA. 2002. Growth at puberty. *J. Adolesc. Health* 31(Suppl. 6):192–200.

Takeda S, Elefteriou F, Levasseur R, Liu X, Zhao L, Parker KL, Armstrong D, Ducy P, Karenty G. 2002. Leptin regulates bone formation via the sympathetic nervous system. *Cell* 111:305–317.

Chapter 11: Changes in Hormone Levels and Responsiveness During Aging Affect Behavior

Anawalt BD, Merriam GR. 2001. Neuroendocrine aging in men. *Endocrinol. Metab. Clin. North Am.* 30:647–669.

Berkley KJ, Hoffman GE, Murphy AZ, Holdcroft A. 2002. Pain: sex/gender differences. In: Pfaff DW, Arnold AP, Etgen AM, Fahrbach SE, Rubin RT, Eds., *Hormones, Brain and Behavior*, Vol. 5. San Diego, CA: Academic Press, pp. 409–442.

Crowe MJ, Frosling ML, Rolls BJ, Phillips PA, Ledingham JG, Smith R. 1987. Altered water excretion in healthy elderly men. *Age Ageing* 16(5):285–293.

Dohanich G. 2002. Gonadal steroids, learning, and memory. In: Pfaff DW, Arnold AP, Etgen AM, Fahrbach SE, Rubin RT, Eds., *Hormones, Brain and Behavior*, Vol. 2. San Diego, CA: Academic Press, pp. 265–327.

Dubal DB, Pettigrew LC, Kashon M, Ren JM, Finklestein SP, Rau SW, Wise PM. 1998. Estradiol protects against ischemia-induced brain injury. *J. Cerebral Blood Flow Metab.* 18:1253–1258.

Dubal DB, Shughrue PJ, Wilson ME, Merchenthaler I, Wise PM. 1999. Estradiol modulates bcl-2 in cerebral ischemia: a potential role for estrogen receptors. *J. Neurosci.* 19:6385–6393.

Dubal DB, Zhu H, Yu J, Rau SW, Shughrue PJ, Merchenthaler I, Kindy MS, Wise PM. 2001. Estrogen receptor α, not β, is a critical link in estradiol-mediated protection against brain injury. *Proc. Natl. Acad. Sci. USA* 98:1952–1957.

Endocrine Society. 2001. 2nd Annual Andropause Consensus Meeting, Beverly Hills, CA.

Fletcher SW, Colditz GA. 2002. Failure of estrogen plus progestin therapy for prevention [editorial]. *JAMA* 288:366–367.

Foster TC. 1999. Involvement of hippocampal synaptic plasticity in age-related memory decline. *Brain Res. Rev.* 30:236–249.

Golden GA, Mason RP, Tulenko TN, Zubenko GS, Rubin RT. 1999. Rapid and opposite effects of cortisol and estradiol on human erythrocyte Na^+, K^+ ATPase activity: relationship to steroid intercalation into the cell membrane. *Life Sci.* 65:1247–1255.

Henderson VW, Reynolds DW. 2002. Protective effects of estrogen on aging and damaged neural systems. In: Pfaff DW, Arnold AP, Etgen AM, Fahrbach SE, Rubin RT, Eds., *Hormones, Brain and Behavior*, Vol. 4. San Diego, CA: Academic Press, pp. 821–837.

Kumar A, Foster TC 2002. 17β-estradiol benzoate decreases the AHP amplitude in CA1 pyramidal neurons. *J. Neurophysiol.* 88:621–626.

Odell WD, Burger HG. 2001. Menopause and hormone replacement. In: DeGroot LJ, Jameson JL, Eds., *Endocrinology*, 4th ed. Philadelphia, PA: Saunders, pp. 2153–2162.

Phillips PA, Bretherton M, Johnston CI, Gray L. 1991. Reduced osmotic thirst in healthy elderly men. *Am. J. Physiol.* 261(1, Pt. 2):R166–R171.

Phillips PA, Bretherton M, Risvanis J, Casley D, Johnston C, Gray L. 1993. Effects of drinking on thirst and vasopressin in dehydrated elderly men. *Am. J. Physiol.* 264(5, Pt. 2):R877–R881.

Phillips PA, Hodsman GP, Johnston CI. 1991. Neuroendocrine mechanisms and cardiovascular homeostasis in the elderly. *Cardiovasc. Drugs Ther.* 6(Suppl.): 1209–1213.

Phillips PA, Johnston CI, Gray L. 1993. Disturbed fluid and electrolyte homeostasis following dehydration in elderly people. *Age Ageing* 22(1):S26–S33.

Raskind MA, Wilkinson CW, Peskind ER. 2002. Aging and Alzheimer's disease. In: Pfaff DW, Arnold AP, Etgen AM, Fahrbach SE, Rubin RT, Eds., *Hormones, Brain and Behavior*, Vol. 5. San Diego, CA: Academic Press, pp. 637–664.

Roth J, Koch CA, Rother KI. 2001. Aging, endocrinology, and the elderly patient. In: DeGroot LJ, Jameson JL, Eds., *Endocrinology*, 4th ed. Philadelphia, PA: Saunders, pp. 529–555.

Rubinow DR, Schmidt PJ, Roca CA, Daly RC. 2002. Gonadal hormones and behavior in women: concentrations versus context. In: Pfaff DW, Arnold AP, Etgen AM, Fahrbach SE, Rubin RT, Eds., *Hormones, Brain and Behavior*, Vol. 5. San Diego, CA: Academic Press, pp. 37–73.

Sadow TF, Rubin RT. 1992. Effects of hypothalamic peptides on the aging brain. *Psychoneuroendocrinology* 17:293–314.

Sapolsky, RM. 2000. Glucocorticoids and hippocampal atrophy in neuropsychiatric disorders. *Arch. Gen. Psychiatry* 57(10):925–935.

Schaaf MJ, Cidlowski JA. 2002. Molecular mechanisms of glucocorticoid action and resistance. *J. Steroid Biochem. Mol. Biol.* 83(1–5):37–48.

Sharrow KM, Kumar A, Foster TC. 2002. Calcineurin as a potential contributor in estradiol regulation of hippocampal synaptic function. *Neuroscience* 113:89–97.

Sherwin BB. 2003. Estrogen and cognitive functioning in women. *Endocrine Rev.* 24:133–151.

Simon NG. 2002. Hormonal processes in the development and expression of aggressive behavior. In: Pfaff DW, Arnold AP, Etgen AM, Fahrbach SE, Rubin RT, Eds., *Hormones, Brain and Behavior*, Vol. 1. San Diego, CA: Academic Press, pp. 339–392.

Snyder PJ. 2001. Effect of age on testicular function and consequences of testosterone treatment. *J. Clin. Endocrinol. Metab.* 86:2369–2372.

Van Coevorden A, Mockel J, Laurent E, Kerkhofs M, L'Hermite-Balériaux M, Decoster C, Neve P, Van Cauter E. 1991. Neuroendocrine rhythms and sleep in aging men. *Am. J. Physiol.* 260:E651–E661.

Vermeulen A. 2001. Androgen replacement therapy in the aging male: a critical evaluation. *J. Clin. Endocrinol. Metab.* 86:2380–2390.

Wilson ME, Dubal DB, Wise PM. 2000. Estradiol protects injury-induced cell death in cortical explant cultures: a role for estrogen receptors. *Brain Res.* 873:235–242.

Wilson ME, Liu Y, Wise PM. 2002. Estradiol enhances Akt activation in cortical explant cultures following neuronal injury. *Mol. Brain Res.* 102:88–94.

Wise PM, Dubal D, Wilson ME, Rau SW, Liu Y. 2001. Estradiol: a trophic and protective factor in the adult brain. *Front. Neuroendocrinol.* 22:33–66.

Wise PM. 2000. Neuroendocrine correlates of aging. In: Conn PM, Freeman ME, Eds., *Neuroendocrinology in Physiology and Medicine*. Totowa, NJ: Humana Press.

Chapter 12: Duration of Hormone Exposure Can Make a Big Difference: In Some Cases Longer is Better; In Other Cases Brief Pulses are Optimal for Behavioral Effects

Parsons B, MacLusky NJ, Krieger MS, McEwen BS, Pfaff DW. 1979. The effects of long-term estrogen exposure on the induction of sexual behavior and measurements of brain estrogen and progestin receptors in the female rat. *Hormones Behav.* 13:301–313.

Parsons B, McEwen BS, Pfaff DW. 1982. A discontinuous schedule of estradiol treatment is sufficient to activate progesterone-facilitated feminine sexual behavior and to increase cytosol receptors for progestins in the hypothalamus of the rat. *Endocrinology* 110:613–619.

Parsons B, Rainbow TC, Pfaff DW, McEwen BS. 1981. Oestradiol, sexual receptivity and cytosol progestin receptors in rat hypothalamus. *Nature* 292:58–59.

Parsons B, Rainbow TC, Pfaff DW, McEwen BS. 1982. Hypothalamic protein synthesis essential for the activation of the lordosis reflex in the female rat. *Endocrinology* 110:620–624.

Reaves PY, Gelband CH, Wang H, Yang H, Lu D, Berecek KH, Katovich MJ, Raizada MK. 1999. Permanent cardiovascular protection from hypertension by the AT(1) receptor antisense gene therapy in hypertensive rat offspring. *Circ. Res.* 85(10):44–50.

Santoro N, Filicori M *et al.* 1986. Hypogonadotropic disorders in men and women: diagnosis and therapy with pulsatile gonadotropin-releasing hormone. *Endocrine Rev.* 7(1):11–23.

Seminara, Hayes, Crowley, 2001. *Endocrine Rev.*
Seminara SB, Hayes FJ. *et al.* 1998. Gonadotropin-releasing hormone deficiency in the human (idiopathic hypogonadotropic hypogonadism and Kallmann's syndrome): pathophysiological and genetic considerations. *Endocrine Rev.* 19(5):521–539.
Tannenbaum GS, Epelbaum J, Bowers CY. 2002. Ghrelin and the growth hormone neuroendocrine axis. In: Kordon *et al.*, Eds., *Brain Somatic Cross-Talk and the Central Control of Metabolism.* Berlin: Springer-Verlag.
Valk TW, Corley KP *et al.* 1980. Hypogonadotropic hypogonadism: hormonal responses to low dose pulsatile administration of gonadotropin-releasing hormone. *J. Clin. Endocrinol. Metab.* 51(4):730–738.
Wilson KM, Margargal W, Berecek KH. 1988. Long-term captopril treatment: angiotensin II receptors and responses. *Hypertension* 11(2, Pt. 2):148–152.

Chapter 13: Hormonal Secretions and Responses are Affected by Biological Clocks

Asakawa A, Inui A, Kaga T, Yuzuriha H, Nagata T, Ueno N, Makino S, Fujimiya M, Niijima A, Fujino MA, Kasuga M. 2001. Ghrelin is an appetite-stimulatory signal from stomach with structural resemblance to motilin. *Gastroenterology* 120(2): 337–345.
Bagnasco M, Dube MG, Kalra PS, Kalra SP. 2002a. Evidence for the existence of distinct central appetite, energy expenditure, and ghrelin stimulation pathways as revealed by hypothalamic site-specific leptin gene therapy. *Endocrinology* 143(11): 4409–4421.
Bagnasco M, Kalra PS, Kalra SP. 2002b. Plasma leptin levels are pulsatile in adult rats: effects of gonadectomy. *Neuroendocrinology* 75(4):257–263.
Balsalobre A, Brown SA, Marcacci L, Tronche F, Kellendonk C, Reichardt HM, Eschutz G, Schibler U. 2000. Resetting of circadian time in peripheral tissues by glucorticoid signaling. *Science* 289:2344–2347.
Cohen P, Zhao C, Cai X, Montez JM, Rohani SC, Feinstein P, Mombaerts P, Friedman JM. 2001. 108(8):1113–1121.
Czeisler CA, Klerman EB. 1999. Circadian and sleep-dependent regulation of hormone release in humans. *Rec. Prog. Horm. Res.* 54:97–132.
Damiola F, LeMinh N, Preitner N, Kornmann B, Fleury-Olela F, Schibler U. 2000. Restricted feeding uncouples circadian oscillators in peripheral tissues from the central pacemaker in the suprachiasmatic nucleus. *Genes Dev.* 14(23): 2950–2961.
Darlington TK, Wager-Smith K *et al.* 1998. Closing the circadian loop: CLOCK-induced transcription of its own inhibitors *per* and *tim*. *Science* 280(5369): 1599–1603.
Ferkin MH, Zucker I. 1991. Seasonal control of odour preferences of meadow voles (*Microtus pennsylvanicus*) by photoperiod and ovarian hormones. *J. Reprod. Fertil.* 92(2):433–441.
Friedman JM. 2002. The function of leptin in nutrition, weight, physiology. *Nutr Rev.* 60(10, Pt. 2):S1–S14; discussion: S68–S87.

Herzog ED, Takahashi JS *et al.* 1998. Clock controls circadian period in isolated suprachiasmatic nucleus neurons. *Nat. Neurosci.* 1(8):708–713.

Kalra SP, Bagnasco M, Otukonyong EE, Dube MG, Kalra PS. 2003. Rhythmic, reciprocal ghrelin and leptin signaling: new insight in the development of obesity. *Regul. Pept.* 111(1–3):1–11.

Kalra SP, Dube MG, Sahu A, Phelps CP, Kalra PS. 1991. Neuropeptide Y secretion increases in the paraventricular nucleus in association with increased appetite for food. *Proc. Natl. Acad. Sci.* 88(23):10931–10935.

LeMinh N, Damiola F, Tronche F, Schutz G, Schibler U. 2001. Glucorticoid hormones inhibit food-induced phase-shifting of peripheral circadian oscillators. *EMBRO J.* 20(24):7128–7136.

Low-Zeddies SS, Takahashi JS. 2001. Chimera analysis of the *Clock* mutation in mice shows that complex cellular integration determines circadian behavior. *Cell* 105(1): 25–42.

Schibler U, Juergen A, Ripperger JA, Brown SA. 2001. Circadian rhythms: chronobiology-reducing time. *Science* 293:437–438.

Song SH, McIntyre SS, Shah H, Veldhuis JD, Hayes PC, Butler PC. 2000. Direct measurement of pulsalile insulin secretion from the portal vein in human subjects. *J. Clin. Endocrinol. Metab.* 85(12):4491–4499.

Toh KL, Jones CR *et al.* 2001. An hPer2 phosphorylation site mutation in familial advanced sleep phase syndrome. *Science* 291(5506):1040–1043.

Vitaterna MH, King, DP *et al.* 1994. Mutagenesis and mapping of a mouse gene, *Clock*, essential for circadian behavior. *Science* 264(5159):719–25.

Xu B, Kalra PS, Farmerie WG, Kalra SP. 1999. Daily changes in hypothalamic gene expression of neuropeptide Y, galanin, proopiomelanocortin, and adipocyte leptin gene expression and secretion: effects of food restriction. *Endocrinology* 140(6): 2668–2675.

Young MW. 2002. Big ben rings in a lesson on biological clocks. *Neuron* 36(6): 1001–1005.

Young MW, Kay SA. 2001. Time zones: a comparative genetics of circadian clocks. *Nat. Rev. Genet.* 2(9):702–715.

Zucker I. 2002. In: Pfaff DW, Arnold AP, Etgen AM, Fahrbach SE, Rubin RT, Eds., *Hormones, Brain and Behavior*, Vol. San Diego, CA: Academic Press.

Chapter 14: Effects of a Given Hormone Can be Widespread Across the Body; Central Effects Consonant with Peripheral Effects Form Coordinated, Unified Mechanisms

McEwen BS. 2002. *The End of Stress As We Know It*, Washington, D.C.: Joseph Henry Press.

Moss R. and McCann SM. 1973. Induction of mating behavior by LRF. *Science* 181:177–179.

Palovcik RA, Phillips MI, Kappy MS, Raizada MK. 1984. Insulin inhibits pyramidal neurons in hippocampal slices. *Brain Res.* 309(1):187–191.

Pfaff DW. 1973. Luteinizing hormone releasing factor (LRF) potentiates lordosis behavior in hypophysectomized ovariectomized female rats. *Science* 182:1148–1149.

Schulkin J. 2003. *Rethinking Homeostasis : Allostatic Regulation in Physiology and Pathophysiology*. Cambridge, MA: MIT Press.

Schwanzel Fukuda M, Pfaff DW. 1989. Origin of luteinizing hormone-releasing hormone neurons. *Nature* 338:161–164, 1989.

Schwanzel Fukuda M, Bick D, Pfaff DW. 1989. Luteinizing hormone-releasing hormone (LHRH)-expressing cells do not migrate normally in an inherited hypogonadal (Kallmann) syndrome. *Mol. Brain Res.* 6:311–326.

Schwanzel-Fukuda M, Crossin KL, Pfaff DW, Bouloux PM, Hardelin J-P, Petit C. 1996. Migration of LHRH neurons in early human embryos: association with neural cell adhesion molecules. *J. Comp. Neurol.* 366:547–557.

Chapter 15: Hormones Can Act at All Levels of the Neuraxis to Exert Behavioral Effects; The Nature of the Behavioral Effect Depends on the Site of Action

Allen AM, MacGregor DP, McKinley MJ, Mendelsohn FA. 1999. Angiotensin II receptors in the human brain. *Regul. Pept.* 79(1):1–7.

Anke J, Van Eekelen M, Phillips MI 1988. Plasma angiotensin II levels at moment of drinking during angiotensin II intravenous infusion. *Am. J. Physiol.* 255(3, Pt. 2): R500–R506.

Bailey CJ. 2001. New pharmacologic agents for diabetes. *Curr. Diab. Rep.* 1(2): 119–126.

Bodnar R, Commons K, Pfaff DW. 2002. *Central Neural States Relating Sex and Pain.* Baltimore, MD: The Johns Hopkins University Press.

Buggy J, Hoffman WE, Phillips MI, Fisher AE, Johnson AK. 1979. Osmosensitivity of rat third ventricle and interactions with angiotensin. *Am. J. Physiol.* 236(1): R75–R82.

Hoffman WE, Phillips MI, Wilson E, Schmid PG. 1977. A pressor response with drinking in rats. *Proc. Soc. Biol. Med.* 154(1):121–124.

Hogarty DC, Speakman EA, Puig V, Phillips MI. 1992. The role of angiotensin, AT1 and AT2 receptors in the pressor, drinking and vasopressin responses to central angiotensin. *Brain Res.* 586(2):289–294.

Hogarty DC, Tran DN, Phillips MI. 1994. Involvement of angiotensin receptor subtypes in osmotically induced release of vasopressin. *Brain Res.* 6371(1–2): 126–132.

Jordan J, Shannon JR, Black BK, Ali Y, Farley M, Costa F, Diedrich A, Robertson RM, Biaggioni I, Robertson D. 2000. The pressor response to water drinking in humans: a sympathetic reflex? *Circulation* 101:504–509.

Kuriyama R. 1996. Angiotensin converting enzyme inhibitor induced anemia in a kidney transplant recipient. *Transplant Proc.* 28:1635.

Maghnie M. 2003. Diabetes insipidus. *Horm. Res.* 59(Suppl. 1):42–54.

McKinley MJ, Gerstberger R, Mathai ML, Oldfield BJ, Schmid H. 1999. The lamina terminalis and its role in fluid and electrolyte homeostasis. *J. Clin. Neurosci.* 6(4): 289–301.

Phillips MI, Hoffman WE, Bealer SL. 1982. Dehydration and fluid balance: central effects of angiotensin. *Fed. Proc.* 41(9):2520–2527.

Phillips MI, Quilan JT, Weyhenmeyer J. 1980. An angiotensin-like peptide in the brain. *Life Sci.* 27(25–26):2589–2594.

Quinlan JT, Phillips MI. 1981. Immunoreactivity for an angiotensin II-like peptide in human brain. *Brain Res.* 205(1):212–218.

Schroeder C, Bush VE, Norcliffe LJ, Luft FC, Tank J, Jordan J, Hainsworth R. 2002. Water drinking acutely improves orthostatic tolerance in health subjects. *Circulation* 106:2806–2811.

Shannon JR, Diedrich A, Biaggioni I, Tank J, Robertson RM, Robertson D, Jordan J. 2002. Water drinking as a treatment for orthostatic syndromes. *Am. J. Med.* 112:355–360.

Sunn N, Egli M, Burazin T, Colvill C, Davern P, Denton DA, Oldfield BJ, Weisinger RS, Rauch M, Schmid HA, McKinley MJ. 2002. Circulating relaxin acts on subfornical organ neurons to stimulate water drinking in the rat. *Proc. Natl. Acad. Sci. USA* 99(3):1701–1706.

Tamura R, Norgren R. 2003. Intracranial renin alters gustatory neural responses in the nucleus of the solitary tract of rats. *Am. J. Physiol. Regul. Integr. Comp. Physiol.* 284:R1108–R1118.

Wong NL, Tsui JK. 2003. Angiotensin II upregulates the expression of vasopressin V2 mRNA in the medullary collecting duct of the rat. *Metabolism* 52(3): 290–295.

Chapter 16: In Responsive Neurons, Rapid Hormone Effects Can Facilitate Later Genomic Actions

Aikey JL, Nyby JG, Anmuth DN, James PJ. 2002. Testosterone rapidly reduces anxiety in male house mice (*Mus musculus*), *Hormones Behav.* In press.

Chambliss KL, Yuhanna IS *et al.* 2002. ERβ has nongenomic action in caveolae. *Mol. Endocrinol.* 16(5):938–946.

Frye CA, Bayon LE, Vongher JM. 2000. Intravenous progesterone elicits a more rapid induction of lordosis in rats than does SKF38393. *Psychobiology* 28(1): 99–109.

Frye CA, Vongher JM. 1999. Progestins' rapid facilitation of lordosis when applied to the ventral tegmentum corresponds to efficacy at enhancing GABA(A)receptor activity. *J. Neuroendocrinol.* 11(11):829–837.

James PJ, Nyby JG. 2002. Testosterone rapidly affects the expression of copulatory behavior in house mice (*Mus musculus*), *Physiol. Behav.* 75:287–294.

Vasudevan N, Kow L-M, Pfaff DW. 2001. Early membrane estrogenic effects required for full expression of slower genomic actions in a nerve cell line. *PNAS* 98(21): 12267–12271.

Chapter 17: Gene Duplication and Splicing Products for Hormone Receptors in the CNS Often Have Different Behavioral Effects

Bain DL, Franden MA *et al.* 2001. The N-terminal region of human progesterone B-receptors: biophysical and biochemical comparison to A-receptors. *J. Biol. Chem.* 276(26):23825–23831.

Burson JM, Aguilera G, Gross KW, Sigmund CD. 1994. Differential expression of angiotensin receptor 1A and 1B in mouse. *Am. J. Physiol.* 267(2, Pt. 1):E260–E267.

Dellovade TL, Chan J, Vennstrom B, Forrest D, Pfaff DW. 2000. The two thyroid hormone receptor genes have opposite effects on estrogen stimulated sex behaviors. *Nat. Neurosci.* 3(5):472–475.

Kakar SS, Riel KK, Neill JD. 1992. Differential expression of angiotensin II receptor subtype mRNAs (AT-1A and AT-1B) in the brain. *Biochem. Biophys. Res. Commun.* 185(2):688–692.

Scott RE, Wu-Peng XS *et al.* 2002. Regulation and expression of progesterone receptor mRNA isoforms A and B in the male and female rat hypothalamus and pituitary following oestrogen treatment. *J. Neuroendocrinol.* 14(3):175–183.

Tung L, Mohamed MK *et al.* 1993. Antagonist-occupied human progesterone B-receptors activate transcription without binding to progesterone response elements and are dominantly inhibited by A-receptors. *Mol. Endocrinol.* 7(10): 1256–1265.

Vegeto E, Shahbaz MM *et al.* 1993. Human progesterone receptor A form is a cell- and promoter-specific repressor of human progesterone receptor B function. *Mol. Endocrinol.* 7(10):1244–1255.

Wen DX, Xu YF *et al.* 1994. The A and B isoforms of the human progesterone receptor operate through distinct signaling pathways within target cells. *Mol. Cell. Biol.* 14(12):8356–8364.

Yudt MR, Cidlowski JA. 2001. Molecular identification and characterization of A and B forms of the glucocorticoid receptor. *Mol. Endocrinol.* 15(7):1093–1103.

Yudt MR, Cidlowski JA. 2002. The glucocorticoid receptor: coding a diversity of proteins and responses through a single gene. *Mol. Endocrinol.* 16(8): 1719–1726.

Chapter 18: Hormone Receptors and Other Nuclear Proteins Influence Hormone Responsiveness

Apostolakis EM, Garai J *et al.* 2000. Epidermal growth factor activates reproductive behavior independent of ovarian steroids in female rodents. *Mol. Endocrinol.* 14(7): 1086–1098.

Apostolakis EM, Ramamurphy, M *et al.* 2002. Acute disruption of select steroid receptor coactivators prevents reproductive behavior in rats and unmasks genetic adaptation in knockout mice. *Mol. Endocrinol.* 16(7):1511–1523

Auger AP, Perrot-Sinal TS *et al.* 2002. Expression of the nuclear receptor coactivator, cAMP response element-binding protein, is sexually dimorphic and modulates sexual differentiation of neonatal rat brain. *Endocrinology* 143(8): 3009–3016.

Holter E, Kotaja N *et al.* 2002. Inhibition of androgen receptor (AR) function by the reproductive orphan nuclear receptor DAX-1. *Mol. Endocrinol.* 16(3): 515–528.

McKenna NJ, O'Malley B. 2000. From ligand to response: generating diversity in nuclear receptor coregulator function. *J. Steroid Biochem. Mol. Biol.* 74(5): 351–356.

Mong JA, Krebs C *et al.* 2002. Perspective: microarrays and differential display PCR-tools for studying transcript levels of genes in neuroendocrine systems. *Endocrinology* 143(6):2002–2006.

Mong JA, Devidze N *et al.* 2003. Estradiol regulation of lipocalin-type prostaglandin D synthase transcript levels in the rodent brain: evidence from high density oligonucleotide arrays and *in situ* hybridization. *PNAS* 100(1):318–323.

Mong, J *et al.* 2004. Hormonal symphony: neural and genetic modules serving sex hormone effects in brain. *Molec. Psychiatry*, in press.

Pfaff DW. 1999. *Drive; Neurobiological and Molecular Mechanisms of Sexual Motivation*. Cambridge, MA: MIT Press.

Pohlenz J, Weiss RE, Macchia PE, Pannain S, Lau IT, Ho H, Refetoff S. 1999. 5 new families with resistance to thyroid hormone not caused by mutations in the thyroid hormone receptor β gene. *J. Clin. Endocrinol. Metab.* 84:3919–3928.

Rowan BG, O'Malley BW. 2000. Progesterone receptor coactivators. *Steroids* 65(10–11):545–549.

Smith CL, O'Malley BW. 1999. Evolving concepts of selective estrogen receptor action: from basic science to clinical applications. *Trends Endocrinol. Metab.* 10(8): 299–300.

Weiss RE, Hayashi Y *et al.* 1996. Dominant inheritance of resistance to thyroid hormone not linked to defects in the thyroid hormone receptor alpha or beta genes may be due to a defective cofactor. *J. Clin. Endocrinol. Metab.* 81(12): 4196–4203.

Chapter 19: Hormone Effects on Behavior Depend Upon Context

Daniel JM, Dohanich GP. 2001. Acetylcholine mediates the estrogen-induced increase in NMDA receptor binding in CA1 of the hippocampus and the associated improvement in working memory. *J. Neurosci.* 21:6949–6956.

Devine J *et al.* 2004. Scientific foundations for the prevention of youth violence. *Ann. N.Y. Acad. Sci.* In press.

Krukoff TL, Khalili P. 1997. Stress-induced activation of nitric oxide producing neurons in the rat brain. *J. Comp. Neurol.* 377(4):509–519.

Lam T, Leranth C. 2002. Locally Administered Estradiol into the MSDB of OVX Rats Affects the Density of CA1 Area Pyramidal Cell Spine Synapses and Glial Processes,

Program No. 444.8, *2002 Abstract Viewer/Itinerary Planner*. Washington, D.C.: Society for Neuroscience (available online).

Leranth C, Shanabrough M, Horvath TL. 2000. Hormonal regulation of hippocampal spine synapse density involves subcortical mediation. *Neuroscience* 101(2): 349–356.

Morgan MA, Pfaff DW. 2001. Effects of estrogen on activity and fear-related behaviors in mice. *Hormones Behav*. 40(4):472–482.

Morgan MA, Pfaff DW. 2002. Estrogen's effects on activity, anxiety, and fear in two mouse strains. *Behav. Brain Res*. 132(1):85–93.

Murphy DD, Cole NB, Greenberger V, Segal M. 1998. Estradiol increases dendritic spine density by reducing GABA neurotransmission in hippocampal neurons. *J. Neurosci*. 18:2550–2559.

Murphy CA, Stacey NE. 2002. Methyl-testosterone induces male-typical ventilatory behavior in response to putative steroidal pheromones in female round gobies (*Neogobius melanostomus*). *Hormones Behav*. 42(2):109–115.

Rudick CN, Woolley CS. 2001. Estrogen regulates functional inhibition of hippocampal CA1 pyramidal cells in the adult female rat. *J. Neurosci*. 21(17): 6532–6543.

Rudick CN, Gibbs RB, Woolley CS. 2002. Estrogen-Induced Disinhibition of Hippocampal CA1 Pyramidal Cells Depends on Basal Forebrain Cholinergic Neurons, Program No. 740.6, *2002 Abstract Viewer/Itinerary Planner*. Washington, D.C.: Society for Neuroscience (available online).

Sandstrom NJ, Williams CL. 2001. Memory retention is modulated by acute estradiol and progesterone replacement. *Behav. Neurosci*. 115:384–393.

Schmidt PJ, Nieman LK, Danaceau MA, Adams LF, Rubinow, DR. 1998. Differential behavioral effects of gonadal steroids in women with and in those without premenstrual syndrome. *New Engl. J. Med*. 338:209–216.

Shors TJ, Chua C, Falduto J. 2001. Sex differences and opposite effects of stress on dendritic spine density in the male versus female hippocampus. *J. Neurosci*. 21(16): 6292–6297.

Woolley CS, McEwen BS. 1992. Estradiol mediates fluctuation in hippocampal synapse density during the estrous cycle in the adult rat. *J. Neurosci*. 12: 2549–2554.

Woolley CS, McEwen BS. 1993. Roles of estradiol and progesterone in regulation of hippocampal dendritic spine density during the estrous cycle in the rat. *J. Comp. Neurol*. 336:293–306.

Woolley CS, McEwen BS. 1994. Estradiol regulates hippocampal dendritic spine density via an NMDA receptor-dependent mechanism. *J. Neurosci*. 14:7680–7687.

Woolley CS, Gould E, Frankfurt M, McEwen BS. 1990. Naturally occurring fluctuation in dendritic spine density on adult hippocampal pyramidal neurons. *J. Neurosci*. 10:4035–4039.

Woolley CS, Weiland NG, McEwen BS, Schwartzkroin PA. 1997. Estradiol increases the sensitivity of hippocampal CA1 pyramidal cells to NMDA receptor-mediated synaptic input: correlation with dendritic spine density. *J. Neurosci*. 17:1848–1859.

Chapter 20: Behavioral/Environmental Context also Alters Hormone Release

Brady JV. 1967. Ulcers in executive monkeys. In: McGaugh JL, Weinberger NW, Whalen RE, Eds., *The Biological Bases of Behavior*. San Francisco, CA: Freeman, pp. 189–192.

Carroll BJ, Curtis GC, Mendels J. 1976. Neuroendocrine regulation in depression. I. Limbic system-adrenocortical dysfunction. *Arch. Gen. Psychiatry* 33:1041–1044.

Carroll BJ, Feinberg M, Greden JF, Tarika J, Albala AA, Haskett RF, James NM, Kronfol Z, Lohr N, Steiner M, de Vigne JP, Young E. 1981. A specific laboratory test for the diagnosis of melancholia: standardization, validation, and clinical utility. *Arch. Gen. Psychiatry* 38:15–22.

Graber JA, Brooks-Gunn J *et al.* 1995. The antecedents of menarcheal age: heredity, family environment, and stressful life events. *Child Dev.* 66(2):346–359.

Helmreich DL, Mattern LG, Cameron JL. 1993. Lack of a role of the hypothalamic-pituitary-adrenalaxis in the fasting-induced suppression of luteinizing hormone secretion in adult male rhesus monkeys (*Macaca Mulatta*). *Endocrinology* 132(6):2427–2437.

Mason JW. 1968. Organization of the multiple endocrine responses to avoidance in the monkey. *Psychosom. Med.* 30:774–790.

McEwen BS. 2002. *The End of Stress As We Know It*. Washington, D.C.: Joseph Henry Press.

Miller RG, Rubin RT, Clark BR, Crawford WR, Arthur RJ. 1970. The stress of aircraft carrier landings. I. Corticosteroid responses in naval aviators. *Psychosom. Med.* 32:581–588.

Pacak K, Palkovits M. 2001. Stressor specificity of central neuroendocrine responses: implications for stress-related disorders. *Endocrine Rev.* 22:502–548.

Rubin RT, Rahe RH, Arthur RJ, Clark BR. 1969. Adrenal cortical activity changes during underwater demolition team training. *Psychosom. Med.* 31:553–564.

Warren MP, Fried JL. 2001. Hypothalamic amenorrhea: the effects of environmental stresses on the reproductive system—a central effect of the central nervous system. *Endocrinol. Metab. Clin. North Am.* 30(3):611–629.

Warren MP, Perlroth NE. 2001. The effects of intense exercise on the female reproductive system. *J. Endocrinol.* 170(1):3–11.

Chapter 21: Neuroendocrine Mechanisms Have Been Conserved to Provide Biologically Adaptive Body/Brain/Behavior Coordination

Bachman ES, Dhillon H, Zhang C-Y, Cinti S, Bianco AC, Kobilka BK, Lowell BB. B-AR signaling required for diet-induced thermogenesis and obesity resistance.

Bates SH, Stearns WH, Dundon TA, Schubert M, Tso AW, Wang Y, Banks AS, Lavery HJ, Haq AK, Maratos-Flier E, Neel BG, Schwartz MW, Myers MG, Jr. 2003. STAT3 signalling is required for leptin regulation of energy balance but not reproduction. *Nature* 421(6925):856–859.

Cerda-Reverter JM, Larhammar D. 2000. Neuropeptide Y family of peptides: structure, anatomical expression, function, and molecular evolution. *Biochem. Cell Biol.* 78(3):371–392.

Conlon JM. 2002. The origin and evolution of peptide YY (PYY) and pancreatic polypeptide (PP). *Peptides* 23(2):269–278.

Elmquist JK, Flier JS. 2004. Neuroscience. The fat-brain axis enters a new dimension. *Science* 304(5667):108–110.

Davisson RL, Oliverio MI, Coffman TM, Sigmund CD. 2000. Divergent functions of angiotensin II receptor isoforms in the brain. *J. Clin. Invest.* 106(1):103–106.

Ducy P, Karsenty G. 2000. The family of bone morphogenetic proteins. *Kidney Intl.* 57:2207–2214.

Galli SM, Phillips MI. 1996. Interactions of angiotensin II and atrial natriuretic peptide in the brain: fish to rodent. *Proc. Soc. Exp. Biol.* 213:128–137.

Garofalo RS. 2002. Genetic analysis of insulin signaling in *Drosophilia. Trends Endocrinol. Metab.* 13(4):156–162.

Manzon LA. 2002. The role of prolactin in fish osmoregulation: a review. *Gen. Comp. Endocrinol.* 125(2):291–310.

Nasokin IO, Alikasifoglu A, Barrette T, Cheng MM, Thomas PM, Nikitin AG. 2002. Cloning, characterization, and embryonic expression analysis of the *Drosophilia melanogaster* gene encoding insulin/relaxin-like peptide. *Biochem. Biophys. Res. Commun.* 295(2):312–318.

Niswender KD, Gallis B, Blevins JE, Corson MA, Schwartz MW, Baskin DG. 2003. Immunocytochemical detection of phosphatidylinositol 3-kinase activation by insulin and leptin. *J. Histochem. Cytochem.* 51(3):275–283.

Niswender KD, Morrison CD, Clegg DJ, Olson R, Baskin DG, Myers MG, Jr, Seeley RJ, Schwartz MW. 2003. Insulin activation of phosphatidylinositol 3-kinase in the hypothalamic arcuate nucleus: a key mediator of insulin-induced anorexia. *Diabetes* 52(2):227–231.

Takahashi A, Yasuda A, Sullivan CV, Kawauchi H. 2003. Identification of proopiomelanocortin-related peptides in the rostral pars distalis of the pituitary in coelacanth: evolutional implications. *Gen. Comp. Endocrinol.* 130(3):340–349.

ACRONYMS

A

ABR — auditory brainstem response
ACE — angiotensin-converting enzyme
ACTH — adrenocorticotropic hormone
ADAM — androgen decline in aging men
ADH — antidiuretic hormone
AGRP — agouti-related peptide
AGT — angiotensinogen
AN — anorexia nervosa
ANP — atrial natriuretic peptide
ARC — arcuate nucleus
ArgAP — aminopeptidase B
ARs — androgen receptors
AspAP — aminopeptidase A
AT_1 — angiotensin receptor
AT_1R — angiotensin type 1 receptor
AT_1R or AT_2R — angiotensin II receptors
AT_2R — angiotensin type 2 receptor
ATP — adenosine triphosphate
AV3V — anterior ventral third ventricle
AVP — arginine vasopressin

B

BDNF — brain-derived neurotrophic factor
BNP — brain natriuretic peptide

C

CAH — congenital adrenal hyperplasia
cAMP — cyclic adenosine monophosphate
CART — cocaine- and amphetamine-regulated transcript
CBG — corticosteroid binding globulin
CBP — cAMP-response-element-binding protein
CCK — cholecystokinin
cGMP — cyclic guanosine monophosphate
CHP — cyclo-histidine-proline
CLIP — corticotropin-like intermediate lobe peptide
CLIP; ACTH18-39 — corticotropin-like intermediate lobe peptide
CNS — central nervous system
CRH — corticotropin-releasing hormone
CSF — cerebrospinal fluid
CVOs — circumventricular organs

D

DAYQUALS — daytime aircraft carrier landing practice
DHEA — dehydroepiandrosterone
DHEA-S — dehydroepiandrosterone-sulfate
DHT — dihydrotestosterone
DI — diabetes insipidus
DOCA — deoxycorticosterone acetate
DRIPs — vitamin D receptor interacting proteins
DST — dexamethasone suppression test

E

E2 — estradiol
ER-α — estrogen receptor-α
ER-β — estrogen receptor-β

F

FFAs — free fatty acids
FSH — follicle-stimulating hormone

G

GABA — γ-aminobutyric acid
GAPPS — growth, amplification, preparation, permissive actions, synchronization
GH — growth hormone

GnRH — gonadotropin-releasing hormone
GRs — glucocorticoid receptors
GR-α — glucocorticoid receptor-alpha

H

HAL — high attack latency
HBM — high bone mass
HHPA — hippocampal-hypothalamic-pituitary-adrenal corticoid
HPA — hypothalamic-pituitary-adrenal cortical
HPA — hypothalamo-pituitary-adrenal cortical
HPG — hypothalamic-pituitary-gonadal
HPG — hypothalamo-pituitary-gonadal
HPT — hypothalamo-pituitary-thyroid
11-β-HSD-1 — 11-β-hydroxysteroid dehydrogenase-1
5HT — 5′-hydroxytryptamine

I

IGF — insulin-like growth factor
IGF-1 — insulin-like growth factor 1
IGF-I — insulin growth factor I
IL-1 — interleukin-1
IL-6 — interleukin-6
IRS — insulin receptor substrate

J

Jak — Janus kinase
JG — juxta glomerulus
JGA — juxtaglomerular apparatus
JP — joining peptide

K

K_{ATP} — ATP-sensitive K channel

L

LAL — low attack latency
L-DOPA — dopamine precursor
LepR — leptin receptor
LH — lateral hypothalamic nucleus
LH — luteinizing hormone
LIF — leukemia inhibitory factor
β-LPH — beta-lipotropin
LTP — long-term potentiation

M

MAOIs — monoamine oxidase inhibitors
MAP-K — mitogen-activated protein kinase
Mc4Rs — melanocortin-4 receptors
MCH — melanin-concentrating hormone
McR — melanocortin receptor
MK801 — dizocilpine
MLP — mirror landing practice
MNPO — median preoptic
mRNA — messenger RNA
MRs — mineralocorticoid receptors
MSG — monosodium glutamate
MSH — melanocyte-stimulating hormone
α-MSH — alpha-melanocyte-stimulating hormone
α-MSH; ACTH1-13 — alpha-melanocyte-stimulating hormone

N

N-CoR — nuclear co-repressor
NE — norepinephrine
NEP — neutral endopeptidase
NITEQUALS — nighttime carrier landing practice
NMDA — *N*-methyl-D-aspartate
N-POC — N-proopiocortin
NPY — neuropeptide Y
NTS — nucleus tractus solitarius

O

Ob/Ob — obese mice
OCD — obsessive-compulsive disorder
OT — oxytocin
OVLT — organum vasculosum of the laminae terminalis

P

PA-I — plasminogen-activating factor inhibitor
PAI-1 — plasminogen activator inhibitor type 1
PD — postnatal day
PgDS — prostaglandin D synthetase
PHYSO — physostigmine
PI-3K — phosphatidylinositol-3-kinase
PKA — protein kinase A
PKC — protein kinase C
PKU — phenylketonuria
PMDD — premenstrual dysphoric disorder
POMC — proopiomelanocortin
PPAR — peroxisome-proliferator-activated receptor
PR — progesterone receptor

PRL — prolactin
PTSD — posttraumatic stress disorder
PVN — paraventricular nucleus
PWS — Prader-Willi syndrome

R

RIO — radar intercept officer
RTH — resistance to thyroid hormone
RVLM — rostral ventral lateral medulla

S

SCN — suprachiasmatic nucleus
SCUBA — self-contained underwater breathing apparatus
SERMs — selective estrogen receptor modulators
SFO — subfornical organ
SHR — spontaneously hypertensive rat
SIADH — syndrome of inappropriate antidiuretic hormone
SON — supraoptic nucleus
SRC-1 — steroid receptor co-activator-1
SSRIs — selective serotonin reuptake inhibitors
STAT — signal transducer and activator of transcription

T

T2DM — diabetes mellitus type 2
T3 — thyroxine (three iodine atoms)
T3 — triiodothyronine
T4 — thyroxine (four iodine atoms)
TGF-β — transforming growth factor-beta
TGF-α — transforming growth factor-alpha
TK — tyrosine kinase
TNF — tumor necrosis factor
TR — thyroid hormone receptor
TRAP-220 — thyroid-receptor-associated protein with an apparent molecular weight of about 220 Da
TRH — thyrotropin-releasing hormone
TR-β — thyroid hormone receptor-beta
TSH — thyroid-stimulating hormone

U

UDT — underwater demolition team

V

VMH — ventral medial hypothalamus
VMH — ventromedial hypothalamic nucleus

W

Y1Rs — Y1 receptors

INDEX

B

C

D